Global Spencerism

Cultural Dynamics of Science

VOLUME 1

The titles published in this series are listed at *brill.com/cds*

Global Spencerism

The Communication and Appropriation of a British Evolutionist

Edited by

Bernard Lightman

BRILL

LEIDEN | BOSTON

Cover illustration: Herbert Spencer contemplates the universe in his study. An oil painting by John McLure Hamilton, circa 1895. NPG 4092. © National Portrait Gallery, London.

Library of Congress Cataloging-in-Publication Data

Global Spencerism : the communication and appropriation of a British evolutionist / edited by Bernard Lightman.
pages cm. -- (Cultural Dynamics of Science, ISSN 2351-9932 ; VOLUME 1)
Includes bibliographical references and index.
ISBN 978-90-04-26399-4 (hardback : alk. paper) -- ISBN 978-90-04-26400-7 (e-book) 1. Spencer, Herbert, 1820-1903. 2. Evolution. I. Lightman, Bernard V., 1950- editor.

B1657.G56 2015
192--dc23

2015032011

This publication has been typeset in the multilingual "Brill" typeface. With over 5,100 characters covering Latin, IPA, Greek, and Cyrillic, this typeface is especially suitable for use in the humanities. For more information, please see www.brill.com/brill-typeface.

ISSN 2351-9932
ISBN 978-90-04-26399-4 (hardback)
ISBN 978-90-04-26400-7 (e-book)

This book is printed on acid-free paper.

Contents

Acknowledgements

The idea for this edited volume came out of discussions with Daw-hwan Wang (Academia Sinica, Taipei, Taiwan) during a conference we both attended on "Darwin in Communication," which took place at Peking University in 2010. The volume came one step closer to reality when we held a symposium on "Global Spencerism" at the International Congress of the History of Science, Technology and Medicine in Manchester in 2013. All of the chapters in this volume were first given as papers at this symposium.

Finding the right contributors was not easy since I needed to recruit scholars who were working in areas with which I was not familiar. I depended on suggestions from a number of colleagues, including Pietro Corsi, Thomas Glick, Nanna Kaalund, Ronald Numbers, Ignacio Silva, and Haiyan Yang. They gave me excellent advice. The contributors were a joy to work with and I learned a lot from them. I am indebted to Michiel Thijssen for suggesting that I submit this volume to Brill, and for his enthusiasm once I began to work with the Press. Sabine Steenbeek and Wendel Scholma, Brill publishing assistants, provided invaluable help. I am grateful for the suggestions of two anonymous referees and for the encouragement of the editors of the series, Lissa Roberts, Agustí Nieto-Galan, and Oliver Hochadel.

I dedicate this volume to Professor Sydney Eisen, who first introduced me to Spencer and the world of Victorian science when I was a timid third year student at York University. Taking Syd's wonderful course on science and religion in Victorian England immersed me in that world and I don't think that I've come up for air ever since. Syd's inspiration has lasted a lifetime.

Bernard Lightman
April 5, 2015

List of Figures, Graphs and Tables

Figures

Graphs

Tables

List of Contributors

Li Bin

is associate professor at the University of Chinese Academy of Sciences. He is editor of *Kexue Jishu yu Shehui Yanjiu* (*Research of Science, Technology and Society*, with Hu Zhiqiang), and managing editor of the *Journal of Dialectics of Nature*. His research focuses on the history of the Industrial Revolution and history of modern Chinese science and technology. Currently he is working on the industrial enlightenment.

Gowan Dawson

is Professor of Victorian Literature and Culture at the University of Leicester. He is the author of *Show Me the Bone: Reconstructing Prehistoric Monsters in Nineteenth-Century Britain and America* (University of Chicago Press, 2016) and *Darwin, Literature and Victorian Respectability* (Cambridge University Press, 2007), and co-author of *Science in the Nineteenth-Century Periodical: Reading the Magazine of Nature* (Cambridge University Press, 2004).

Heloisa Maria Bertol Domingues

is a Full Researcher at the Museum of Astronomy (MAST/MCTI). She is also Professor at the Post-Graduate Program History from UNIRIO and the Post-Graduate Program, Philosophy and History of Science from UFBA. She works on the history of the natural sciences and anthropology in Brazil, and the history of scientific policy and theories, particulary on the history of Darwinism in Brazil.

Marwa Elshakry

is Associate Professor in the History Department at Columbia University. She is the author of *Reading Darwin in Arabic, 1860–1950* (University of Chicago Press, 2013), and co-editor, with Sujit Sivasundaram, of *Science, Race and Imperialism*, volume 6 of *Victorian Literature and Science* (Pickering and Chatto, 2012).

Mark Francis

is Professor of Political Science at the University of Canterbury, Christchurch, New Zealand. He has been Fowler Hamilton Senior Research Fellow at Christ Church, Oxford, and Rutherford Scholar at Trinity College, Cambridge. His publications include *Herbert Spencer and the Invention of Modern Life* (Acumen, 2008) and "Herbert Spencer" in *The Wiley-Blackwell Companion to Major Social Theorists* (2011).

G. Clinton Godart
is an assistant professor at the department of history at the University of Southern California. He received his Ph.D. from the University of Chicago, and was a postdoctoral researcher at the University of Cambridge. His work focuses on the intellectual history of modern Japan, and he is currently working on a book on the history of evolutionary theory and religion in Japan.

Michael D. Gordin
is Rosengarten Professor of Modern and Contemporary History at Princeton University. He is the author of several monographs, including *A Well-Ordered Thing: Dmitrii Mendeleev and the Shadow of the Periodic Table* (Basic Books, 2004) and *Scientific Babel: How Science Was Done before and after Global English* (University of Chicago Press, 2015).

Paola Govoni
is Assistant Professor at the University of Bologna. She is the author of books and articles on science and society, and on gender and science in the 19th and 20th centuries. She recently edited *Writing about Lives in Science: (Auto)Biography, Gender, and Genre* (Göttingen, 2014, with Z. Franceschi).

Hans Henrik Hjermitslev
is associate professor at University College South Denmark in Aabenraa. He has contributed to *The Reception of Charles Darwin in Europe* (Vol. 1, Continuum, 2008; Vol. 3, Bloomsbury. 2014), and co-edited *Creationism in Europe* (Johns Hopkins University Press, 2014). Currently, Hjermitslev is investigating the circulation of scientific knowledge among Grundtvigians in late nineteenth- and early twentieth-century Denmark.

Bernard Lightman
is Professor of Humanities at York University. Lightman's most recent publications include *Victorian Scientific Naturalism* (co-edited with Gowan Dawson), *Evolution and Victorian Culture* (co-edited with Bennett Zon), and *The Age of Scientific Naturalism* (co-edited with Michael Reidy). He is currently working on a biography of John Tyndall and is one of the editors of the John Tyndall Correspondence Project.

Ricardo Noguera Solano
is a lecturer and researcher in history of biology at the School of Sciences, UNAM, Mexico. His publications include "Darwin and Inheritance: The Influence of Prosper Lucas" (*Journal of the History of Biology*, 2009) and "The Metaphor of the Architect in Darwin: Chance and Free Will" (*Zygon*, 2013). He is currently part of UNAM's project on morality and evolution.

Adriana Novoa
is Associate Professor of History at the University of South Florida. Her work deals with the reception of evolutionary ideas in Latin America. She is currently completing a manuscript on evolutionism and its relationship to gender and race titled *From Virile to Sterile: Darwinism and Masculinity in Argentina, 1870–1930*.

Gregory Radick
is Professor of History and Philosophy of Science at the University of Leeds. His books include *The Simian Tongue: The Long Debate about Animal Language* (Chicago, 2007) and, as co-editor, *The Cambridge Companion to Darwin* (Cambridge, 2nd edition 2009). He is completing a book about the early-twentieth-century controversy over Mendel.

Nathalie Richard
is Professor in Modern History at Le Mans University (France). She is a member of the Centre de recherches historiques de l'Ouest (CERHIO, CNRS UMR 6258). Her research themes are the history of human and social sciences and the history of archaeology in XIXth Century France. She has recently published: *Hippolyte Taine. Histoire, psychologie, littérature* (Garnier, 2013) and *La* Vie de Jésus *de Renan: La Fabrique d'un best-seller* (PUR, 2015).

Juan Manuel Rodriguez Caso
holds a PhD in HPS from the University of Leeds, UK. He is currently an independent scholar. His main research interests include the work of Alfred Russel Wallace, the history of Victorian anthropology, and issues related to science and religion.

Rosaura Ruiz Gutierrez
holds a PhD in biology from UNAM, Mexico, where she is a full-time lecturer in history of biology at the School of Sciences. She is the author of over seventy academic papers, and her books include *The Reception of Darwinism in the Iberian World* (Kluwer, 2001).

Ke Zunke
is assistant professor at the University of Chinese Academy of Sciences. His research interests include the social history of Victorian science, the history of modern Chinese science, and the global history of science. He is currently working on the early development of the Institute of High Energy Physics, Chinese Academy of Sciences.

Introduction

Bernard Lightman

The year 2009 marked the 150th anniversary of the publication of Charles Darwin's *Origin of Species* (1859) and the 200th anniversary of Darwin's birth. The coinciding, twin anniversaries generated over 300 celebratory events, held across the globe.[1] In that same year scores of publications were produced on all things Darwin.[2] Three years after the Darwin Year in 2009 another anniversary in the history of evolution took place. 2012 marked the 150th anniversary of the publication of Herbert Spencer's *First Principles* (1862), the first volume of his ambitious *System of Synthetic Philosophy* (1862–1896). In Spencer's mind the theory of evolution was the key to his entire system. In ten dense volumes on biology, sociology, ethics, psychology and religion, Spencer presented a new intellectual synthesis by interpreting all phenomena according to the law of evolution. But in contrast to the extensive Darwin celebrations, 2012 passed with no fanfare. No major events were held to draw attention to Spencer's role in the dissemination of evolutionary theory. Instead of an avalanche of publications in one year, as in Darwin's case, only a handful of books on Spencer have appeared over the last decade, including Jones and Peel's edited collection *Herbert Spencer: The Intellectual Legacy*, Francis's *Herbert Spencer and the Invention of Modern Life*, and Taylor's *The Philosophy of Herbert Spencer*.[3] Both the non-event of the 2012 anniversary and the paucity of recent scholarship speak volumes about Spencer's reputation in the early twenty-first century. He is entirely overshadowed by Darwin.

And yet, in the latter part of the nineteenth century, Spencer's vision of cosmic evolution was extremely appealing to his contemporaries. Peter Bowler

1 For a long list see: "Darwin Online" <http://www.Darwin-online.org.uk/2009.html>. Accessed July 13, 2014.

2 Some of the more important books were Adrian Desmond and James Moore, *Darwin's Sacred Cause: How a Hatred of Slavery Shaped Darwin's Views on Human Evolution* (Boston: Houghton Mifflin Harcourt, 2009); Diana Donald and Jane Munro, ed., *Endless Forms: Charles Darwin, Natural Science and the Visual Arts* (New Haven: Yale University Press, 2009); John Holmes, *Darwin's Bards: British and American Poetry in the Age of Evolution* (Edinburgh: University of Edinburgh, 2009).

3 Greta Jones and Robert A. Peel, ed., *Herbert Spencer: The Intellectual Legacy* (London: The Galton Institute, 2004); Mark Francis, *Herbert Spencer and the Invention of Modern Life* (Ithaca, New York: Cornell University Press, 2007), 8; Michael W. Taylor, *The Philosophy of Herbert Spencer* (London: Continuum, 2007).

 | DOI 10.1163/9789004264007_002

has asserted, "Spencer's writings were far more effective than Darwin's in spreading the general gospel of evolutionism."[4] Similarly, Mark Francis has pointed out that Spencer became "the world philosopher of the late-nineteenth century," while Michael Taylor has referred to Spencer as "the first international public intellectual" who was read by a large global audience.[5] In addition to admiring readers around the world, Spencer also had devoted British and Americans disciples, like John Fiske, Grant Allen, and Edward Youmans, who magnified his influence by praising his accomplishments and actively disseminating his ideas. Through a donation of £1,000, Shyamaji Krishnavarma, one of Spencer's Indian disciples, was responsible for the founding of the Herbert Spencer lectures at Oxford University. Spencer's Chinese admirer, Yan Fu, translated Spencer's *The Study of Sociology* and *Principles of Sociology* into Chinese.[6]

In his counterfactual history of evolutionism, Peter Bowler has tried to imagine what the world would be like if Darwin had died after being swept overboard while on the *Beagle* voyage. If the *Origin of Species* had never been written, where would Spencer fit into the story? Bowler argues that "a hybrid of liberal Christianity and Spencerism would have been the dominant form of evolutionary thinking during the 1860s and 1870s, and it would have remained influential in America long after that."[7] Without the Darwinian theory of natural selection, Bowler asserts, Spencer's religiously tinged vision of struggle as a stimulus to individual and racial self-improvement would have been even more attractive to liberal Christian thinkers. Spencer's connection with "Darwinian" materialism would not have been a barrier to acceptance. But putting counterfactual history to the side, Bowler maintains that Spencer, in his day, enjoyed significant support, especially from liberal Christians.[8]

Bowler's "what if" story is not the only way to re-evaluate Spencer's place in the history of evolutionism. It is not necessary to kill off the *Beagle*'s most famous passenger to remind us that Spencer didn't stand in Darwin's shadow in the latter part of the nineteenth century. Another way of exploring Spencer's crucial role is to draw on recent scholarly developments in the areas of print culture and global history of science. If we pay more attention to the modes

4 Peter Bowler, *Darwin Deleted: Imagining a World Without Darwin* (Chicago and London: University of Chicago Press, 2013), 245.

5 Francis, *Herbert Spencer and the Invention of Modern Life*, 8; Taylor, *Philosophy of Herbert Spencer*, 2.

6 Taylor, *Philosophy of Herbert Spencer*, 3.

7 Bowler, *Darwin Deleted*, 225.

8 Ibid., 224–225.

of communication in different national contexts, as well as the circulation of knowledge worldwide, then we can more clearly see how Spencer became – for a time – a significant global figure.

Understandably, even Darwin scholarship has not treated Darwin in this light until recently. Thomas Glick's pioneering edited collection, *The Comparative Reception of Darwinism* (1972), examined the impact of Darwinism in various national contexts.[9] Glick himself acknowledged the problems in omitting national traditions from Asia, and the majority of the chapters focused on Europe and the United States.[10] There was little attention to how evolutionary ideas were communicated and the emphasis on "reception" tended to obscure the active appropriation of Darwin's ideas by those claiming to be for, and against, evolution. *Disseminating Darwinism* (1999), co-edited by Ronald Numbers and John Stenhouse, aimed to complement previous work by drawing attention to previously neglected factors, such as gender and race.[11] Some of the chapters offered revisionist accounts of the role of religion in responses to evolution. Another strength of the collection was the incorporation of insights from the new scholarship on the role of space and place in the circulation of scientific knowledge, as well as covering such places as Australia, New Zealand, and Canada. But there was little use of insights drawn from the study of print culture and global history of science.

More recently, James Secord's chapter on "Global Darwin" (2010) does attempt to integrate the recent scholarship on print culture and global history.[12] Secord begins by pointing out that an examination of Darwin's extensive international correspondence shows that "even the most intimate aspects of intellectual work were embedded in networks of cross-cultural exchange."[13] He also reminds us that Darwin's career in science was originally forged during his trip on the *Beagle*, an imperial voyage approved by the British Admiralty to chart the southern coasts of South America and to compile a chronometrical

9 Thomas F. Glick, ed., *The Comparative Reception of Darwinism* (Austin and London: University of Texas Press, 1972). See also David Kohn, ed., *The Darwinian Heritage* (Princeton: Princeton University Press, 1985), chapters 22–24, for more examples of early scholarship on the reception of Darwin in diverse national contexts.

10 Ibid., ix. The more recent edited collection by Engels and Glick focuses on the reception of Darwin in Europe. See: Eve-Marie Engels and Thomas F. Glick, eds., *The Reception of Charles Darwin in Europe*, 2 vols (London: Continuum, 2008).

11 Ronald L. Numbers and John Stenhouse, ed., *Disseminating Darwinism: The Role of Place, Race, Religion, and Gender* (Cambridge: Cambridge University Press, 1992).

12 James A. Secord, "Global Darwin," in *Darwin*, ed. William Brown and Andrew C. Fabian (Cambridge: Cambridge University Press, 2010), 31–57.

13 Ibid., 32.

survey of the globe to help determine longitude for trading vessels.[14] But, as Secord points out, understanding the "emergence of Darwinism throughout the world" is more difficult than analyzing the imperial networks and global perspectives underlying Darwin's work. The previous studies of the reception of Darwin, organized around the nation state, seem to present a similar pattern. Initially there are pro-Darwinians, who are opposed by a traditionalist group; this early stage gives way to a process of "accommodation and compromise."[15] Secord's solution for finding a way out of this repetitive mode of analysis is to

> ... approach evolutionary texts in their concrete material forms: as books, journals, letters, songs, scientific articles, cartoons, illustrated newspapers, public lectures and reports of conversation. We need to think about the availability of these works and their audiences: about reviews, translations, textbooks and elementary redactions.[16]

Secord suggests that the best way to make sense of Darwin and Darwinism is to start with how evolutionary ideas were communicated. This means treating Darwin, those who called themselves Darwinians, and those who considered themselves anti-Darwinians, as authors who were part of a commercial system of print, and not as disembodied thinkers.

The idea behind this collection is to try to follow out Secord's suggestion for Herbert Spencer, rather than Darwin, while incorporating more of the insights from global history of science. This means taking into account recent work on postcolonial and transnational studies, on go-betweens, and on centre-periphery issues. The chapters will focus on how Spencerian ideas were disseminated, communicated, and appropriated around the globe, primarily from the 1860s to the 1920s.[17] It is possible to separate Spencerism from Darwinism, although contemporaries sometimes conflated them. Spencer's evolutionary theory, cosmic in nature, and based as it was on Lamarckian modes of thought, was quite different from Darwin's. Piers Hale, for example, has pointed to the

14 Ibid., 34.

15 Ibid., 35–36.

16 Ibid., 37.

17 Recent Spencer studies tend to focus on Spencer as an important intellectual figure. The edited collection by Jones and Peel does contain a chapter on how Spencer was interpreted in France and Italy, but the subtitle of the book is "the intellectual legacy." Francis and Taylor are primarily interested in Spencer as scientific, political, sociological, ethical, and religious thinker. There is nothing on Spencer comparable to Glick or Numbers and Stenhouse's edited collections on Darwin.

importance of Darwin's debt to Thomas Malthus, in contrast to Spencer's sympathy with William Godwin. He argues that this had profound implications for their political views and those of their followers, separating liberal radicals into two camps.[18] Since Darwin's theory of natural selection did not become fully accepted until the 1920s, Spencer's cosmic, Lamarckian form of evolution still had scientific legitimacy into the early decades of the twentieth century. In the last few decades of the nineteenth century Spencer was read across the globe. For different reasons, different audiences were fascinated with Spencer's story of cosmic evolution.

However, ironically, in subjecting Spencer to modes of analysis developed by scholars working on the global history of science, it turns out that his evolutionary thought was not what necessarily appealed to many of his readers. For Spencer, the idea of evolution tied together all of his investigations into the disparate fields of knowledge covered in his system of synthetic philosophy. But Spencer's audiences had their own sense of his identity and importance. They often focused on a specific area of his thought – his work as psychologist, or sociologist, or philosopher, or religious theorist, or educationalist – and dispensed with the rest. They could even ignore the evolutionary structure underlying his system. Scholars, by focusing primarily on Spencer the evolutionist, have tended to concentrate their attention on a rather narrow view of him that has come out of Anglo-American appropriations of his thought. This view of Spencer does not match his earlier identities and reputation across the globe.

In addition to recovering the varied experiences of readers who encountered Spencer, this collection aims to understand how his ideas were transmitted *within* a variety of national contexts. This means examining the state of publishing, which varied considerably from country to country. Britain is generally considered to have been in the forefront of publishing in the nineteenth century. In the second quarter of the century the mode of communicating science underwent a significant transformation. Far more cheap works on science began to be produced. Starting in the 1830s, British publishers could reach out to a growing reading audience as new segments of the population became literate, especially members of the middle class and the wealthier working class. If the nature and size of the British reading audience was evolving in the first half of the nineteenth century, so was the technology of publishers. The new steam powered publishing technologies under development would eventually allow publishers to reach the rapidly growing reading audience through

18 Piers J. Hale, *Political Descent: Malthus, Mutualism, and the Politics of Evolution in Victorian England* (Chicago and London: University of Chicago Press, 2014), 68.

books and periodicals. The growth of the British railway system provided a better transportation system for publishers and the reduction of the cost of paper and the "taxes on knowledge," an intricate system created to prevent radicals from publishing, made it even more possible to produce cheap publications. James Secord, who has written extensively about Robert Chambers and the *Vestiges of the Natural History of Creation*, has argued persuasively that all of these developments combined to create a communication revolution. He asserts that this revolution represented the "greatest transformation in human communication since the Renaissance' that led to 'opening the floodgates to a vastly increased reading public."[19] Although Secord may be right about when and how this communications revolution took place in Britain, we cannot assume that the same process took place in other nations around the world. The way that Spencer's ideas were circulated through books and periodicals within individual nation-states very much depended on the state of publishing in each case.

Many of Spencer's readers did not read English, so translations are crucial to the story. This brings us to how Spencer's ideas were transmitted *between* national contexts – the transnational dimension of global Spencerism. Recent scholarship has emphasized the important role of go-betweens in the construction of the modern world. Brokers, missionaries, spies, entrepreneurs, messengers, and translators played, as Shaffer, Roberts, Raj and Delbourgo declare, a "crucial role in making sustained encounter and interaction across different cultures possible throughout history." All go-betweens mediated between disparate cultures "by being able to translate between them."[20] This is especially true of translators, of course. We have tried to include detailed information on the translations of Spencer's work that were available to different audiences. Once we understand the crucial role of translators in establishing new networks of knowledge it becomes clear that a rigid distinction between center and periphery is profoundly misleading. As Sivasundarum has argued, each locality can potentially become the central node of its own circuit of information.[21] By paying close attention to the role of translators we can remap

19 James A. Secord, *Victorian: Sensation: The Extraordinary Publication, Reception, and Secret Authorship of* Vestiges of the Natural History of Creation (Chicago: University of Chicago Press, 2000), 2.

20 Simon Schaffer, Lissa Roberts, Kapil Raj, and James Delbourgo, "Introduction," in *The Brokered World: Go-Betweens and Global Intelligence, 1770–1820*, ed. Simon Schaffer, Lissa Roberts, Kapil Raj, and James Delbourgo (Sagamore Beach: Science History Publications, 2009), xi, xiv.

21 Sujit Sivasundarum, "Sciences and the Global: On Methods, Questions, and Theory," *Isis* 101, no. 1 (March 2010), 158.

the networks of knowledge, circulation, and exchange in last third of the nineteenth century.

Finally, understanding the global dissemination, communication, and appropriation of Spencer's work requires careful attention to the scientific, cultural, social, and political contexts in each region. There are a series of questions that can be raised at the outset about each national context. Did a well-defined scientific community exist that could evaluate Spencer's work and possibly disseminate it? Or did other groups within the intellectual elite, such as journalists or novelists, for example, take on this role? How did previous cultural traditions shape how Spencer was interpreted? What role did social and political discontinuity play in how Spencer was appropriated? Was it primarily liberals, who welcomed a break with the past, who looked to Spencer as providing authority for their plans for social and political reform? How did colonialism factor into the way Spencer was taken up in different social and political contexts? In the Middle East Spencer's emphasis on gradual progressive change appealed to the emerging elite of colonial intellectuals. Spencer's theories also provided an uncontroversial way for nineteenth-century Western science to be absorbed by Indian intellectuals.[22] Was this a common response to Spencer in areas that were part of a colonial empire?

There are twelve chapters and an afterword in this collection. The chapters are organized roughly by global region. The first four chapters deal with Russia, the Middle East, Japan and China. Gordin's chapter on the hostile appropriation of Spencer in Imperial Russia confronts us immediately with a seeming contradiction. Spencer was widely read in Russia, where a new, vibrant world of print existed in the wake of the Great Reforms. But his Russian readers reacted negatively to his ideas. Both radical populists like Mikhailovskii, and conservative liberals like Chicherin, treated Spencer as a straw man to be shred to pieces. Even if it seems that the Russian intelligentsia vehemently rejected Spencer, Gordin argues that his ideas became a template for alternate social visions. Gordin's point is that the process of appropriation is seldom straightforward.

Elshakry's study of Spencer's Arabic readers, focusing on the journal *Al-Muqtataf*, underscores the crucial role of periodicals. The editors of *Al-Muqtataf* presented Spencer as the preeminent philosopher of evolution who was more important than Comte, Mill, and even Darwin. Endorsing his attempt to reconcile science and religion helped them to avoid charges of materialism; his holism resonated with traditional Arabic philosophies; and they were attracted to his emphasis on gradual, progressive change. Since only *Education* and

22 Shruti Kapila, "The Enchantment of Science in India," *Isis* 101, no. 1 (March 2010), 131.

Philosophy of Style were translated into Arabic at this time, most readers came to Spencer through summaries of his ideas in journals like *Al-Muqtataf*. Elshakry argues that Spencer's "transcendental positivism" made him popular with readers of the early twentieth century as well. In sum, Spencer appealed to several generations of Arabic readers from the late nineteenth and early twentieth century because he offered a broad, synthetic social and natural philosophy of progress and of the cosmos.

In his chapter, Godart discusses how Spencer's ideas flourished in Japan in the 1870s and 1880s, after the Tokugawa Shogunate had been overthrown and the Meiji era had begun with the imperial restoration. In his desire to modernize the country rapidly, the Emperor reversed the older policy of closing Japan to the outside world. The result was the dramatic expansion of journals and newspapers, the adoption of new printing technologies, the formation of a new national, compulsory education system, and an increase in literacy rates. This led to a boom in translations of Spencer's works, though some intellectuals had already read him in English. Spencer appealed to Japanese liberals who desired a political system that embraced popular representation. But Spencer was also important in debates concerning psychology, biology, the position of women, education, and religion. Godart also analyses the "Spencer bust" from the 1890s onwards. In this period Spencer's evolutionary pattern from militant to industrial was belied by later historical events, which seemed to discredit Spencerian pacifism. Also, when capitalism and industry took off, Spencer's promised moral improvement did not take place.

Ke Zunke and Li Bin draw our attention to the fascinating two-part story of Spencer's appropriation in China in their chapter. In general, Spencer's educational theory interested Chinese intellectuals more than his relationship with Darwin and evolution. In the early 1880s Yen Yung Kiung, a Chinese minister of the American Episcopal Church in China and a professor of natural science, translated Spencer's education essays as part of his argument for educational reform. After the Sino-Japanese War in 1894, a publishing revolution occurred that quickly changed the character of Chinese culture. The explosion of journals had a major impact on how Spencer's ideas were appropriated. After 1894, Yan Fu, an important translator and writer, played a key role in the diffusion of Spencer's thought. He believed that Spencer's sociological and educational ideas showed China how it could survive in the future. Like Yen Yung Kiung, Yan Fu asserted that traditional education had to be reformed and that science had to be given a much larger role. Eventually, in 1912, Yan Fu's endorsement of Spencer's theories shaped educational reform and led to the development of a cadre of scientists in China. The Chinese story reminds us of the importance of translators and periodicals.

Francis's chapter on James, Royce, and Dewey signals the beginning of a section containing five studies of Spencer and the Americas. Francis focuses on the fight between two groups each representing themselves as the legitimate heirs to Spencer's synthesis of philosophy and science. One group, which included Fiske and Youmans, claimed to be the "orthodox" group and had the virtue of being recognized by Spencer. The other group, which included James, Royce and Dewey, were the revisionists, who rewrote Spencer's philosophy and psychology so that they would be metaphysically sound. Rather than just being hostile towards Spencer, as some scholars have claimed, the revisionists were held together by the common belief that evolutionary psychology posed the chief challenge for American philosophy and by their desire to rewrite Spencer's psychology in order to blunt its materialist qualities. But, as Francis points out, the relationship between the three reformers and Spencer was complicated. They were both critics and appropriators.

In my chapter I deal with the other group of American Spencerians, composed of Fiske and Youmans. They were among Spencer's closest friends in the United States. Initially a scientific lecturer and writer, and later an advisor to the Appleton publishing house, Youmans became Spencer's literary agent in the United States. The younger Fiske, lecturer, philosopher and historian, wrote a two-volume outline of Spencer's evolutionary system. Both of them first encountered Spencer during the tensions leading up to the American Civil War (1861–1865). To Youmans, Spencer's analysis of first principles in the cause of mental emancipation was exactly what was needed to heal a nation on the brink of destruction. For Fiske, Spencer's evolutionary theory could be applied to American history to understand that God's destiny for the United States was to become the world's leading national power. Both Fiske and Youmans believed that emphasizing the religious dimensions of Spencer's thought was the best way for him to win a large American readership.

Ruiz Gutiérrez, Noguera Solano, and Rodríguez Caso explore the appropriation of Spencer in Mexico during the presidency of Porfirio Díaz. They argue that Spencer's ideas were promoted primarily in the press and that they strengthened an already existent national scientific culture based on positivism. References to Spencer in the Mexican press appeared relatively late, in 1875. Though Spencer's books were available previously in French, it wasn't until 1894 that Spencer was translated into Spanish. Once again, a periodical plays an important role in the dissemination of Spencer's views. *La Libertad* (1878–1884), a liberal newspaper, looked to Spencer for inspiration. It contained articles arguing that it was crucial to know the laws of nature in order to establish the proper development of a social organism like Mexican society. Gutiérrez, Solano, and Caso point out that Justo Sierra, one of the founders of

La Libertad, should be considered the driving force behind Spencerian positivism in Mexico. From 1881 onwards Sierra was actively involved in the quest to found a free and secular, public university. This was finally achieved in 1910 when the National University of Mexico was inaugurated.

The story of Spencer's appropriation in Argentina, as told by Adriana Novoa, is quite different. Novoa presents a detailed account of the arc that goes from the quick adoption of Spencer as a thinker who perpetuated treasured European ideas on progress, morality, education, and the importance of civilization, to the dismissal of his work as a betrayal of true enlightened ideas. After the military triumph of the liberals in 1852, printing houses and bookstores were opened to signal the importance of a reading culture. The expansion of the printing business, which ran parallel to an education campaign that increased literacy by the 1870s, both amplified the interest in Darwin and Spencer's ideas. But Spencer's synthetic philosophy was used to correct the perceived materialism of Darwin's evolutionary theory. Spencer's adoption of Lamarckian ideas on inheritance, and his concept of moral progress, also appealed to those who wanted to stress the notion that humans could shape their evolutionary future through social and educational reform. In the 1880s the influence of Spencer's work was widely acknowledged. The importance of his work was connected to a growing number of science publications, which helped spread his ideas. But in the 1890s there was a growing discomfort with Spencer's work. In Argentina, the theory of natural selection became dominant among scientists. Moreover, Spencer was criticized for being too materialistic and too individualistic.

In late nineteenth-century Brazil, as in Argentina during the seventies and eighties, Spencer was preferred over Darwin. Heloisa Maria Bertol Domingues demonstrates that more of Spencer's books were translated than Darwin's. Moreover, Spencer's books appear in larger numbers in Brazilian libraries. Taking Silvia Romero and Capistrano de Abreu as the chief representatives of Spencerism, Domingues shows that both drew on Spencer to articulate a notion of national identity that resonated with their readers, based on an evolutionary interpretation of Brazilian history. The political context makes it clear why such appropriations of Spencer would have been attractive. It was a time of transition from an imperial regime to a republican one, which produced a crisis between Church and State. Later, when the social and political context had changed, the relevance of Spencer to Brazil was soundly rejected.

The final three chapters on Italy, Scandinava, and France take us back across the Atlantic to Europe. Spencer reached Italy late as the first translation, from *Education*, was published in 1876. In the wake of Italian unification in 1861, and the tensions between the Catholic Church and liberal elites, Spencer's writings became a useful resource for the latter. Paola Govoni's study of the "Spencer effect" relies on the metaphor of a two-wave pattern to describe how Spencer's theories were appropriated in Italy. The first wave of success occurred in the 1870s, when scientists supported Spencer's ideas. Later, in the 1880s, economists and politicians, who saw themselves as part of a liberal, positivist culture, also enthusiastically endorsed Spencer. Spencer was more widely read in Italy than Darwin. Politicians, economists, writers, journalists and members of the well educated public were attracted to Spencer's politically reassuring system of thought, to his utilitarianism, and to the way the concept of personal liberty was so central to his politics. After a decline in Spencer's reputation in the 1890s, a second wave of success took place from around 1900 to 1920, when lay audiences continued to read his books, though scholarly interest was still low. When fascism became influential, Spencer disappeared from the publishers' catalogues.

Hans Henrik Hjermitslev's chapter provides an examination of how Spencer was appropriated in Sweden, Denmark and Norway from 1870 to 1920. The reception of Spencer's ideas in Scandinavia, Hjermitslev argues, was characterized by remarkable heterogeneity due to the different national contexts even though all three countries witnessed a gradual modernization. The most important sites for understanding Spencer's appropriation were the national universities, where philosophy was compulsory for most students. But the philosophical traditions at these universities differed. In Denmark, where radical-liberal literary critic George Brandes was influential, philosophers with positivist sympathies paved the way for Spencer's ideas from the 1870s. Brandes's "Modern Breakthrough" advocated naturalism in literature, positivism in science and liberalism in politics. Declaring a clear break with tradition, Brandes's followers embraced Spencer and Darwin as providing the principal ideological material for their revolution. But in Norway and Sweden philosophy departments remained strongholds for idealistic and conservative trains of thought well into the twentieth century, preventing Spencer's philosophy from gaining ground in the universities. It was left to radical students and modernist authors to popularize Spencer's philosophy in Norway and Sweden.

In the final chapter Nathalie Richard discusses Spencer and France. At first, Richard asserts, Spencer was seen as an ally to the spiritualists. But after the founding of the Third Republic in 1870, Spencer was enlisted to stabilize the fragile new government. The political context of the time fostered and directed the reception of Spencer's philosophy. Republican supporters believed that Spencer's system supplied a new political philosophy based on the social and moral sciences that did not fall into materialism. A new generation of academics who began their careers under the Third Republic also became interested in

Spencer's *Principles of Psychology* and *Principles of Sociology*. French publishers like Baillière, an anticlerical republican, were important for Spencer's academic reception. So were translators, like Cazelles, a republican who translated Spencer for Baillière. All of Spencer's major books were translated into French by the end of the 1870s. The years 1880 to 1885 marked the peak of Spencer's influence in France. Spencer's name, even more than Darwin's, stood for British liberal and evolutionary thought. In passing the Freedom of the Press Act in 1881, the republicans encouraged the growth of new periodicals and books, which benefited Spencer's reputation. In the 1880s Spencer's name was introduced in crucial controversies over secularization, civic morality and educational reform. His philosophy was often referenced during parliamentary debates, the popular press, and in novels. Spencer's success faded in France by the end of the 1880s. A series of crises showed that civil rights, secularization and education reform were not sufficient to ensure the future of the Republic. Simultaneously, scientific and philosophical agendas had changed due to a revival of interest in spiritualism and the development of more specialized social sciences

The diversity of these stories about Spencer's appropriation around the world makes it clear that the older notion of "reception" is no longer adequate. Examining the process of appropriation brings to the fore the agency of writers who created new networks of knowledge that challenge the traditional notion of a center and a periphery. It also highlights the role of interpreters as go-betweens. By focusing on the relation of Spencer's readership, rather than Darwin's, to revolutions in communication within a global context we gain fresh insights into the circulation of evolutionary theory in the second half of the nineteenth century and the early twentieth century. But it also points to the way that aspects of Spencer's thought, divorced from his evolutionary theory, also appealed to a significant portion of his audience. There is much to learn from the largely untold story of Global Spencerism.

CHAPTER 1

"What a Go-a-Head People They Are!": The Hostile Appropriation of Herbert Spencer in Imperial Russia

Michael D. Gordin

There was no reception of Herbert Spencer in Imperial Russia. Certainly, the English sociologist, philosopher, and universal theoretician never set foot on lands governed by the Romanov Tsars, and he even refused his one chance to be received by Emperor Alexander II in May 1874 at the British Foreign Office, ostensibly because he disliked wearing formal dress.[2] Hence, no literal reception. But also no metaphorical reception in the sense used by intellectual historians and historians of science of yesteryear, a sense that has been largely superseded, with good reason. According to this older model, intellectual creations such as scientific theories were born in one place and then distributed around the globe like so many parasols, unchanged by the pressures of transit and unaffected by the beliefs and commitments of those who received them in Kuala Lumpur or San Francisco. The picture we now have is richer and more multifaceted: a scientific theory is not passively received; it is actively appropriated. That is, specific books are selected to be read or translated because of interests or beliefs of the consumers, and the way the theories are understood depends strongly on prior training and the intellectual practices by which they are adopted and adapted.[3] Herbert Spencer's work – or, rather, a subset of his

1 I would like to thank Bernie Lightman, Greg Radick, and Hans Hjermitslev for helpful comments on the framing of this project. Transliteration for all names and terms except common ones (like *intelligentsia*) follow the modified standard Library of Congress system. All unattributed translations are mine.

2 David Duncan, *Life and Letters of Herbert Spencer*, 2 vols (New York: D. Appleton, 1908), I: 246.

3 That this is true even within the same national context has been amply demonstrated, for example, in the case of Britain by James A. Secord, *Victorian Sensation: The Extraordinary Publication, Reception, and Secret Authorship of* Vestiges of the Natural History of Creation (Chicago: University of Chicago Press, 2000), and Andrew Warwick, *Masters of Theory: Cambridge and the Rise of Mathematical Physics* (Chicago: University of Chicago Press, 2003). The varieties of appropriation become even more complex and multifaceted when the text/theory/object in question travels beyond national and/or linguistic borders.

 | DOI 10.1163/9789004264007_003

work, for who could hope to encompass the entirety of that gargantuan corpus? – met no simple reception by the Russians.

Neither, however, was it simply appropriated. I do not mean that Spencer was not read by the *intelligentsia* of St. Petersburg, Moscow, and places further afield. On the contrary, Herbert Spencer was literally *everywhere* on the pages of the "thick journals" (*tolstye zhurnaly*) that comprised the dominant form of intellectual interchange in the late nineteenth-century Russian Empire, and essentially everything Spencer wrote was translated into Russian, often earlier and more extensively than in any other European or non-European language (see Table 1.1 at the end of this chapter). But Spencer's ubiquity did not necessarily imply endorsement, or even popularity. Rather, one of the most striking things about the engagement with Spencer in the 1860s and 1870s – the decades after his first translation into Russian, but before the advent of Marxism in Russian space unmanageably complicates the picture – is how negative, even hostile, those intellectual appropriations were.

In this essay, I use the case of Imperial Russian Herbert Spencer (an oxymoron of truly Spencerian grandeur) to explore the phenomenon I call "hostile appropriation." There are many ways of appropriating a new idea. The simplest, which one used to call "reception," is to endorse the ideas lock, stock, and barrel, and become a devout "Spencerian." This did not happen in Petersburg's dominion, even though *intelligentsia* practices could be enormously faddish, and there were times when you could not (metaphorically) swing a cat without hitting a Left Hegelian, a Right Hegelian, a Darwinist, and a Proudhonist. Nor am I merely referring to what we might think of as "negative appropriation," whereby individuals read Spencer and found him not to their tastes and set him aside, perhaps with an acerbic review or two. "Hostile appropriation" goes a step farther. In such cases, the new idea was not simply rejected, but the arguments for rejection became a template upon which an alternative vision was built. For many thinkers in Imperial Russia, Herbert Spencer was neither accepted nor rejected, but critiqued so thoroughly that he became the – sometimes unacknowledged – photographic negative of new intellectual visions. Spencer's work was so copious, and so amenable to different interpretations, that he was hostilely appropriated for diametrically opposite purposes. Here, after a brief description of Spencer's involvement with Imperial Russia and his initial translation into Russian,[4] I will focus on two quite different cases: radical

4 In this essay, I restrict the story to the Russian-language engagement with Herbert Spencer. Imperial Russia (much like the Soviet Union and the present-day Russian Federation) was a multilingual state, and there are surely interesting and divergent stories to be gained from examining the publication history and reactions in Polish, Yiddish, or other languages within the Empire.

populist Nikolai Konstantinovich Mikhailovskii (1842–1904) and conservative liberal Boris Nikolaevich Chicherin (1828–1904). Both used their scorched-earth attacks on Herbert Spencer as springboards to develop alternative visions for Russia's future, burying him in the process. Spencer's greatest legacy in this period of Russian history is to be found where he was most absent.

From Herbert to Gerbert

Herbert Spencer did not know much about Imperial Russia, but that did not prevent him from writing about it with great vigor. His sources were mostly indirect, drawing from scholarly literature or travelers' accounts (usually English or French), with some limited conversations with individuals who had traveled to Russia or (rarely) the occasional Russian acquaintance. (For example, he seems to have adopted without attribution many of the views of Vladimir Ivanovich Kovalevskii on the origins of private property in Russia.)[5] Russia entered most of his publications as a convenient contemporary example of a peasant society undergoing modernization and industrialization, a prominent topic of Western conversations about the country after the February 1861 Emancipation of the serfs. Russia, in Spencer's account, was an exemplar of a "civilized" polity but of the "militant" type, in some ways a fossil representing the past of other European states such as Prussia, preserved through the vagaries of history for first-hand observation by Western travelers.[6]

Politically, Spencer found the Russian autocracy distasteful, and it is hard to avoid the suspicion that some of his reluctance to attend on the Tsar in 1874 stemmed from these views. His unease, to be sure, reflected his general views about political structure enunciated in many works, but he also opposed specific policies and lent his name to their reversal. Toward the turn of the century, for example, he opposed the intensification of Imperial control over the Grand Duchy of Finland (since 1809 an autonomous state within the Russian

5 Alexander Vucinich, *Social Thought in Tsarist Russia: The Quest for a General Science of Society, 1861–1917* (Chicago: University of Chicago Press, 1976), 154. On the complex history of property rights in late Imperial Russia, see Ekaterina Pravilova, *A Public Empire: Property and the Quest for the Common Good in Imperial Russia* (Princeton: Princeton University Press, 2014).

6 See, for example, Herbert Spencer, *The Principles of Sociology*, 3 vols (London: Williams and Norgate, 1882–1898), III: 433, 458, 481, 483, and many other references in this crucial work. Spencer's view of Russia has generated passing comments from several later scholars, as in Hugh Elliot, *Herbert Spencer* (New York: Henry Holt and Company, 1917), 161; and David Wiltshire, *The Social and Political Thought of Herbert Spencer* (Oxford: Oxford University Press, 1978), 210.

Empire).[7] More controversially, he noted in a letter to G.J. Holyoke in September 1894, "[a]t the instigation of some Jewish periodical I expressed my detestation of the persecutions in Russia, thoughtlessly supposing that my letter would have no further circulation."[8] As might be expected, the letter was widely distributed and incited angry objections from loyalists, sparking yet another wave of anti-Spencerian comment in Russia.

Not that Spencer really knew about the *earlier* waves. Spencer's major awareness of his own status as a thinker within Imperial Russia drew from his knowledge of his shockingly persistent translatability into the Russian language. The enthusiasm of Russian translators for Spencer's copious output arrived to the author as "an agreeable surprise," as he wrote in one of the last letters to his father, dated 27 March 1866. The gratifying news of a new edition in a foreign clime was spiced by the frisson of hearing that the translator/editor, Nikolai L'vovich Tiblen (1825-after 1869), had been arrested due to his involvement with the work. (Tiblen asked that Spencer put a paragraph about it in the *Times* if the case proceeded. The charges were dismissed.)[9] Spencer understood the legal case as attaching to his *Essays*, specifically to his refutation of the divine right of monarchs, but actually Tiblen's difficulties were more of the mundane sort: financial arrears and bankruptcy, culminating in the hapless entrepreneur fleeing abroad.[10] (Tiblen was instrumental in distributing a good deal of British intellectual life among the Russians; he had organized translations of Thomas Babington Macaulay in 1863 and Henry Thomas Buckle the following year.) Occasionally, stereotypical stories of Russian repression reached Spencer's receptive ears. The *Times* of 28 July 1903 reported that a student had been charged with holding socialist views during his interview for admission to university based on his having been observed, when he was fifteen, with a copy of Spencer's *Sociology*.[11] This might have happened, of course, but it is difficult to ascertain the reliability of such reports. Their historical value is more to illustrate Spencer's imagined understanding of the vitality of his thought in an autocratic society. (Despite his objections to socialism,

7 Spencer to James Sully, 10 December 1899, reproduced in Duncan, *Life and Letters of Herbert Spencer*, II: 152.

8 Spencer to G.J. Holyoke, 17 September 1894, reproduced in Duncan, *Life and Letters of Herbert Spencer*, II: 66.

9 Duncan, *Life and Letters of Herbert Spencer*, I: 169.

10 James H. Billington, *Mikhailovsky and Russian Populism* (New York: Oxford University Press, 1958), 189n. The conventional story, drawn from Spencer's writings, is understandably widespread in the Western historiography, e.g., in Wiltshire, *The Social and Political Thought of Herbert Spencer*, 77.

11 Duncan, *Life and Letters of Herbert Spencer*, II: 206.

Spencer did not comment on the political allegiances of the ostensible martyrs to Spencerianism.)

When Russians read Spencer, they read him in the context of a vibrant and flourishing world of print, one with characteristics distinct from those of other European polities. Rapid discontinuities hit Russian printers and readers in the 1860s, due to the series of administrative measures known as the Great Reforms. With the Emancipation of the serfs in 1861, both administrative decrees and primary schooling extended into the countryside, producing a small but significant expansion of readership. University reforms in 1863 enabled many individuals from lower ranks of the hierarchical Russian social order to participate in higher education, expanding the pool of readers, authors, and (importantly) translators of foreign works. Changes in the civil service, military service, and governance also promoted literacy and expanded readership; coupled with an easing of the censorship regime from preliminary to punitive – that is, allowing periodicals to publish articles without pre-clearance and fining promoters of seditious or libelous material only afterward – in 1865, this generated a sudden expansion of the number of readers and the kinds of books and especially periodicals that could be directed at them.[12]

The periodicals demand particular attention, because the great majority of the Russian readership encountered intellectual life through them. The primary vehicle here were the "thick journals," fat monthly periodicals (usually circa 300 pages an issue) which published political commentary, book reviews, historical essays, philosophy articles, and serialized fiction. (With the exception of Ivan Turgenev's *Fathers and Sons* and Nikolai Gogol's *Dead Souls*, every major novel of the century was first published in installments in thick journals.) People subscribed to (often several) thick journals based on aesthetic preferences and political leanings, and in the process were routinely exposed to a wide array of genres and ideas across a gigantic country with very limited booksellers but a relatively robust postal system. The system continued to flourish after 1865 while facing stiffer competition from an expanding newspaper market. Editors' need for content to fill each issue summoned into being a Grub Street of students, radicals, and moonlighting civil servants who wrote reviews or translated Western European fashions into Russian, and discussions of contemporary science were certainly included. Building on the demand for new intellectual stimulation triggered by thick journals, publishers generated books

12 On these transformations, see especially Jeffrey Brooks, *When Russia Learned to Read: Literacy and Popular Literature, 1861–1917* (Princeton: Princeton University Press, 1985); and Charles A. Ruud, *Fighting Words: Imperial Censorship and the Russian Press, 1804–1906* (Toronto: University of Toronto Press, 2009 [1982]).

from the content that first appeared in them, and then were reviewed in the same, creating a tight-knit culture of reading and writing.[13]

Tiblen's edition appeared as stand-alone volumes, but it had been primed by mentions of Spencer in the thick journals, and in the end represented only the beginning of a wave of translations of almost everything Spencer wrote. Tiblen intended to publish seven volumes of Spencer's collected works, appearing serially in 1866, but only the first two appeared (a common enough outcome for such ventures, especially for the unlucky Tiblen). The contents indicate both the diversity of Spencer's oeuvre as well as the chaos that appeared before his Russian readers: the first volume contained Spencer's shorter essays ("Progress: Its Law and Cause," 1857; "The Philosophy of Style," 1857; "The Ultimate Laws of Physiology," 1857; "The Origin and Function of Music," 1857; "Use and Beauty," 1854; "The Development Hypothesis," 1852–4; "The Sources of Architectural Types," 1852–4; "The Theory of Tears and Laughter," 1852; "Gracefulness," 1852–4; "The Valuation of Evidence," 1853; "Personal Beauty," 1852–4; "The Use of Anthropomorphism," 1852; "Railway Morals and Railway Policy," 1854; "The Genesis of Science," 1854; "Manners and Fashion," 1854; and "The Social Organism," 1860); the second volume included "The Physiology of Laughter," 1860; "Bain on the Emotions and the Will," 1860; "The Morals of Trade," 1860; "State-Tamperings with Money and Banks," 1858; "Prison Ethics," 1860; and so on. Also available in Russian were *First Principles*, 1862; *The Principles of Biology*, 1864–7, and *Education: Intellectual, Moral and Physical*, 1861 (the translation of which included *The Nebular Hypothesis* and *Illogical Geology*).[14] In 1864, *The Classification of Sciences* and *Reasons for Dissenting from the Philosophy of M. Comte* had been published as separate pamphlets. Spencermania continued into the 1870s. As Spencer wrote to his American acolyte Edward L. Youmans on 28 December 1875: "A professor at Kiev" – Spencer couldn't quite make out the name – "proposes, in conjunction with his colleagues and pupils, to translate the *Descriptive Sociology*. He tells me, to my surprise, that all my books have now been translated into Russian with the exception of the *Descriptive Sociology*, which will thus soon be added to the list."[15] Two years later, having apparently

---

13 On these journals, see Deborah A. Martinsen, ed., *Literary Journals in Imperial Russia* (Cambridge: Cambridge University Press, 1997). On seriality and literature in the Russian context, see the classic study by William Mills Todd III, *Fiction and Society in the Age of Pushkin: Ideology, Institutions, and Narrative* (Cambridge, MA: Harvard University Press, 1986).

14 Gerbert Spenser, *Sobranie sochinenii v 7-mi tomakh. Nauchnye, politicheskie i filosofskie opyty*, t. 1, perevod pod redaktsiei N.L. Tiblena (St Petersburg: Tip. N. Tiblena i ko., 1866).

15 Reproduced in Herbert Spencer, *An Autobiography*, 2 vols (New York: D. Appleton, 1904), II: 338.

completely forgotten about the offer, he wrote again to Youmans with news that the translation had arrived. It was all very impressive: "What a go-a-head people they are!"[16]

And so Herbert Spencer became widely known as Gerbert Spenser (there is no "h" in Russian). Of course, no book ever arrives shorn of context, as readers encounter the work in a penumbra of other books they consider to be related, in addition to the cultural milieu which shapes how they interpret the book. Translations, being books, are no different, and "Gerbert Spenser" was read within a matrix of writings which decisively affected his hostile appropriation. For while I maintain that Spencer was almost universally cited and interpreted as an (only slightly exaggerated) strawman, it matters very much that these Russian thinkers picked *Spencer* and not someone else as the figure against which they built up their own narratives about politics and society.

Two characteristics of Spencer facilitated this appropriation. The first was the unassailable fact that he was English. English thought was decisively *à la mode* among the *intelligentsia* and bureaucrats in the wake of the Russian defeat by the British in the Crimean War (1854–1856), the proximate impetus for the Great Reforms, and Spencer also fit nicely into pre-existing excitement about Henry Buckle's *History of Civilization in England* and the growing hubbub in the 1860s about the evolutionary thought of Charles Darwin. As has been well described by historian Daniel Todes, Darwin was read by Russians voraciously but with moderate skepticism, since the overly fecund tropics that provided the bulk of his evidence for the Malthusian trap of natural selection seemed ill-suited to the geography of the underpopulated steppes.[17] The point, however, is the deep *engagement* with Darwin's ideas, as well as the respectful appropriation; Spencer would be different. Mention of Darwin brings up the second feature of Spencer's thought that picked him out for Russian commentators: his thoroughgoing articulation of a theory of "progress." As mentioned earlier, in 1865, as part of the Great Reforms, Russian censorship was slightly eased, but not to the point where thoroughgoing political reform could be openly discussed. As a result, conversations about such topics took place through what has been termed "Aesopian language," whereby coded metaphors

16 Spencer to Youmans, reproduced in Spencer, *An Autobiography*, II: 362.

17 On interest in the English, see Esther Kingston-Mann, "In Search of the True West: Western Economic Models and Russian Rural Development," *Journal of Historical Sociology* 3 (1990): 23–49. Darwin's translation into Russian and the Russian expansions on his theory of natural selection have been well studied by historians of science. See Daniel P. Todes, *Darwin without Malthus: The Struggle for Existence in Russian Evolutionary Thought* (New York: Oxford University Press, 1989); and Alexander Vucinich, *Darwin in Russian Thought* (Berkeley: University of California Press, 1988).

and allusions allowed censors (who obviously understood the code) to authorize interesting works while protecting themselves from repercussions should the approved texts later be interpreted by police as seditious. "Progress" had, since the longstanding Russian engagement with G.F.W. Hegel's thought (dating from at least the 1830s), served as precisely such an Aesopian code word, soon joined by "evolution." Spencer – sometimes alongside Darwin, and sometimes in his stead – represented for the growing but still largely elite and urbanized reading public an excellent foil to frame discussions of the most sensitive topics of the day.

One final point about how the Russians read Spencer before turning to our two examples: Spencer was uniformly interpreted within the positivist tradition, although the man himself vehemently objected to this characterization when it cropped up in Western contexts. For Russians, "positivism" meant, almost interchangeably, Auguste Comte, John Stuart Mill, and Herbert Spencer.[18] The collapsing of Spencer especially into Comte was exacerbated by the fact that, during the Imperial period, the Tiblen edition of Spencer's earlier works and the later translations of the British thinker were the main access to any stadial theory of history based on purportedly scientific observations. Although Spencer protested that his evolutionism was distinct from Comte's three phases (theological, metaphysical, positive), the second-hand reports of Comte arrived in Russia filtered through a public that had already read the more widely available Spencer (Comte's *Cours de philosophie positive* was not translated until 1899 – probably a consequence of the fact that Russian *intelligenty* were more able to handle the French original than anything written in English).[19] Officially, the Holy Synod, Russia's established Orthodox Church, opposed positivism, understood as a manifestation of the European crisis of materialism, although it is likely that ecclesiastical objections intensified interest in Spencer rather than warning readers away.[20]

18 N.F. Utkina, *Pozitivizm, antropologicheskii materializm i nauka v Rossii* (*vtoraia polovina XIX veka*) (Moscow: Nauka, 1975), 12. Sometimes, "Darwin" was added to the triumvirate, but this was relatively rare. Robert E. MacMaster, "In the Russian Manner: Thought as Incipient Action," in Hugh Maclean, Martin E. Malia, and George Fischer, eds., *Russian Thought and Politics* (Cambridge, MA: Harvard University Press, 1957): 299.

19 Utkina, *Pozitivizm, antropologicheskii materializm i nauka v Rossii*, 7. On Comte's impact in Russia, see James H. Billington, "The Intelligentsia and the Religion of Humanity," *American Historical Review* 65 (1960): 807–821; and Esther Kingston-Mann, "In the Light and Shadow of the West: The Impact of Western Economics in Pre-Emancipation Russia," *Comparative Studies in Society and History* 33 (1991): 100–101.

20 P.S. Shkurinov, *Pozitivizm v Rossii XIX veka* (Moscow: Izd. Moskovskogo universiteta, 1980), 162.

And so Spencer was read, in almost all ideological circles of the tiny stratum of educated Russians, as being both a representative of English thought and an apostle of one variant of "progress," and therefore, necessarily, a positivist. Yet the striking fact remains that, according to historian of Russian science Alexander Vucinich, "The Spencerian theory of the survival of the fittest as a key mechanism of social evolution did not have a single supporter among the leading Russian sociologists."[21] It is not enough to point to the absence of explicit advocates; one must go further and explore the complex topography of the rejection. As one Soviet intellectual historian observed: "[I]t is not easy to judge about [Spencer's] real influence, about how far his readers went in the pursuit of his fundamental ideas, whether they shared these ideas or selected from his work whatever corresponded to their own moods."[22] Such a selective reading is the essence of all appropriation, given in this instance an additional wrinkle in that Spencer – unlike Comte and especially unlike Darwin – was predominantly read in the negative-template mode.

Spencer among the Populists

Although essentially unknown to intellectual historians outside of specialists on late Imperial Russia, Nikolai Mikhailovskii was one of the most important and consequential social theorists of the late nineteenth century (see Fig. 1.1). He set out in St. Petersburg to study to be a mining engineer in the early 1860s, only to be swept up by the intellectual ferment surrounding the atmosphere of the Great Reforms. Drawn to political and social commentary, he began work as a bookbinder, which cemented connections with the publishing world, enabled him to read more widely in French and German, and connected him as a tutor to wealthy young foreigners. He soon became a regular contributor to, and editor of, several thick journals, especially *Notes of the Fatherland* (*Otechestvennye Zapiski*) and the *Northern Courier* (*Severnii Vestnik*).[23]

Mikhailovskii's reputation as a leading theorist of the radical movement known as the *narodniki* – often translated as "Populists" in English from the Russian word *narod*, meaning "people," but not to be confused with American or Western European electoral populism – was made by a series of essays he composed in the late 1860s and early 1870s to articulate his "subjective method" in sociology and a voluntarist conception of political transformation, including

21 Vucinich, *Social Thought in Tsarist Russia*, 236.

22 Utkina, *Pozitivizm, antropologicheskii materializm i nauka v Rossii*, 52.

23 Billington, *Mikhailovsky and Russian Populism*, 20.

FIGURE 1.1
Photograph of Nikolai K. Mikhailovskii.
SOURCE: A.J. SACK, *THE BIRTH OF RUSSIAN DEMOCRACY* (NEW YORK: RUSSIAN INFORMATION BUREAU, 1918), 79.

"Darwin's Theory and Social Science," "The Analogical Method in Social Science," and "Louis Blanc's Philosophy of History."[24] The first of these pieces, published in 1869 in *Fatherland Notes*, was entitled "What Is Progress?" It was a massive and detailed review of volume one of the Tiblen edition of Herbert Spencer's essays.

Mikhailovskii aggressively and uncompromisingly panned it, but in the process "made the name of Herbert Spencer a household word in the Russian intellectual community."[25] Mikhailovskii began with an exposition and then demolition of Spencer's aesthetic theory – a seemingly innocuous choice in other contexts, but not in Russia, where literary critic Vissarion Belinskii (1811–1848) had transformed aesthetic criticism into a powerful Aesopian vehicle for political commentary. From this example, Mikhailovskii set out the task of his enormous essay:

24 Vucinich, *Social Thought in Tsarist Russia*, 21.
25 Vucinich, *Social Thought in Tsarist Russia*, 44.

> The reader sees that the entirety of Spencer's argumentation does not sustain even the slightest criticism, and that here one cannot even talk about the evaluation of his theoretical principles, because their logical supports give way all by themselves. And if, in the ten fascicles of his collected works which have so far appeared in Russian, there was nothing besides such proofs and propositions, then I, it stands to reason, would not think it necessary to speak about him. But I intend to speak, and to speak at length.[26]

Spencer was worth attacking because, despite his misconceptions, he was "a mind of great caliber; one of those all-encompassing synthetic minds, which from time to time introduce a spirit of unity and life in the disparate facts achieved by several generations of less gifted and even completely ungifted toilers of science."[27] Mikhailovskii then proceeded to outline Spencer's sociological theory, "the Achilles heel of his philosophy,"[28] with a lengthy detour through an exposition of Comte's framework (as well as Mill, Darwin, and Buckle), in order to articulate his "subjective sociology," whose "negative frame of reference was provided by the sociological and historical conceptions of Spencer, first of all his theory of progress," as noted by the most prominent intellectual historian of Russian Populism.[29]

Mikhailovskii's engagement with Spencer stemmed from a concern he had with Darwinism's implications for the boundary between biology and sociology. He appreciated that Darwin's picture of biological evolution required randomness; at the same time, he thought humans' capacity to develop intentional plans – and in particular the goal of *progress* in those plans – meant that non-directed evolution could not account for social change. Therefore, he felt compelled to engage with Spencer's pre-Darwinian articulations of purposive evolution, and especially Spencer's dominant metaphor of society as an organism – which represented "progress" as increasing levels of social differentiation – and posit an alternative. Out of his detailed and sensitive reading of Spencer, Mikhailovskii argued that the social sciences functioned on a completely

---

26 N.K. Mikhailovskii, "Chto takoe progress?," in N.K. Mikhailovskii, *Polnoe sobranie sochinenii*, volume 1 (St. Petersburg: Obshchestvennaia pol'za, 1911): 5.

27 Mikhailovskii, "Chto takoe progress?," 6–7.

28 Mikhailovskii, "Chto takoe progress?," 16.

29 Andrzej Walicki, *The Controversy over Capitalism: Studies in the Social Philosophy of the Russian Populists* (Notre Dame, IN: University of Notre Dame Press, 1989 [1969]), 46. See also Billington, *Mikhailovsky and Russian Populism*, 70; and E.E. Mikhailova, *Russkii pozitivizm na rubezhe XIX-XX vekov: Problema sotsiokul'turnogo razvitiia* (Moscow: Izd. MGOU, 2004), 31.

different plane from the natural sciences, and one could not be reduced to the other. (Later, the Marxist philosophy of science known as dialectical materialism would argue something similar from different principles.) Further, whereas Spencer thought sociologists should study social change with the same level of disinterestedness as physicists study molecules, Mikhailovskii contended that moral values could not be dissociated from facts, the social sciences were necessarily incapable of being approached objectively because of the omnipresence of unconscious emotions and conscious ideologies, and that individuals could override the putative "laws of history."[30]

These notions formed the intellectual core of Populism, which later became associated with political terrorism and assassinations in late Imperial Russia. Its roots, however, lay in Mikhailovskii's school of social criticism designed to move beyond mechanistic Western European sociology and to harness the tremendous energy of peasants' willpower (*volia*). Along with other important thinkers, like Nikolai Chernyshevskii and Petr Tkachev, Mikhailovskii maintained that the social conditions of Russia, especially the peasant commune, endowed Russia with the sociological infrastructure to leapfrog over capitalism (with all its horrors of proletarianization and mass impoverishment) directly into socialism.[31] If progress were defined as greater social differentiation (*à la* Spencer's evolutionary organismal analogy), then exploitative capitalism was inevitable and specialization of labor was to be welcomed. If progress could be defined through subjective value judgments that emphasized the greater *integration* of society, rather than its atomization, then a harmonious social utopia might be achieved without a capitalist phase. The radicals, in other words, used the critique of Spencer in order to delink Darwin from social change, and thus maintain their progressive credentials as scientifically-minded Darwinists without having to compromise on their voluntarist political program. An attack on Spencer could serve multiple different ends.

30 See the succinct exposition in Walicki, *The Controversy over Capitalism*, 31–32; Vucinich, *Social Thought in Tsarist Russia*, 23–24.

31 On Tkachev's more qualified critique of Spencer, see Walicki, *The Controversy over Capitalism*, 40, 246. Chernyshevskii – who translated some of Spencer in order to support himself – was violently repelled by objective sociology. See Franco Venturi, *Roots of Revolution: A History of Populist and Socialist Movements in Nineteenth-Century Russia*, tr. Francis Haskell (Chicago: University of Chicago Press, 1960 [1952]), 186.

Mr. Chicherin Objects

One of those ends, in the hands of Boris Chicherin (see Fig. 1.2), was the *anti*-revolutionary goal of liberal change through the conservative channels of the extant Russian state. Born in 1828, Chicherin was a generation older than Mikhailovskii and – in the classic Turgenevesque conflict between "fathers" and "children" – belonged without question to a cohort raised on debates about Hegelian necessity and the laws of history rather than the willful scientism of these "men of the sixties" (*shestidesiatniki*).[32] Chicherin is one of those Zelig-like figures who shows up in almost every episode in the intellectual history of Imperial Russia, and he published broadly in an astonishing array of fields, ranging from theology to chemistry (he postulated a mathematized formulation of the periodic system of chemical elements). Here are some of his publications: *State Institutions of Russia in the 17th Century* (1856), *On Popular Government* (1866), *History of Political Doctrines* (5 volumes, 1869–1902), *Science and Religion* (1879), *Mysticism and Science* (1880), *Property and the State* (2 volumes, 1882–1883), *The Positive Philosophy and the Unity of Science* (1892), *The Foundations of Logic and Metaphysics* (1894). It is a range reminiscent of someone like Herbert Spencer. The connection is not, I insist, accidental, although it has long gone unnoticed as one of the most sustained instances of hostile appropriation of Spencer in Imperial Russia.

Chicherin had moved to Moscow in 1844 and quickly became part of a classic circle of "Westernizers," thinkers who saw the future of Russia lying in its transformation into a Western European state as opposed to the so-called "Slavophiles" – both monikers were originally terms of abuse adopted gleefully by the intended targets – who advocated a future built out of particularly Slavic traditions of the Russian peasantry. Despite their mutual sniping, both groups focused their intellectual efforts on the same goal: the end of serfdom. For the Westernizers, serfdom held Russia back in a medieval stupor; for the Slavophiles, it oppressed the peasants and dampened the invigorating Slavic spirit. Chicherin drank deeply of the Hegelian brew of the epoch, and quickly became known as a pioneering legal thinker and historian.[33]

32 See, for example, the hagiographic obituary pamphlet by Evgenii N. Trubetskoi, *Boris Nikolaevich Chicherin kak pobornik pravdy v pravie* (St. Petersburg: Senatskaia tip., 1904). More biographical details can be found in Georges Gurvitch, "Chicherin, Boris Nikolaeyevich," in Edwin R.A. Seligman, ed., *Encyclopedia of the Social Sciences*, vol. 3 (New York: Macmillan, 1930): 372.

33 On the significance of Hegel, see Aileen Kelly, "'What Is Real Is Rational': The Political Philosophy of B.N. Chicherin," *Cahiers du monde russe et soviétique* 18 (1977): 195–222; and Andrzej Walicki, *A History of Russian Thought: From the Enlightenment to Marxism*, tr.

FIGURE 1.2 *Portrait of Boris N. Chicherin by Vladimir O. Sherwood, late nineteenth century.*
SOURCE: <http://commons.wikimedia.org/wiki/File:Sherwood_PtChicherina.jpg>.

His historical work has attracted the greatest scholarly attention, especially his 1858 thesis that the peasant commune (*obshchina*), which both Slavophiles and later Populists held to be the seed of a renovated Russia, was not an autochthonous Slavonic institution but was rather imposed by the Muscovite state in the seventeenth century in an effort to more efficiently collect taxes.[34] Immensely controversial at the time, Chicherin's general thesis is mostly accepted today, and it remains a landmark in the intellectual development of the "State School" of Russian historiography (so dubbed because it attributes most historical developments to top-down actions of the state). Chicherin continued his scholarship at the law faculty of Moscow University, eventually resigning with several colleagues in 1868 in a dispute over state intervention against academic freedom. (He also served briefly as mayor of Moscow in 1882, resigning for similar reasons.) His early career as a self-professed "liberal" and interlocutor of such giants of Russian thought as Alexander Herzen has focused commentators on his writings of the 1850s and early 1860s, when he theorized the legal structures of peasant Emancipation (although without participating in it directly, opting instead to spend the early years of the Great Reforms on a European tour). As a liberal (however conservative in approach), Chicherin stood almost alone for a political position of moderate reform in an intellectual hothouse dominated by revolutionaries and reactionaries.[35] It was a lonely spot.

Hilda Andres-Rusiecka (Stanford: Stanford University Press, 1979 [1973]), 150. On Chicherin's mentor in Hegelianism, see Priscilla Reynolds Roosevelt, *Apostle of Russian Liberalism: Timofei Granovsky* (Newtonville, MA: Oriental Research Partners, 1986); and E.V. Mironov, *Filosofskie i sotsial'no-politicheskie vozzreniia B.N. Chicherina* (Cheliabinsk: Izd. IuUrGU, 2008).

34 Gary M. Hamburg, *Boris Chicherin and Early Russian Liberalism, 1828–1866* (Stanford: Stanford University Press, 1992); Hamburg, "Peasant Emancipation and Russian Social Thought: The Case of Boris N. Chicherin," *Slavic Review* 50 (1991): 890–904; David Saunders, "The Political Ideas of Russian Historians," *Historical Journal* 27 (1984): 757–771; Walicki, *The Controversy over Capitalism*, 104; and G.S. Krinitskaia, *Istoricheskaia kontseptsiia B.N. Chicherina* (Tomsk: Tomskii gosudarstvennyi universitet, 2001).

35 On the difficulties of being "liberal" in the Russian context, see Isaiah Berlin, "Fathers and Children: Turgenev and the Liberal Predicament," in Berlin, *Russian Thinkers* (New York: Penguin, 1994): 264–305; Marc Raeff, "Some Reflections on Russian Liberalism," *Russian Review* 18 (1959): 221. On Chicherin as an archetypal liberal, see Sumner Benson, "The Conservative Liberalism of Boris Chicherin," *Forschungen zur osteuropäischen Geschichte* 21 (1975): 17–114; Leonard Schapiro, *Russian Studies*, ed. Ellen Dahrendorf (New York: Elisabeth Sifton Books, 1986), 59; and Schapiro, *Rationalism and Nationalism in Nineteenth-Century Political Thought* (New Haven: Yale University Press, 1967), 89–97.

What the scholarly focus on the "early Chicherin" ignores is his persistent, decades-long engagement with the works of Herbert Spencer, perhaps the second greatest vehicle for the development of his thought next to Hegel – although this time as a photographic negative.[36] Starting in the 1860s, Chicherin began to refer to Spencer in his many publications, but always as a strawman, an alternative view that he would proceed to demolish before erecting his own picture.[37] For example, in his 1879 book *Science and Religion*, Chicherin devoted a crucial segment to a systematic dismantling of aspects of Spencer's project of synthesizing the world under a principle of the unity of force, dismissing such an approach as devolving into "pure materialism" and resulting in "a scandalous misuse of the experimental method."[38] Chicherin also believed Spencer's notion of progress through differentiation justified a caste-bound society as the highest form of human development, concluding that "[b]oth logic and experience equally come out as losers in this theory, which presents a brave attempt to unify the world's phenomena without any firm point of support, both in terms of speculation and in its proposed data."[39]

That work was focused on the intellectual and theological bankruptcy of positivism, considered from an idealist Hegelian standpoint, and it is not surprising that Spencer would be a target for Chicherin's ire. More intriguing is that even in the area of political science and sociology, Chicherin – no friend of the Populists' subjectivism – point-by-point constructed his own interpretation against Herbert Spencer's. "Only by means of the most strained analogies and omission of the most essential indications," he noted in his 1882–3 textbook *Property and the State*, "is it possible to forcibly bring together both sorts of laws into one category. That is precisely what we find in Spencer."[40] Chicherin devoted most of his critique to Spencer's 1857 article "Progress: Its Law and Cause." After articulating his opponent's views, in a classic instance of hostile appropriation, he proceeded to build up his own in mirror-image: "Such is Spencer's theory. Upon a superficial glance it appears rather well-rounded and consistent. But if we look at it a little more closely, we see that it contains nothing besides entirely arbitrarily selected facts and conclusions, in which one

36 For example, V.D. Zor'kin mentions Spencer only once in passing in his biography of Chicherin, in the context of Mill, Adam Smith, and Wilhelm von Humboldt. Zor'kin, *Chicherin* (Moscow: Iuridicheskaia literatura, 1984), 41.

37 There was, for example, no reference to Spencer in Chicherin's earlier essays, collected in *Neskol'ko sovremennykh voprosov* (Moscow: GPIB, 2002).

38 B.N. Chicherin, *Nauka i religiia* (Moscow: Tip. Martynova i Ko., 1879), 494–495.

39 Chicherin, *Nauka i religiia*, 497.

40 B.N. Chicherin, *Sobstvennost' i gosudarstvo* (St. Petersburg: Izd. RKhGA, 2005 [1882–3]), 301.

can find everything except logic. Let us follow it step by step."[41] We end up with a perspective that rejected many of the foundational postulates of English liberalism in favor of greater state involvement in managing the economy and the polity – yet still "liberal" for its Russian context. A similar account also appears in Chicherin's 1900 textbook *Philosophy of Law.* (Textbooks are an excellent place to observe the kind of hostile appropriation I am describing, since these are systematic expositions of an entire field which necessarily have to introduce countervailing views as well as the author's own.) Here, Spencer was addressed not through one sustained engagement, but by repeated jabs, in part a response to the increased translation of Spencer's work in the 1880s and 1890s within the proliferating universe of professional legal journals (wherein the influence of Mill and Comte was markedly downplayed, and that of the neo-Kantians emphasized).[42]

Conclusion

As can be seen from this brief survey, while Chicherin could not disagree more vocally with the specific projects of Mikhailovskii and his ilk, the two camps were of one accord in wanting to be as anti-Spencerian as possible. Given the copiousness of Spencer's oeuvre and his occasional tendency to stake out inconsistent positions, being "non-Spencer" could mean a wide variety of things. That said, the emphasis on how much some Russians hated Spencer might make it sound like he found no sympathetic ears, and I do not want to give that impression. Nonetheless, much like his detractors, those rare supporters of Spencer were highly selective in what they read and praised, and even then their approbation was hardly sufficient to merit calling them "Spencerians."

The only significant work of Spencer's which seems to have met with a consistent degree of admiration in Russia was his 1861 *Education: Intellectual, Moral, and Physical* (which appeared in a staggering *five* different translations before 1900; see Table 1.1). This was hardly his most important work from either a sociological or evolutionary point of view, but writers as disparate as Fedor Dostoevskii and Anton Chekhov wrote positive assessments in their notes concerning Spencer's advocacy of co-education and his opposition to rote

41 Chicherin, *Sobstvennost' i gosudarstvo*, 751.

42 E.g., B.N. Chicherin, *Filosofiia prava* (St. Petersburg: Nauka, 1998 [1900]), 64–65, 72. On the legal context, see Shkurinov, *Pozitivizm v Rossii XIX veka*, 329–330.

learning.[43] (Leo Tolstoy, on the other hand, remained resolutely anti-Spencerian.)[44] Spencer's thought had other brief renaissances as well. In 1885 Aleksandr S. Lappo-Danilevskii, soon to become a leading historian, devoted a whole year along with the Scientific Literary Society at St. Petersburg University to intensive study of Spencer's sociology and psychology.[45] Later still, prominent Constitutional Democrat Pavel Miliukov on several occasions referred to both Spencer's and Comte's ideas in articulating his commitment to democracy and social reform after the 1905 Revolution.[46] None of these invocations of Spencer had the same duration and intellectual intensity as the hostile appropriation of the 1860s and 1870s.

Had Spencer known about them, two final positive interactions with his work on the part of Russian authors would have gratified him deeply. First, in the dispute between Spencer and August Weismann in 1893 about the inheritance of acquired characteristics, Russian naturalists and *intelligenty* tended to side with Spencer, who argued that there was some evidence of neo-Lamarckian heredity (or at least that Weismann had not proved their absence).[47] This is a matter of some interest given the later intense propagation of neo-Lamarckian ideas in the Soviet Union under the auspices of Trofim Lysenko. But perhaps the gentlest and least controversial was the proliferation of sympathetic obituaries that heralded the Englishman's passing in December 1903. Bromides and platitudes about the encyclopedism of Spencer's thought and the ambition of his synthesis were to be found in several intellectual journals, but without critical engagement with (or even exposition of) his central ideas.[48] In the ensuing months, commentary dwindled to a trickle. The legacies of Chicherin's conservative liberalism on the one hand, and much more Mikhailovskii's Populism on the other, would be of greater import. Hostile appropriation prevailed.

---

43 Irene Zohrab, "Dostoevsky and Herbert Spencer," *Dostoevsky Studies* 7 (1986): 45–72; Shoshana Knapp, "Herbert Spencer in Čexov's 'Skučnaja Istorija' and 'Duél': The Love of Science and the Science of Love," *Slavic and East European Journal* 29, no. 3 (Autumn 1985): 279–296; John Tulloch, *Chekhov: A Structuralist Study* (New York: Harper & Row, 1980), 94–95, 151, 154.

44 Shkurinov, *Pozitivizm v Rossii XIX veka*, 237.

45 Vucinich, *Social Thought in Tsarist Russia*, 111.

46 George Fischer, *Russian Liberalism: From Gentry to Intelligentsia* (Cambridge, MA: Harvard University Press, 1958), 70–71; Mikhailova, *Russkii pozitivizm na rubezhe XIX-XX vekov*, 95.

47 Vucinich, *Darwin in Russian Thought*, 151, 167–169.

48 Much is made of these obituaries in V.A. Bazhanov, *Vospriiatie Britanskoi sotsial'no-filosofskoi mysli v Rossii (XIX-nachalo XX vv.)* (Ul'ianovsk, UlGU, 2005), 29–40, although the publications invoked were relatively marginal to the dominant intellectual conversations in contemporary Russia.

Given the apparent ubiquity of Spencer's thought in the latter half of the nineteenth century in the *intelligentsia* publications of Petersburg and Moscow, it is striking that the topic has attracted so little attention from historians. At times, Spencer even seems invisible. For example, in a 1978 history book entitled *The Positivist Theory of Law in Russia*, Herbert Spencer was not referenced even once – despite the evidence of significant citations we have seen on Chicherin's part (who played a large role in the study).[49] As with all historical questions about absences, one can only speculate, but it seems that there are two dominant explanations for the blinders concerning Spencer's appropriation in Imperial Russia. The first is that to the extent that historians have examined the kinds of questions to which Spencer devoted the lion's share of his corpus – evolution, social organization, progress – they have interpreted Russian intellectuals' engagement with those questions as creative readings of either Charles Darwin or Karl Marx. In this, they are no doubt at least partly correct, but it seems to me highly likely that Russians' readings of Spencer colored their engagement with those better-studied authors, and also certainly possible that elliptical references that have been taken as relating to Darwin or Marx were sometimes references to Spencer. Spencer's very proximity to these two giants obscures his centrality.

The second and final point concerns the challenges of writing histories of "hostile appropriations." If individuals had waved banners proclaiming "I am a Spencerian," it would be much easier to locate them and tell the story of an active and vigorous positive reception of Herbert Spencer's corpus. When a scholar is appropriated largely in the context of being a universal punching bag, however, the only way to see his or her impact is to read the footnotes carefully, and see who is cited and in what context. These appropriations do not usually show up in indices or bibliographic searches, as some of the citations can be allusive. But we should look harder for these negative appropriations, because it seems entirely reasonable that not only is the phenomenon of building one's own theories on vanquished rubble of an interlocutor a common practice, but it might even be the *more* frequent form of appropriation of intellectual legacies. Spencer, even when honored more in the breach than the observance, has much to teach us still.

---

49 V.D. Zor'kin, *Pozitivistskaia teoriia prava v Rossii* (Moscow: Izd. Moskovskogo universiteta, 1978).

TABLE 1.1 *Translations of Herbert Spencer's works into Russian, 1866–1900 (excluding summaries and simplified expositions, and reprints)*

Years	Original	Title	Publisher	Translator (if known)	Comments
1866–9	Various essays	*Sobranie sochinenii v 7 t.*	N.L. Tiblen	N.L. Tiblen (supervisor)	Advertised as with Spencer's consent; 7 volumes
1870	*Principles of Biology*	*Osnovaniia biologii*	N.P. Poliakov	Al. Gerd	2 volumes
1872	*Social Statics*	*Sotsial'naia statika*	N.P. Poliakov		
1874–5	*The Study of Sociology*	*Izuchenie sotsiologii*	Znanie	I.A. Gol'dsmit (supervisor)	2 volumes
1876	*Principles of Psychology*	*Osnovaniia psikhologii*	I.I. Bilibin		From 2nd English edition, 4 volumes; 2nd ed. (1897–8), published by I.D. Sytin
1876–7	*Principles of Sociology*	*Osnovaniia sotsiologii Gerberta Spensera*	I.I. Bilibin		2 volumes
1877	*Education: Intellectual, Moral, and Physical*	*Vospitanie umstvennoe, nravstvennoe i fizicheskoe*	N.P. Karbasnikov		2nd ed. (1878); 3rd ed. (1880)
1878	*Descriptive Sociology*	*Opisatel'naia sotsiologiia*	M.P. Frits	I.V. Luchitskii (supervisor)	
1880	*Principles of Sociology: Ceremonial Institutions*	*Nachala sotsiologii: Obriadovye uchrezhdeniia*	E. Ia. Fedorov	I.V. Luchitskii (supervisor)	
1880	*Principles of Ethics*	*Osnovaniia nauki o nravstvennosti*	I.I. Bilibin		
1880	*Data of Ethics*	*Dannye nauki o nravstvennosti*	Mysl'		Republished from the journal *Mysl'*
1881	*Education: Intellectual, Moral, and Physical*	*Vospitanie umstvennoe, nravstvennoe i fizicheskoe*	Rodnik	E. Sysoevaia	Republished from the children's journal *Rodnik*; 2nd ed. (1883); 3rd ed. (1889)

Years	Original	Title	Publisher	Translator (if known)	Comments
1881	"On the Physiology of Laughter"	*Fiziologiia smekha*	A.S. Suvorina		
1882	"Political Institutions"	*Razvitie politicheskikh uchrezhdenii*	Mysl'		Republished from the journal *Mysl;'* with preface by L.E. Obolenskii
1884	"The Coming Slavery"	*Griadushchee rabstvo*	A.E. Riabenko		
1886	*First Principles*	*Osnovnye nachala*	F. Ioganson	L. Alekseev	
1894	*Education: Intellectual, Moral, and Physical*	*Vospitanie umstvennoe, nravstvennoe i fizicheskoe*	Dom prizreniia malolet. bednykh	N. Fedorova	
1894	"The Inadequacy of 'Natural Selection'"	*Nedostatochnost' estestvennogo podbora*	Iu. Ia. Riman		Republished from journal *Nauchnoe obozrenie*
1894	*From Freedom to Bondage*	*Ot svobody k rabstvu: Dovody protiv sotsializma*	Sokolovskii i Minsker		
1895	Selection from *Education*	*O nravstvennom vospitanii*	G. Beilenson i I. Iurovskii		no. 33 in the series "Mezh-dunarodnaia bibliioteka"
1896	*Principles of Ethics*	*Nauchnye osnovaniia nravstvennosti*	I. Perevoznikov i I. Ivanov		
1896	*The Study of Sociology*	*Sotsiologiia, kak predmet ucheniia*	A. Iakobson nasl.		
1897	*Justice*	*Spravedlivost'*	A.A. Porokhov-shchikov	M. Filippov (supervisor)	
1897	*Classification of the Sciences*	*Klassifikatsiia nauk*	Kn. delo	N.N. Spiridionov	
1897	*First Principles*	*Osnovnye nachala*	L.F. Panteleev		
1898	*Education: Intellectual, Moral, Physical*	*Vospitanie umstvennoe, nravstvennoe i fizicheskoe*	V.A. Tikhanov	M.A. Lazareva	
1898	Extract from *Principles of Ethics*	*Prava zhen-shchiny*	Vladim	L. Gol'dmer-shtein	On rights of women

TABLE 1.1 Cont.

Years	Original	Title	Publisher	Translator (if known)	Comments
1898	"Chastity," "Marriage," "Parenthood"	*Tselomudrie, brak, roditel'stvo*	F.A. Ioganson	A. Samoilov	
1898	"Chastity," "Marriage," "Parenthood"	*Tselomudrie, brak i roditel'stvo*	L.A. Zolotarev	L.A. Zolotarev	2nd ed. (1900)
1898	["History of the Visit"]	*Istoriia vizita*	P.P. Soikin		
1898	"The Genesis of Science"	*Proiskhozhdenie nauki*	O.N. Popov		
1898	"Gracefulness"	*Slezy, smekh i gratsioznost'*	Dom prizreniia malolet. bednykh		
1899	*Principles of Biology*	*Osnovaniia biologii*	Izdatel'	N.A. Rubakin (editor)	From the 1898 edition
1899	Various works	*Sochineniia Gerberta Spensera*	Iuzh.-rus. kn-vo	N.A. Rubakin (editor)	New translation, not a reprint of Tiblen
1899	["The Mysterious World"]	*Tainstvennyi mir*	V.I. Gubinskii	I.P. Rapgof	
1899	["The Development of Sexual Relations"]	*Razvitie polovykh otnoshenii*	F.I. Mitiurnikov	N. d'Andre	
1899	"Good and Bad Conduct"	*Khoroshee i durnoe povedenie*	A.L. Turnov i Ko.		
1900	"The Rights of Women"	*Prava Zhenshchin*	Vladimirskaia tip. Mordukhovskogo	N. d'Andre	
1900	*Education: Intellectual, Moral, and Physical*	*Vospitanie umstvennoe, nravstvennoe i fizicheskoe*	A. Ia. Panafidin	M. Glovatskaia	
1900	["The Forms of Sexual Relations"]	*Formy polovykh otnoshenii*	Borisenko i Breslin		

CHAPTER 2

Spencer's Arabic Readers

Marwa Elshakry

In the summer of 1903, the British traveler, essayist, and activist Wilfrid Scawen Blunt introduced the eighty-three-year-old Herbert Spencer to an Egyptian admirer of his, the Grand Mufti of Egypt, Muhammad 'Abduh. 'Abduh was just one of Spencer's many worldwide followers who similarly went to the ageing English philosopher's home to pay homage to him. Indian sociologists, Confucian thinkers, and Japanese reformers had similarly made the pilgrimage, and most of them had been also drawn to his vision of a unified science of the cosmos, the state, and society. The Mufti had been a fan of Spencer's *Education: Intellectual, Moral and Physical* in particular, his most widely translated book; 'Abduh himself tried his hand at translating it. He also referred to Spencer repeatedly in his lectures on the exegesis of the Qur'an (*tafsir*). Indeed, on one occasion he described Spencer as "the greatest living philosopher."

At the meeting between Spencer and 'Abduh, Blunt was to "act as interpreter." Without one, he feared "much exchange of ideas is impossible" since Spencer "understood little more French than is needed for traveling purposes." Spencer would die just three months later, and no mention of the encounter is to be found in his writings. Blunt, however, wrote vividly about it in his diaries. The aging Spencer was in a frail condition. Nevertheless, during their morning meeting, the bedridden Spencer offered Blunt and the Mufti his views on politics and prophesied the future decline of Europe. Their host lamented the disappearance of "right," denounced the Transvaal war as "an outrage on humanity," and spoke of a coming "reign of *force* in the world" when "there will be again a general war for mastery, when every kind of brutality will be practiced."

After lunch, Spencer turned to matters of philosophy and asked the Mufti "whether it was true that thought in the East was developing on the same lines as the thought of Europe." 'Abduh replied (according to Blunt) that "what the East was learning from the West was the evil rather than the good, but that still the best and most enlightened of thought was the same." This opaque answer left Spencer dissatisfied:

> "To go to the bottom of things," said Spencer, "I suppose that the conception of the underlying force of the world, what you call Allah and we call God, is not very different?" In his reply the Mufti made a distinction that struck Spencer as new. "We believe," said Abdu, "that God is a *Being*, not a *Person*." Spencer was pleased at this, but said the distinction was rather difficult to grasp. "At any rate," he said, "it is clear that you are Agnostics of the same kind as our agnosticism in Europe."

We can only surmise what Egypt's Grand Mufti would have made of being called an agnostic, for the meeting seems to have ended there. Spencer, who had recently had a stroke, would consent only to brief conversations lasting a few minutes.

Blunt, however, pursued the matter with 'Abduh as they walked back to the station:

> I questioned the Mufti more closely on the point. *I.* "Do you believe that God has consciousness, that he knows that you exist and I exist, and is not such knowledge personality?" *The Mufti.* "He knows." *I.* "If he knows, he knows that you are good and I am bad." He agreed. *I.* "And he is pleased with you and displeased with me?" *The Mufti.* "He approves and disapproves." *I.* "And he approves to-day because your actions are good, and he disapproves to-morrow because your actions have become bad. Is not this change from approval to disapproval characteristic of personality? How then has God no person?" *The Mufti.* "God knows all things at all times; to him there is no to-day and no to-morrow, and therefore in him is no change; His is an eternal unchanging consciousness of all things. This I call Being, not Personality." *I.* "And Matter? Is not Matter eternal, too, or did God create it? If he created it he made a change?" *The Mufti.* "Matter, too, is eternal as God is eternal." Here evidently is the foundation of Abdu's thought, and we agreed that our ideas are the same.

In translation, however, the rendition was perhaps less startling. In Arabic, the statement "*in al-madda aazaliya kama aydan Allah aazali*" (matter is eternal as God is eternal), for instance, might have struck Muslim readers at the time as a rather familiar, even if not universally held, formulation. But it did not hold the specific materialistic implications that it might have had for the likes of Blunt or even Spencer.[1] The debate over the eternity of God and of matter was a

1 Blunt's recollections were later published in Arabic and the dialogue was excerpted and republished as part of a biography of 'Abduh in the 1930s by Muhammad Rashid Rida, whom

familiar philosophical debate for most Arabic readers; its connection to a materialist view of the cosmos would not have made sense on several counts: indeed, in the Arabic version, the discussion of the Transvaal war is described in translation as evidence of the dangerous rise and spread of a "materialistic spirit" in Europe and the world. Translations, we know, hold as many shades of meaning as the possibilities of synonymy allow between any two languages, and reading these two accounts of the Mufti's conversation with Spencer offers a fascinating glimpse into the ways in which ideas can be carried across languages, like conversations, in multiple ways.

'Abduh once described as the "transcriber of his thoughts." (Rida compiled and personally published much of the work written under 'Abduh's name.) But for Rida writing later, the implications of his mentor's encounter with Spencer were quite different from those that led the mufti to visit the English writer in the first place. The Middle East between the world wars was a hotbed of anticolonial foment, and presenting the late grand mufti as the admirer of a British writer – one not quite so eminent as he had been a generation before – was a risky move. Rida's version, however, gave no cause for worry, and reading the Arabic version against the original, one notices a number of interesting omissions and additions. Whereas in Blunt's recollections, Spencer's lamentations had concerned imperialism and the Europeans' love of war, in Rida's version, he also identified the rise of *materialism* (*al-maddiya*) as the cause of England's – and Europe's – moral decline. Rida's concern with materialism continues, and he has 'Abduh reckoning that ultimately "truth and the good will triumph over materialist thought." Spencer agrees and says that the "materialistic tendency will run its course still in Europe." When the Englishman explains to the Mufti that "truth for Europeans now is power," 'Abduh replies that the fascination with power has also gripped Easterners, who imitate Europeans. This exchange about truth and power was absent in the original. Other additions further alter the conversation, guiding it into familiar theological channels: "*SPENCER*: What do Islam's *'ulama* say of the Creator – does he exist in or outside the world? *AL-IMAM*: The Companions say that God is above all things and beyond the world. The *mutikallimun* [theologians] say that he is neither in nor outside the world. And the Sufis, who profess a unity in all existence, say that all the manifestations of the world are a manifestation of his existence. And we believe that God is a *Being* and not a *Personality*." Blunt's account was thus made by Rida to fit easily within a standard explication of Muslim theological and cosmological thought, touching on themes that would have seemed familiar: the eternity of matter, and God's transcendence and being in the world. Not surprisingly, no mention is made of agnosticism in the Arabic version. Rather, at the end, Spencer is quoted as saying "In any case it is clear that you are profound thinkers, like the profound thinkers among us Europeans," Rida diplomatically substituting "profound thinkers" for "agnostics" in the original. The following and above account are taken from my *Reading Darwin in Arabic, 1860–1950* (Chicago: University of Chicago Press, 2013).

The Mufti's conversation with the aged English philosopher offers one reading of Spencer's many polyglot incarnations.[2] As the Arabic readings of Spencer in the late nineteenth and early twentieth century demonstrate, and as I hope to show in this chapter, Spencer was often read against a background of contexts, and against a series of "metatextual" references, as well as intellectual and institutional backdrops. As even this brief meeting with the Mufti highlights, Spencer's noted anti-imperial views were clearly a key feature of his appeal for many of his readers in the newly minted global presses of the nineteenth century. He offered new credence – and credentials – for a specifically natural theory of society. This too appealed to many of his readers in the popular monthlies and periodicals of the growing late nineteenth and early twentieth century global presses. Spencer's philosophy represented a broad, social and natural, philosophy of man and the state, if not the cosmos (and for some, even God). It was evolutionary in the widest sense: his synthetic philosophy encompassed the formation of the cosmos from matter and force to the evolution of humanity, with all moral customs, political systems and social institutions seemingly accounted for. Combined with a unified vision of the sciences, it offered the perfect framework for colonial intellectuals who found themselves concerned with the opportunities and problems of state reform projects that similarly viewed progress in social scientific terms. This was what made him so interesting to international readers of the time. He provided a new, scientific language for the study of "society," itself a key nineteenth century neologism. A variety of 'social questions' gained urgency for the new states and empires of the nineteenth century and after and Spencer's Arabic readers were often highly conscious of this political backdrop.

Yet, Spencer offered his polyglot readers something else too: a "synthetic philosophy" of the evolution of both man and the cosmos. For many, this was both a positivist and a metaphysical philosophy, a transcendent ontology as much as a new epistemology (for those who pushed his abstruse declarations on the "unknowable" into this territory, in any case). His idea of the unknowable solved what many felt were the incompatibilities between positivism and materialism. (Many modern Arabic readers versed in Arabic logic and speculative theology, were also quick to point out that they were incompatible, since first causes were by definition always beyond positive, empirical verification.) As the Arabic version of Spencer's conversation with the Mufti also highlights, his philosophy left open a back door for faith in a transcendent, albeit 'unknow-

2 On tracing the ideas of a polyglot author, see Patrick O'Neill, *Polyglot Joyce: Fictions of Translation* (Toronto: University of Toronto Press, 2005), Introduction.

able' force in nature; for some, like the Mufti, this also left room for the possibility of an eternal, whether material or immaterial, being in the world.

Yet the Mufti's interest in the philosopher extended beyond the bounds of even this broad if somewhat ambiguous conversation. The Mufti had in fact just finished his translatation of parts of Spencer's *Education: Moral, Intellectual and Physical* (though he never ended up publishing it). That this was by far the most important among Spencer's translated works worldwide is perhaps not surprising. The question of education was a growing nineteenth-century concern, for scholars and statesmen alike. Spencer was among the first to write popularly on the subject and his vision of education was unique in its time in that it offered a social and moral program for uplifting society (rooted above all in the family). It also presented a broadly liberal view of the self, society, and the state – and education was key for the proper progress of all three. For the liberal Englishman, it was proper pedagogy which might shape the state, not the other way around. For colonial bureaucrats like 'Abduh, this was appealing on a number of counts: it offered a path to progress that could rest on moral as much as intellectual or material grounds and it emphasized the work of political and social reform through the education of ordinary citizens and kinsmen. For someone like 'Abduh, this vision of education could be said to be broadly in line with someone of his position: despite the official title, he was as keen to preserve the autonomy of the *'ulama* against both the growing interventions of state reformers and imperial overseers. (This was a risky strategy, however, for he did in fact rely on British support when he needed a buffer against Azhari or Khedival contenders, an irony, many of his detractors pointed out at the time.) Yet 'Abduh's reading of Spencer formed only one part of a much broader metatextual landscape for the Mufti. He was above all a pedagogue, intent on carving out a new social role for the *'ulama* through education. Indeed, one could scarcely understand the Mufti's own writings on education generally without attending to his own reading of classical theories of *tarbiya* ("education" broadly, or a *paidagōgeō*, or method of leading, cultivating or rearing, children). As I hope this introduction shows, reading and translating Spencer in Arabic allowed for a creative interplay of ideas, old and new.

As an internationally best-selling author, Herbert Spencer was a figure who captured many of the contradictory impulses of his place and era. A solitary figure, he was nevertheless an astute observer of human behavior, and while he lived a life of routine, he subscribed to a philosophy of cosmic change. He was an evolutionist who cared more about a cosmic evolutionary process than the specifics of species change, and though he subscribed to a positivist philosophy of science, he also left room (or so many thought) for metaphysics; and, finally, despite his interest in the natural and physical sciences, it was as spokesman of a new social (and even ethical) science that he gained his reputation: that he saw the evolution of morality as the both the cause and effect of human progression also appealed to many, and was recalled creatively, by many around the world. An arm-chair naturalist and social commentator of his day, he rarely ventured abroad, spending most of his time at his home in Brighton, even while nevertheless entertaining numerous visitors from as far afield as Cairo, Bombay, and Buenos Aries, who, like 'Abduh, came to pay their respects to him. For while he seemingly spoke to the concerns of late Victorian society, his ideas circulated among and captured the imagination of people the world over.

So how did this somewhat reclusive if not parochial Victorian thinker become a global icon during his own lifetime (and for some time after his death in 1904)? This chapter considers some of the reasons behind Spencer's appeal, while focusing on the example of several generations of Spencer's Arabic readers. Looking at how Arabic authors appropriated, digested, summarized, or otherwise referenced Spencer, it examines how Spencer's own creative, if at times ambiguous, resolution of numerous polarities – such as transcendentalism and positivism, individualism and collectivism, among others – lead to equally creative and highly varied interpretations in Arabic from the late nineteenth to early twentieth centuries.

Following Spencer's Arabic readers, even schematically, as this chapter does, also means tracing, to some extent, the different communities of readers that formed around his works, as well as the varied metatextual landscapes that they inhabited. It means paying attention to translators and readers alike, and to the institutional networks and sociopolitical contexts that they were a part of, while at the same time listening to the various textual echoes that shaped their readings.

Focusing on a reception of Spencer's complete *oeuvres* as such would not take us very far. His works were only partially translated – his *Education*, as mentioned, was one of two of his complete works available in Arabic for much of the twentieth century. Yet there were partial summaries and translations, as well as numerous discussions of and references to his works long before the turn of the century and well after. These were creative works in their own right, formed on the basis of selective appropriations or, at times, on singular interpretations of minor works or even tangential ideas, and all composed against a deep set of historical and contemporary, philosophical and political debates.

This article attends to some of these readings and references by looking, somewhat thematically and somewhat schematically, at the broad category of ideas that Spencer's works inspired among some of his Arabic readers. These include, first and foremost, the appeal to Spencer's broad, synthetic philosophy. For some of his readers, this was one that could bridge a commitment to both a positivist science and a philosophical humanism as well as a transcendent ontology, metaphysics, or philosophy. Indeed, it was really the very language, at once universal and historicist, at once metaphysical and positivist, of Spencer's "synthetic philosophy" that appealed to his many Arabic readers. This was one reason why so many of them thought his ideas resonated with older or more familiar natural, moral and political theories. Another common theme could be found in the enthusiasm for Spencer's vision of social science, and his views on education and role of the state, as we saw already. Coming both during the heyday of empire and evolution, it was Spencer's evolutionary science of society and the state that appealed to so many of his disparate colonial readers. After the 1880s, following the wake of the "scramble for Africa" and the French and British occupation of Tunisia and Egypt respectively, the question of "social evolution," and the "struggle for life between nations," became prominent in the minds of many. Spencer's anti-imperial views and his progressive theory of the state also proved salutary in this setting. Yet Spencer appealed as much to liberal intellectuals as he did to early Arab socialists and collectivists –two groups one would otherwise imagine might inhabit more separate literary or at least ideological realms. Following Spencer's Arabic readers along these broad thematic and metatextual themes, it is precisely this intellectual dexterity and ideological ambiguity that rested behind so many of the readings and translations of Spencer that this chapter aims to explore.

"A Synthetic Philosophy"

For many lay readers of the late nineteenth and early twentieth century, Spencer was the key *fin-de-siècle* philosopher of the evolution of life. Indeed, his writings on evolution were vast and cosmic in both their scale and content. Unlike Darwin (whose name often accompanied mention of Spencer's), he did not shy from speaking outright about the development, or more specifically the progress, of humanity. As a key international spokesman for both a positivist methodology for the human sciences and for the new theory of evolution, his importance for Arabic, as for many other readers worldwide at that time, outweighed that of Comte, Mill, and perhaps even Darwin himself. By the time of his death in 1904, the editors of an Arabic popular science monthly,

Al-Muqtataf, could write of how they "mourned his death like that of the greatest Egyptian."[3]

What lay behind their curious devotion to him? It was, as they advised their readers, that Spencer's philosophy provided a truly "synthetic philosophy" – a means of discovering the order of nature and the history of humanity in one, in other words. What appealed to them, in particular, was that it provided the basis for uncovering the universal laws of nature and man alike. Indeed, for many Arabic readers, Spencer's philosophy was regarded as most valuable for providing the basic evolutionary principles of good governance. The editors emphasized how both our understanding of the cosmos and our role in it could be revealed through this synthetic and holistic yet still precise and positivist philosophy.[4] Outlining his philosophy of the "Unknown" a few issues later, they noted how his vision combined metaphysics and morality with scientific insight, and how this very commitment to "the advancement of science" was thus to be appreciated for the breadth of its contributions to our own understanding of the plight of humanity itself.

It might have seemed odd then when shortly after this, a reader was quoted as asking them the following: "What are the benefits of Spencer's books?" Of course this curiously vague question makes more sense when we consider how few people could have actually read his books in translation: only his *Education* and *The Philosophy of Style* were available in Arabic at this time, and even those had not yet been widely circulated. Yet people were clearly familiar with his ideas, and his name appears in numerous late nineteenth-century discussions on a variety of topics, including evolution, ethics, education, empire, and the state. In a process of translation and transmission that was characteristic of intellectual life among readers outside Europe and America at this time, most people came to Spencer through summaries of his ideas in popular journals.[5]

To their curious but uninformed reader, *Al-Muqtataf*'s editors replied summarily, emphasizing how the totality of his works could be defined in terms of utility and uplift: Spencer's works, they wrote, helped "strengthen minds." Those who read them, they clarified, were invariably made more informed of the role of modern science in our world and were more insightful and successful (and as evidence they cited the scientific progress enjoyed in Europe and

3 "Al-Faylasuf Hirbirt Spinsir" [The Philosopher Herbert Spencer], *Al-Muqtataf* 29 (1904): 1–9.

4 Ibid.

5 "Spencer's Works," *Al-Muqtataf*: 29 (1904): 96. Among the books it lists are: *First Principles*; *Principles of Biology*; *Principles of Psychology*; *Principles of Sociology*; *Principles of Ethics; The Study of Sociology*; *Education*; *Social Statics*; and *Man versus the State*.

America itself!) Spencer's appeal seemed to rest precisely upon this promise of progress. In a colonial context, where the notions of progress and development were often thought to be synonymous with social uplift, political freedom or national independence, it also carried the hope of a participation in a vastly different future, geopolitical order. Yet what appealed to Spencer's Arabic readers, as with his many colonial readers elsewhere at this time, was that while his view of the evolution of the state and society was implicitly anti-imperial (he famously saw imperialism as a sign of an evolutionarily regressive rather than progressive state), it was also a gradualist one. For the editors of *Al-Muqtataf*, who had begun their monthly "Journal of Science and Industry" under the patronage of the Syrian Protestant College in Beirut in 1876, this emphasis on a gradual progression of state and society, under the proper evolutionary development of a natural social order, was an important theme in their journal.

It was thanks largely to the popularizing summaries and commentaries published in *Al-Muqtataf* (1876–1952), especially during the early decades of its publication, that Spencer was first widely introduced to Arabic readers in fact. Their own translations and transmission of his ideas followed particular routes and references that mobilized or catalyzed his ideas, however, in particular ways. The journal's readings of Spencer were as much influenced by the context of imperial and anti-imperial politics, for instance, as they were by American Protestant theology or by older Arabic traditions of natural history and the philosophy of creation.[6]

Then there was the very specific epistemological resonance of Spencer's approach. In the year of the philosopher's death, when the editors of the monthly had paid tribute to him in the series of short articles that detailed his life and philosophy, they claimed he ascribed "all creation" – from the movement of the stars to human thought itself – to the fundamental and universal operations of matter and force. At the same time, they also emphasized that his was no mere "materialist philosophy."

The charge of materialism – toxic above all in Ottoman Syria and later in the southern Arab provinces since at least the early 1880s – was one they were keen to avoid. Originally they had relied on the works of Protestant theologians, many of whom had similarly utilized Spencer's writings for just this kind of theological reconciliation with natural philosophy. After their break with the Protestant College and their move to Cairo in 1885, however, the

6 For more on the influence of Protestant theology on late nineteenth century readings of Spencer, and on the general background to reading evidence of Arabic natural philosophy into modern ideas of evolution, see *Reading Darwin in Arabic*.

editors of the Arabic monthly turned to Spencer more directly.[7] The spectre of materialism still haunted them and they sometimes wondered whether Spencer himself was the best antidote against it: in their 1898 "Commentary on Spencer," they pointed out that his views on the evolution of faith – from totemism to monotheism – were themselves too suggestive of a materialist leaning, for example.[8]

Yet, when reviewing his works shortly after his death, they highlighted Spencer's definition of the "Unknown" in detail. Citing his *First Principles*, they used this to make the point that his philosophy of the cosmos and of man did not touch upon ultimate causes. This, they reasoned, left room for belief in a higher, "unknowable" *power*.[9] While the "knowable" rested within the domain of science, the object of which was to relate natural laws to one another, the unknowable concerned all ultimate ideas, or those cases where reason offers no guidance, Spencer thought. [10] Neither ultimate religious ideas, nor even ultimate scientific ideas – such as the final cause of matter, motion or force, for example – they concluded, could be positively evaluated by reason alone.[11]

The editors had long had a keen interest in promoting an empirical and positivist methodology of science. In fact, *Al-Muqtataf* had been among the first to use the term "positivism" in Arabic and was, by 1898, defining it as knowledge "established on reason and its proofs," and as distinct from materialism, Epicureanism, and all those schools of thought that "rely on sensation."[12] For them, as for many other global popularizers of science at that time, it was Spencer rather than Comte who served as the main spokesman for a positivist epistemology, and this despite Spencer's ambiguous appeal to an abstruse and decisively "unknown" limit to scientific knowledge itself.

Finding echoes of a vague transcendentalism in Spencer's works was especially appealing to the editors of *Al-Muqtataf* whose entire mission of popularizing science had been one premised on the idea of a possible reconciliation between science and religion. In terms they found appealing, Spencer had argued: "If Religion and Science are to be reconciled, the basis of the

7 See Elshakry, *Reading Darwin in Arabic*, chapters one and two.

8 From *Al-Muqtataf* 22 (1898): 818–821.

9 "Falsafat Spinsir" [Spencer's Philosophy], *Al-Muqtataf* 29 (1904): 281–285.

10 Drawing from Henry Mansel's *Limits of Religious Thought*, he uses the example that we cannot argue in favour of an ultimate, first cause in nature, as it is both infinite and absolute (as the two are mutually contradictory).

11 See also George Smith, "Herbert Spencer's Theory of Causation," *Journal of Libertarian Studies* 5 (1981): 113–52.

12 "Al-madhhab al- falsafi"[The Philosophical School, or Positivism], *Al-Muqtataf* 22 (1898): 106–12.

reconciliation must be the deepest, widest, and most certain of all facts – that the Power which the Universe manifests to us is inscrutable."[13] Indeed, for Spencer this commitment to a positivist demarcation between the known and the unknown, and the shunting aside of ultimate causes, was itself a sign of all sophisticated "cultivated theology of the present day." Yet, the appeal to unity – or *tawhid*, as it was typically translated in Arabic – as both a force in nature and a larger, inscrutable, hidden or unknowable reality (*al-ghayb*) would have held other resonances for Arabic readers, and held multiple philosophical and theological implications too.

Yet at least as important as his ambiguous transcendentalism was another theme the editors had long promoted: the unity of nature. His philosophy, they argued, "has demonstrated how the universe is one complete picture – like parts of a puzzle that fit together." "The laws that apply to its parts, also consist of general laws which apply to the whole." They praised his "synthetic philosophy" for this vision of universal progression – the movement, whether in societies or in organisms – from heterogeneity to homogeneity.[14] In his classic essay on "Progress: Its Law and Cause," Spencer argued that the "advance from the homogenous to the heterogeneous" was evident in geology, biology and history, and he offered an ethnography of the growing complexity of governments, from tribes to nations, of labor, of language and of art, as evidence of this "law of progress." The roots of this desire for holism can surely be traced back in part to the natural theological tradition that was so important for science popularizers everywhere; perhaps too in the case of these Arab intellectuals, ideas of the unity of God's creation dated back centuries and still formed part of a continuous tradition of Arabic translations and publications on the "Marvels of Creation." This fusion of Protestant theologies of nature, Spencerian views of the unknown, and classical Muslim views of metaphysics and existence might seem surprising, until we remember that it was all taking place in presses manned and read by Syrian Christian science enthusiasts, American missionaries and Egyptian Muslim educators, and bureaucrats alike.

For all of them, Spencer's natural philosophy was easily encompassed within a number of older traditions, as well as within a number of growing, contemporary intellectual trends. Spencer's holism was inextricably linked to his skepticism and to his emphasis on the inscrutable: "the sincere man of science," he wrote, discovers that "inward and outward things" are "alike inscrutable in their ultimate genesis and nature." The distinction between

13 Herbert Spencer, *First Principles* (New York: D. Appleton and Co., 1897), 48.

14 On Von Baer's principle of complexity in Spencer's *First Principles*, see "Spencer's Philosophy," part II, *Al-Muqtataf* 29 (1904): 383–6.

transcendental and positive knowledge blurred by the editors of *Al-Muqtataf*, as by many positivists around the world at that time, offered the possibility of a reconciliation between scientific and spiritual knowledge. After all, for Spencer, the "Materialist and Spiritualist controversy is a mere war of words." Faced with the range of experience and all that "transcends experience," the "sincere man of science" also "sees that absolute knowledge is impossible." And "he alone knows that under all things there lies an impenetrable mystery."[15]

Society: 'An Organism'

Discussions of Spencer had followed those of Darwin since the first days of *Al-Muqtataf*'s publication in particular. After the British occupation of Egypt, and the journal's subsequent move south, however, their references to Spencer became more frequent. The question of social evolution – like their interest in the problem of progress itself – became ever more urgent in their minds.

In 1885, as they settled into their new offices in Cairo, the editors of *Al-Muqtataf* announced the "discovery" of a new organism. Relaying the news with characteristic flair, and headlining the column "A Prodigious Animal" (*Hayawan ha'il*), they described it as "an animal so massive that the image of it has not crossed the imagination of any of the Ancients has only now been discovered by the modern philosophers. … We are not speaking figuratively or in riddles. That such an animal exists is quite certain, if we are to believe the modern philosophers. You will ask us: 'And what is this strange animal?' The answer is 'Society' [*al-ijtma' al-insani*, literally the 'human collective']." This organism functioned like others, they noted; it too operated according to natural laws that ensured its growth and evolution or its gradual decay and senescence. Speaking directly to their readers, they reminded them that "you are a member of this globule [*al-karriya*]."[16]

The 1885 article served as part of a commentary on a longer series of articles published earlier on "The Natural History of Societies" (*Tarikh al-ijtima' al-tabi'i*). The term "society" – as a disciplinary category – emerged as a nineteenth-century neologism across many languages, and early Arabic discussions were heavily influenced by Spencer's social scientific program in particular.

15 Herbert Spencer, *Essays: Scientific, Political and Speculative* (New York: D. Appleton and Co., 1888), vol. 1, 61–62, 67.

16 "A Prodigious Animal," *Al-Muqtataf* 10 (1885): 119.

Added to this, was the rising concern with the questions of state and social reforms that emerged in the wake of European imperialism in the region. In this context, it is hardly surprising that more than a few colonial intellectuals saw in Spencer's idea that societies, like species, progressed a powerful diagnostic for the rise and fall of states. Spencer's concern with "society" from the perspective of a unified vision of science was a key factor behind his appeal among Arabic readers.

Spencer, like other emerging nineteenth century social scientists often argued, society, like nature itself, operated according to natural laws.[17] To intellectuals in Egypt under British occupation such ideas were especially resonant: "society" could function as an object of imperial solicitude, a rationale for colonial occupation, or contrarily (and later on), as a vehicle for anti-colonial mobilization. Either way, it marked a common preoccupation among intellectuals in a multi-sovereign and a multi-confessional Ottoman milieu.

These views on the natural laws of social progression took on extra urgency for colonial intellectuals in Egypt. The stagnation, decline, and decay of nations, wrote the editors of *Al-Muqtataf*, following Spencer, and in terms that spoke of the highly charged conditions of their own "national" trajectory, were themselves a condition of progress, for the "very life of the social organism depends upon the death of nations." The rise and fall of civilizations – such as those "glorious Eastern civilizations of the ancient Egyptians, Phoenicians, Babylonians, and Assyrians" – constituted a natural law of universal progress.

Yet, progress for Spencer, they thought, was far from inevitable: it was conditional upon being "unhindered, undisturbed, unrepressed, undistorted, undwarfed, undeformed, uninjured." As they saw it, a "reversion towards the old type" was one possible outcome of a hindered or disturbed society.[18] While Spencer feared a reversion in the case of England from an industrial society to what he dubbed a "militaristic" one, Sarruf and Nimr, by contrast, found the idea of a return to the past consoling in the Arab context. "There is an absolute and invariable law in the universe," they wrote, "and that is that men's ancestral racial traits do not easily disappear. And if they do so partially, it is quite easy to bring them back by a resumption of the necessary conditions." It might be possible, in other words, to return Eastern civilization to its former glory. "The Arab East may once again be on the road to progress, if we attend to those conditions that allow for it." The editors subscribed to a classically organicist view

17 Walter Simon, "Herbert Spencer and the 'Social Organism,'" *Journal of the History of Ideas* 21 (1960): 295.

18 Elshakry, *Reading Darwin in Arabic*, 86.

of society and yet, following Spencer, embedded both an evolutionary and a historicist view of civilizations within it.[19]

These organicist and holistic views had classical resonances, but the novelty of this approach was that it was both nomothetic and historicist. This held particular appeal for the many technocratic bureaucrats and colonial intellectuals that collectively comprised *Al-Muqtataf*'s general authorship and readership. In that sense, it was history, as much natural as national, which brought together evolution and empire for many of that generation. The new "natural history of societies" or of "nations" (the latter was a broad rubric in Arabic that could also imply "races") that Spencer helped in large part to popularize among Arabic readers thus helped to transform conceptions and categories of history too.

Yet the question of a progressive vision of knowledge brought with it new conceptions of the past, present and future. Cultivating a historicist reading of one's past alongside a natural history of societies allowed for a social scientific diagnostic for the future. Caught between Ottoman and European political control, many Arab nationalists of that era could not but reflect on their historical predicament, and the way out. The question of rise and fall was raised because it brought with it new possibilities for the future: indeed, many who took up the banner of evolution and progress were also key figures in the "Renaissance" literary movement or *Nahda* of their time, if not the key instigators and inventors of it. This Renaissance or "awakening" of the East was to follow a broadscale translation movement intended to transfer knowledge from West to East. It was in the course of these discussions that notions of "Arab civilization," and of "Arabic science" – or their later counterparts, Muslim civilization and science – entered the timeline of universal history in Arabic histories themselves.

For many colonial intellectuals of the late nineteenth century and after, uncovering, reclaiming, and even redefining one's own classical or golden age was crucial for claiming a place in the hierarchy of civilizations. Many of the these discussions carried over new notions of science as much as of 'civilization.' The connection between the two was one that occupied many in post-Napoleonic Europe who were concerned with what one French commentator called "that new divinity – civilization." As one early nineteenth-century Arabic writer put it, writing on the idea of civilization: "the more we go back in time, the more you can see the backwardness of people in terms of human skills and civil sciences. Conversely, as you descend and observe the slope of

19 "Al-Faylasuf Hirbirt Spinsir," *Al-Muqtataf* 29 (1904): 1–9.

time (to the present), you generally see the elevation and progress of these."[20] A later generation built upon these formulations. They turned to Mill's discussion of 'Civilization' and to his distinction between moral perfection, on the one hand, and wealth and worldly power, on the other.[21] Meanwhile, translation disarmed the racialized readings of Le Bon's treatment of "the evolution of races" and transformed it into a debate about the perfection of peoples (*umam*). They read Guizot on the importance of religion in cementing social solidarity and recalled Ibn Khaldun, whose much earlier discussion of similar themes helped naturalize the entire debate. As with all these creative readings, this emphasis on social progression explains why so many of Spencer's Arabic readers (like Darwin's) seized upon their ideas as evidence for the evolution of peoples, societies or states. Spencer in particular was called upon to provide evidence of the evolution of humanity. For many Arabic readers in particular, evolution was about much more than biology or even history: after all, the new universal history for which it provided the blueprint was one with the power to recast the future itself.

Empire and Education

We can hardly ignore the connections between Spencer's global appeal and the new imperialism of the late nineteenth century. After all, Spencer's popularity among Arabic readers coincided with the growth of the British Empire and its expansion into the Ottoman Arab provinces. Yet, this was, as we saw, a double-pronged appeal. Spencer was a staunch anti-imperialist, but this did not mean that he advocated for self-governance among a nation of equal citizens: for Spencer social development implied a process of gradual progression. Change required time.

20 Rifa 'a Rafi ' al-Tahtawi, *Takhlis al-Ibriz fi Talkhis Bariz*, translated by Daniel Newman as *An Imam in Paris* (London: Saqi Books, 2014), 108.

21 The editors took for granted (as did Spencer) that Europe was at the pinnacle of the social evolutionary ladder. Yet in their discussion of "civilization's" flip-side, "barbarism," the editors (again, like Spencer) proved more ambiguous. Their article on "Barbarism and Civilization" describes the complex process of social progression on a universal scale. Not all of the nations that Europeans invaded, took over or otherwise colonized were necessarily barbaric, they argued, referring to Spencer's discussion of native Americans to prove the point. Reading arguments about Europe's civilizational superiority led many Arabic readers to refer back to their own once civilizationally glorious past in new terms, and it also explains why they understood evolution to be a doctrine of progress.

This commitment to gradualism, and its anti-imperial and yet still classically liberal and progressivist line suited the proprietors and readers of many science and literary Arabic journals of the time. These were liberal colonial intellectuals who saw technocratic oversight by a newly educated elite as the basis for a true yet gradual progression of the state and its society. We should also bear in mind, however, that for many of them, British rule had also originally implied the possibility of ending what they considered to be the despotism of Ottoman rule; but emancipation from Ottoman control came simultaneously new forms of political oversight. At the same time, Britain – and its intellectual elite – furnished a model for a future emancipation; yet it was a model that would take time to develop. In any case, for this literary elite, there were few alternatives to imagining the progressive evolution of Arab nations other than the adoption of a gradual path to progress.

Education emerged as a key concern in this context. For Spencer, education would lead to self-cultivated and self-sufficient national subjects, thereby contributing to the general progress of its citizens. Building upon Enlightenment theories of education, with their emphasis on human perfectibility, the cultivation of the mind, and the fostering of a reasoned moral sense, Spencer added to these a contemporary fascination with the evolution of the human mind.

In their 1903 article on "The Great English Philosopher Herbert Spencer," the editors of *Al-Muqtataf* described *Education* as his most important book, noting that it was "translated into all the known, civilized languages." By then more than 40 years old, we know that it had long proved to be his most popular and most widely translated work. Reading through the short book, it is not difficult to see how it might have appealed to Arabic readers too. When Spencer complains about the tyranny of the classical curriculum, for example, one could easily imagine how this might have spoken to the similar anxieties among his Egyptian readers, wrestling with a rapidly transforming national school system while also railing against "traditionalism" in the classroom to make room for their own pedagogic reforms.[22] Spencer insisted that education's sole aim was to fulfill the mind's natural potential. Schooling should therefore mimic the natural unfolding or cultivation of the mind, and foster the power of reasoning, the higher intellectual and moral faculties, to create a self-perpetuating

---

22 For instance, on the uselessness of a classical education in England, Spencer wrote: "We are guilty of something like a platitude when we say that throughout his after-career a boy, in nine cases out of ten, applies his Latin and Greek to no practical purposes. ... If we inquire what the real motive is for giving boys a classical education, we find it to be simply conformity to public opinion." See Herbert Spencer, *Education: Intellectual, Moral and Physical* (New York: Appleton, 1861), 23.

and pleasurable curiosity in the nature of things. For Spencer, proper intellectual and moral (like physical) training must progress in tandem.[23]

These ideas paralleled older Arabic notions of *adab* (proper manners, morals, and taste) and ideas of *tarbiya* (education), which included similar themes on proper moral cultivation and child rearing. Classical conceptions of virtue and of knowledge, were extended in this context, however, to include the "social function" of education and its role in the creation of a harmonious social (and national) order. For instance, in his 1881 *Discourse on Eight Words*, Shaykh al-Azhar Husayn al-Marsafi described *tarbiya* as a process of "perfection by degrees toward completion" – or as an innate force for improvement that humans inculcated for the "benefit and aid of a society." As each individual was merely a "limb," or part, of the *umma* (community or state), so he (or she) must be trained in accordance with the natural development of the collectivity. For Marsafi, education referred above all to the "cultivation of morals" (or *tahdhib al-akhlaq*). Physicians of society or the state, educators "serve their country" (or *watan*, another new "keyword" in Marsafi's octet). Given the reinvigoration if not recreation of ideas of education, Spencer's own emphasis on moral education was one reason why his book was popular among Arabic readers. 'Abduh, we might recall, saw education as a moral as much as an intellectual enterprise, a concept he had essentially derived from canonical Muslim models of pedagogy and ethics. Spencer's repeated critiques of a narrow, dogmatic adherence to authority, the problems of rote memorization and classicism would also have struck a chord with the Mufti, whose own reform program would later be hailed as the prototype of "Islamic modernism." Like Spencer, he thought good pedagogy was good politics: questioning authority on scholarly matters set an example for those who wished to question political dogmatism and or the problem of autocracy.

There was also the question of the family. For Spencer, care of the family was the key foundation for good society and therefore for good government. The rational management of the family was the starting point for Spencer's chapter on "Moral Education": "To prepare the young for the duties of life is tacitly admitted by all to be the end which parents and school- masters have in view." But he criticized the current state of affairs: "While many years are spent by a boy gaining knowledge, of which the chief value is that it constitutes 'the education of a gentleman;' and while many years are spent by a girl in those decorative acquirements which fit her for evening parties; not an hour is spent by either of them in preparation for that gravest of all responsibilities – the management of the family."[24] Spencer thought that education should give chil-

23 *Al-Muqtataf* 28 (1903): 896.

24 Spencer, *Education*, 22–23.

dren the skills to understand right and wrong and that a proper moral development would lead them to acquire an innate knowledge of the natural consequences of both. This emphasis on the proper safeguarding of the family suited the reform program of many of Spencer's Arabic readers, including the Mufti. As Qasim Amin, a friend of 'Abduh's and a noted writer and educational reformer, argued in his 1899 *Liberation of Women*, the proper education of boys and girls – rooted in the liberation of women and men, or the mothers and fathers of the nation – was the key to progress.[25]

Metaphysics and Positivism

Spencer's theory of education was a key draw. But it was the appeal to transcendental positivism that gave Spencer an unexpected literary afterlife among Arabic readers later in the twentieth century. This was clearly evident in the reading of Spencer by Isma'il Mazhar, the first translator of Darwin's *Origin of Species.*

In addition to *Origin*, Mazhar also published numerous other works on evolution. He was also a keen translator, translating popular works on science from Draper to George Sarton. Like *Al-Muqtataf*, a journal to which he owed many of his own ideas, Mazhar was committed to a fundamentally positivist vision of science. He too was bothered by materialism, which he saw as fundamentally counterposed to positivism: He argued that one could not square a subscription to materialism with a positive scientific method (and he regarded the latter as the only valid scientific method if not philosophy). The problem, as he saw it, rested in the former's unempirical and therefore ultimately metaphysical claims about the origin of life from matter.

Yet for Mazhar, positivism demanded and did not merely accommodate, still less challenge, a transcendental view of life and its origins. Following Spencer, he combined what we can only call transcendental positivism with a cosmological evolutionism. One of the most systematic of the early Arabic transcendental positivists, he effectively referenced Spencer for his own reconciliation between science and metaphysics.[26] The latter's ideas on the limits of

---

25 See Qasim Amin, *The Liberation of Women*, translated by Samiha Sidhom Peterson (Cairo: American University in Cairo Press, 2000), originally published as *Tahrir al-mar'a* (Cairo, 1899).

26 Later, Mazhar would turn more directly to what we might label transcendental materialism and took a more explicit interest in the merging of matter and mind, as his later translation of Edmund Synott's *Life of Matter* demonstrates.

knowledge proved useful for Mazhar in this context for a number of reasons. For one, Mazhar insisted that the theory of evolution was not concerned with the ultimate nature of things (*ma hayat*), and he spent a considerable amount of time criticizing Arab materialists' use of concepts like "force" and "matter," which he classed as abstract and therefore fundamentally beyond empirical investigation – or, ultimately, "unknowable." Like Spencer (a figure much cited in all of Mazhar's works on evolution), Mazhar felt that a belief in the Unknowable permitted him to sidestep any metaphysical commitments to the nature of reality while also avoiding any subscription to "materialism, determinism, or to any 'ism' that implied knowledge of reality as it exists apart from consciousness." The true nature of space, time, matter, motion, and force were, in Spencer's words, "absolutely incomprehensible." "Ultimate Scientific Ideas, then, are all representations of realities that cannot be comprehended": "in its ultimate nature nothing can be known." The "reality existing behind all appearances is, and must ever be," for Spencer, "unknown."[27]

Like *Al-Muqtataf* before him, and like many late Victorian popularizers who found theistic resonances in Spencer (Mazhar drew on many of these, including John Beattie Crozier's *Civilization and Progress*), Mazhar argued that positivism and a belief in ultimate causes or even a deity were not strictly incompatible. Indeed, Mazhar followed Crozier by outlining several examples of principles that a positive science must utilize though it cannot positively or empirically verify them: the existence of a world outside ourselves, the persistence of force, and the belief in scientific causation chief among them. Crozier's conclusion, which Mazhar followed, was: "If then, the main foundation stones on which our intelligence is built must be believed, although they cannot be known by Science, Comte's first great argument against the existence of a Deity, viz., that He cannot be known by Science, falls to the ground." Mazhar took a similar line regarding the question of ultimate truths and the existence of a "Greater Being."[28]

In the end, however, Mazhar's sympathies were made very clear: he thought that only evolution – and not the doctrine of independent or special creation, as he put it – could explain and bring together all the evidence available from the extant fossil and geological records of his time or from the new disciplines of comparative anatomy, embryology, and ethnology. Only evolution could

27 Cited from George Smith, "Herbert Spencer's Theory of Causation," *Journal of Libertarian Studies* 5 (1981): 116.

28 Isma'il Mazhar, *Malqa al-sabil fi madhhab al-nushu' wa al-irtiqa' wa-atharuhu fi al-fikr al-hadith* [Evolution at the crossroads: The theory of evolution and its impact on modern thought] (Cairo: Al-Matba'a al-'Asriya, 1924), 63–66.

make sense of the distribution, diversification, adaptation, and finally the extinction of the world's flora and fauna. Yet he also thought that this did not necessarily threaten either a spiritual view of creation or even faith in God: after all, Wallace, he reassured his readers, believed in both evolution and the spiritual uniqueness of humanity, while Spencer provided numerous arguments in favor of the compatibility of both. Writing well after Spencer's influence had ceased in much of western Europe, Darwin's Arabic translator still found in his ideas both a remedy for materialism and an argument in favor of his own transcendental positivism.

Spencer's Arabic Readers

Spencer appealed to several generations of Arabic readers from the late nineteenth and early twentieth century. He offered the promise of a broad, synthetic, social and natural philosophy of progress. It was evolutionist in the widest sense and its unified vision of science, society and the state that offered the perfect framework for national reformers and colonial intellectuals concerned with the problem of progress. As we saw, Spencer's emphasis on the "unknowable" in science also lent many the creative license to blend metaphysics with positivism. Behind all this, too, lay the realities of the new imperialism and the varieties of anti-colonial nationalisms that followed in their wake. Spencer was himself bound up with this: a noted anti-imperialist, his remarks on the regression of industrial societies to military ones through colonial expansion no doubt impressed and inspired his audiences abroad. Yet like many Victorian writers of the time, this did not necessarily mean that he thought all colonial territories were as yet capable of self-rule (as his famous letter advising a Japanese constitutional lawyer, quoted in full in *Al-Muqtataf*, suggests). Hence, his developmental and evolutionary philosophy held a strange attraction for these colonial intellectuals searching for their own program of social, material and political reform. If in following his Arabic translators and commentators we see how Spencer came to represent the ambiguities of modern ideas of societies as much as states or empires, we can say much the same for the ways in which people came to understand his subscription to both liberalism and positivism. This chapter has dealt primarily with the latter, and the question of the methods and limits of an empirical enquiry into the natural world. There was an implicit appeal too to a transcendental ontology that this epistemology created. The creative resuscitations that these then inspired, shows how Spencer was referenced for numerous ends. His views on the former were similarly diverse. This is a subject I have taken up elsewhere, but it may be instructive to note in passing here that his ideas worked to both reinforce early individualist, laissez-faire view of society and the state as much as later collectivist, socialist ones.[29] As this chapter suggested at the start, Spencer's own attempt to reconcile the natural and moral philosophies as well as the social and political theories of his time were themselves reflected in the numerous readings, translations and appropriations of his ideas in Arabic: whether in terms of serving as a bridge between a transcendental and positivist philosophy, in terms of providing both an anti-colonial political program and a decisively developmentalist and gradualist reformist one, or in terms of bringing together the role of the individual with the collective when thinking about the nature of the progress of society and the state. It was perhaps precisely because of these embedded and refracted creative ambiguities that he became such an enduring icon of modern science for generations of colonial intellectuals who were themselves caught up in their own web of contradictory individual and collective philosophical and political agendas.

29 See Elshakry, *Reading Darwin in Arabic*, chapter six.

CHAPTER 3

Spencerism in Japan: Boom and Bust of a Theory

G. Clinton Godart

It is well known that Herbert Spencer was widely read in Japan, especially during the "Spencer boom" in the early Meiji period (1870s and 1880s). But it is still not entirely clear what the actual scale and the long-term effects of Spencer's thought upon Japan were. In this chapter, I will introduce the usages of Spencer in Japan and attempt to provide a way to rethink the role of Spencer in Japan. One problem is that the appropriation of Spencer's ideas in Japan has been simultaneously over- and underestimated. Overestimated, because many historians of Japan have assumed the existence of a ubiquitous Spencerian "Social Darwinism" supposedly leading to Japan's colonial adventures, this despite Spencer's own pacifism and anti-imperialism, and little evidence to support such a wide-ranging influence beyond the mid-Meiji period.[1] On the other hand, most actual research on Spencer in Japan has focused on how Spencer's thought was used in political debates among a limited group of thinkers and activists during a very restricted time period, the Spencer Boom of the 1870s and 1880s.[2] Studies have rarely looked into the many Japanese criticisms of Spencer's thought, and the reasons for the surprisingly quick decline of his popularity (in Japan and elsewhere) has gone largely unquestioned. On the other hand, the emphasis on the political dimension has left other elements of Spencer's thought that interested Japanese, such as biology, religion, and education, to large part unexamined. I will thus argue that Spencer's role in modern

1 To give just one example: "These ideas [of Social Darwinism], made popular in Japan in the translations of the European philosopher Herbert Spencer, suggested that just as Charles Darwin's theory of "natural selection" taught that only the stronger, more "evolved" species triumphed over weaker and less-well-adapted species, similar natural laws governed human civilization. Japan wanted to survive and to triumph; the natural way to do this was to prey on the weak." In Louis G. Perez, *The History of Japan* (Westport: Greenwood Press, 1998), 116.

2 Important studies on Herbert Spencer in Japan are the seminal work by Yamashita Shigekazu, *Supensā to nihon kindai* (Ocha no suishobo, 1983); and in English Nagai Michio, "Herbert Spencer in Meiji Japan," *The Far Eastern Quarterly* 14, 1 (1954): 55–64; Julia Thomas, *Reconfiguring Modernity: Concepts of Nature in Japanese Political Ideology* (Berkeley: University of California Press, 2002); David Howell, "Society Reified: Herbert Spencer and Political Theory in Early Meiji Japan," *Comparative Studies in Society and History* 42, 1 (2000), 67–86.

 | DOI 10.1163/9789004264007_005

Japanese thought has been on the one hand more *limited* than has been assumed, but on the other hand much more *diverse* in its effects.

In this chapter, I do not intend to answer the question of Spencer's appropriation in Japan comprehensively. Instead, I will introduce the story of Spencer in the 1870s and 1880s (of which some will be familiar to the intellectual historians of modern Japan), while zooming in on one popular liberal champion of Spencer in Japan to illustrate the reasons for the appeal of Spencer in the early Meiji period. I will also make two new contributions: Firstly, in contrast to the political emphasis, I want to draw attention to how Spencer was appropriated in other areas, beyond the immediately political: in religion, education, and the organization of knowledge itself, hence emphasizing a multi-dimensional picture of Spencer in Japan.

A second objective of this essay is to draw attention to the limits of Spencer's popularity in Japan and especially its swift decline around the turn of the century. I will argue that while there might have been a "Spencer boom," the scale of it should be put in perspective. And importantly, for all the talk about the "Spencer boom" in Japan, few have talked about what we might call the "Spencer bust." While the reasons why Spencer was much read in Japan do actually not require that much explanation – Great Britain was the world's most powerful nation and Spencer one of her most respected philosophers – the reasons why Spencer's philosophy went from such great heights of popularity to declining so quickly (and worldwide) are less clear, but I think that understanding the Spencer bust is crucial to a full understanding of the phenomenon of Global Spencerism. Why was such a widely read and discussed system of thought, with such an overarching theoretical scope, dropped in a relatively short time, by a large majority of intelligentsia? Taking this into consideration, how profound was, really, the appropriation of Spencer's philosophy in Japan?

The "Spencer Boom"

In a period of three decades around thirty translations of Spencer's works appeared in Japanese (see table 3.1). But the bulk of these appeared during one decade, between 1877 and 1886, hence the "Spencer boom." The reasons for the transfer of Spencer's ideas in his period were, as I will explain below, indeed largely political, but we should also note the conditions for the transmission of knowledge in this period. After a decade of internal turmoil and American and European military pressure, the Tokugawa Shogunate was overthrown in 1868. The country was unified under the Emperor (or *Tennō*, "Heavenly Sovereign") as head of state, and was plunged into a breakneck pace of modernization. Important for the story of Spencer in Japan, the period saw a great explosion in information transfer, not only between Japan and the rest of the world, but also domestically. Japan under the Tokugawa was never truly "closed" to the outside world, but information transmission was restricted by Tokugawa policies: Japanese were not allowed to travel abroad, and trade with Asia and the West was very limited. This all changed with the opening of treaty ports in the 1850s, and even more after 1868. This is the background to why Spencer's ideas were imported into Japan somewhat late.

In the Meiji period, Japanese went abroad to study, and foreign experts (*yatoi*) were hired in large numbers, arriving on steamships, and a railroad network, a national postal system, and a telegraph system were built. The late Tokugawa period had seen the development of nation-wide, informal networks of scholars, but this expanded dramatically in the Meiji period when freedom of movement was expanded. The status system was abolished. The press, journals, and newspapers also expanded dramatically during this time, finding a large audience of readers educated in a new national compulsory education system. Estimating literacy rates is a very slippery affair, with great differences among regions, class, and gender, but the usual rough figure given is that, the literacy rate grew from 30 to 70 percent over the course of the Meiji period (1868–1911).[3] The Meiji period also saw a number of changes in publishing, such as a change from woodblocks to moveable type, from hand presses to engine powered rotary presses, and from traditional binding to Western binding, although the early translations of Spencer's books were published in traditional Japanese format.[4]

Translations of course played a crucial role in the transmission of knowledge, the creation of new vocabularies, and also the dissemination of information and ideas to a wider public. But when considering the appropriation of Spencer in Japan, an important caveat in this respect is that we should not confuse dates of translations with important moments of the transmission of its content. Educated Japanese were often fluent in English, and a number of them read Spencer or secondary sources on Spencer in the original, travelled abroad and listened to lectures about Spencer in Japan and abroad. Foreign books were available through, for example, Maruzen Company (*Maruzen shōsha*), founded in 1869 by Yuteki Hayashi (1837–1901), a pupil of the liberal

3 For a nuanced discussion and history of literacy in Japan, see Richard Rubinger, *Popular Literacy in Early Modern Japan* (Honolulu: University of Hawai'i Press, 2007).

4 See the first chapter of Edward Mack, *Manufacturing Modern Japanese Literature: Publishing, Prizes, and the Ascription of Literary Value* (Durham and London: Duke University Press, 2010).

thinker Fukuzawa Yukichi.[5] Maruzen Company became the most important importer of foreign books into Japan, with stores in Yokohama and Tokyo, and later nationwide. Thus, outside of the world of translations, there is a large but unknown X number of transmissions of Spencer to Japan.

The boom in translations of Spencer from the mid-1870s began in earnest when his political philosophy was appropriated by the liberal Freedom and Popular Rights Movement (*Jiyū minken undō*) in the 1870s. Soon, Spencer's ideas also came to be used by their opponents, who argued for a powerful state and rejected, or argued, for a more gradual transition to popular representation. Historians of Japan have seen this as partly reflective of a tension between Spencer's liberalism and individualism on the one hand, and his organic theory of society on the other (although it is open to debate whether this is a correct interpretation of Spencer). We can perhaps call this the "dual use"-theory of Spencer's appropriation in Meiji era Japan.[6] Let us follow the story of Spencer in Japan along these lines for the time being.

The Freedom and Popular Rights Movement was a diverse movement of ex-samurai and others who, for a variety of reasons, appealed to the government for the establishment of popular political representation. "Freedom" (*jiyū*) became one of the neologism buzzwords of the day. These political activists found in Spencer (and others, like J.S. Mill) a scientific spokesman for their cause, and they translated accordingly. Spencer's *Social Statics* (1851) was first translated in abbreviated form in 1877 by the young student Ozaki Yukio (1858–1954), who would become one of the most important politicians in Japan. Another liberal activist, Matsushima Kō (1853–1940), who later became an important educator, subsequently translated *Social Statics* in full, significantly, as *On Equal Rights in Society* in 1882. This translation was widely read, and Itagaki Taisuke (1837–1919), the leader of this liberal movement, called Spencer's *Social Statics* the "textbook of social rights."[7] The intended audiences of many of the translations of Spencer were the members of the movement and other educated readers, including those of the elite, to provide arguments for popular sovereignty and other liberal causes.

Competing with the liberals were more conservative thinkers, often at positions at the University of Tokyo, such as Katō Hiroyuki (1836–1916) and Ariga

---

5 In this chapter, for Japanese names, the Japanese name order is used: family name comes first.

6 See among others Yamashita Shigekazu, *Supensā to nihon kindai* (Tokyo: Ocha no mizu shobō, 1983); Nagai Michio, "Herbert Spencer in Early Meiji Japan," *The Far Eastern Quarterly* 14, 1 (1954): 55–64; and Julia Thomas, *Reconfiguring Modernity: Concepts of Nature in Japanese Political Ideology* (Berkeley: University of California Press, 2002).

7 Yamashita, *Supensā to nihon kindai*, 59.

Nagao (1860–1921), and with closer relations to the Meiji state. Especially Katō Hiroyuki promoted a vision of societies as biological organisms, its individuals as cells, and these organisms locked in a struggle for survival. While invoking Spencer, he developed his own theory of social evolution, and argued "rights" were not natural, but bestowed on the weak by the strong, in their own self-interest. The implication of Katō's theory of social evolution was that Meiji state could decide to grant which rights and when, for the good of the nation. Not surprisingly, this theory of Katō was severely criticized by the liberals.[8]

A word of caution concerning the "Spencer boom" is relevant here. The number of translations of Spencer's works is indeed large, but has been somewhat inflated, since many earlier translations were only partial translations or translations of individual chapters. More importantly, there was a larger boom in translations from all Western things happening at the same time, Spencer being just one. As a comparison, in the period from 1877–1895, at least sixteen book-length translations of John Stuart Mill (who was often mentioned in the same breath as Spencer) appeared and this is just one example. While this is indeed less than Spencer, the difference is not large enough to take the number of translations of Spencer as a sign of a country as being in the grip of Spencer. A large number of translations of Spencer's work is also reflective of his being a prolific writer, and it does not preclude the possibility that one translation of another author might have been more widely *read* (Marx's *Capital* for example come to mind). Darwin's *Origin* was indeed only translated fully for the first time in 1896 (a partial translation of the *Descent* had appeared in 1881), but when discussing "evolutionary theory," Japanese intellectuals tended to refer to Darwin, Spencer, and Haeckel, the usual lineup.

A Spencerian: Tokutomi Sohō

"Are the Japanese not doomed to become traders and producers?" This prediction came from what is the most influential and ambitious "Spencerian" book to appear from the liberal camp during this period: *The Future Japan* (*Shōrai no nihon*), written in 1886 by Tokutomi Sohō (1863–1957).[9] Tokutomi was part of a

---

8 For Katō Hiroyuki, see Winston Davis, *The moral and political naturalism of Baron Kato Hiroyuki* (Berkeley: University of California Press, 1995), and in Japanese see among others Yoshida Kōji, *Katō Hiroyuki no Kenkyū* (Tokyo: Ōhara Shinseisha, 1976).

9 Iichirō Tokutomi, *The Future Japan*, trans. Vinh Sing, Ed. Matsuzawa Hiroaki and Nicholas Wickenden (Edmonton: University of Alberta, 1989), 127. In this article, for readers who cannot read Japanese, I will refer to this English translation.

generation that had come of age in the early years after the Meiji restoration, experiencing its dramatic changes. Tokutomi would become one of Japan's most prolific and important journalists. He had read Mill, Bentham, and others while studying at the Dōshisha, a Christian university in Kyoto. Tokutomi's reading of Spencer's second volume of *The Principles of Sociology* in August 1884, was, according to John D. Pierson's intellectual biography of Tokutomi, a breakthrough in his thought, and it prompted him to write *The Future Japan*. At age twenty-three, he "attempted to develop a comprehensive theory of social progress in history, to analyze world conditions and trends in international relations in the nineteenth century, and, on these bases, to argue for the democratic reform of Japanese society."[10] First serialized in 1886 in the journal *Tokyo Keizai Zasshi*, the book became a major bestseller. Around this time, Tokutomi put himself squarely behind Spencer's ideas, before rejecting much of it a decade later. Since Spencer's own works (as contemporaries complained about at the time) are notoriously laborious to read through, it is worth noting that Tokutomi's version of Spencer's ideas probably reached a vastly wider audience than Spencer's own works.

The Future Japan was, in essence, an exposé of Spencer's vision of social evolution as developing from the "militant" to the "industrial" type. Tokutomi saw these two tendencies or "organs" (*kikan*) as fundamentally opposed to each other and irreconcilable (although the industrial had, through a process of the division of labor developed out of the militant). Human history, and especially the nineteenth century, was therefore a battleground between these two systems, with the industrial winning out. This battle, Tokutomi argued forcefully, was the environment to which Japan had to adapt, or face the fate of "annihilation" by the West, like Burma and Vietnam had. It is interesting to note that Tokutomi did *not* consequently present the "West" as superior, or as a model to follow, since many European nations were showing strong "militant" tendencies, keeping large armies and developing a weapons industry. Germany and Russia were especially guilty of this, and Japan should avoid their predicament and better follow the example of Britain and especially the United States. Tokutomi combined this Spencerian version of social evolution with an enthusiastic endorsement of liberalism, democracy (also relying on J.S. Mill), and the free-market (also inspired by the Manchester school).

It is important to note that Tokutomi strongly endorsed an organic conception of society, which, while not developed very elaborately, held that any change in society involved changes in its other parts as well as the whole. This

10 John Pierson, *Tokutomi Soho, 1863–1957: A Journalist for Modern Japan* (Princeton: Princeton University Press, 1980), 127.

might sound trivial today, but it had a scientific aura at the time. Importantly, with this organic metaphor, Tokutomi warned his fellow Japanese they could not change structures in society, technology, and dress without transforming Japanese culture. For example, Tokutomi predicted that communication technologies would change people's feelings. This use of the organic metaphor is important, since a tendency of the "dual use" interpretation of Spencer in Japan was to see the organic conception of society as something more the prerogative of the conservatives. In Tokutomi's case, as in Spencer's own, the organic conception of society was in fact a component of his liberalism.

Tokutomi also reinterpreted Japan's past in the light of Spencer's theories. The Tokugawa period, he argued, was a typical example of Spencer's "militant" society. "There had never been previously a feudal society like ours. There will never be a feudal society like ours again."[11] What were the fall of the Tokugawa in 1868, the subsequent abolishment of the samurai, and the opening to world trade, if not proof of the Spencerian transition from militant to industrial? The nation's leaders had not planned this transition, but "Extemporaneous measures, designed to meet immediate needs, completely changed our militant society into an industrial society and our aristocratic society into a democratic society."[12] Thus, for contemporary Japanese, looking at Japan through a Tokutomi-Spencerian lens must have seemed to explain many things, such as the bewildering changes, and the strange mix between the old and the new, and where it all would lead.

Tokutomi's book located and explained Japan's transformation as following long-term world historical patterns, but his vision, while seemingly believing in the inevitable victory of democracy and free trade, was not one of inevitable, unilinear progress in "stages" that societies inexorably follow, nor was it truly teleological. On the contrary, individual societies could still retrogress, stagnate, or collapse. In order to avoid this, they do well to adapt to the international *environment*, which did show a general progress towards industry and democracy. *The Future Japan* was therefore also meant as a wake-up call for his fellow Japanese, and was permeated with an anxiety that Japan might lose the momentum of progress in the right direction, if "feudal" and militaristic tendencies, regional rivalries, the illegal amassing of wealth, and increasing authoritarianism held Japan back or drove it in the wrong direction. Concretely, Tokutomi warned that the tendencies of the Meiji state to develop a large army and navy, increase armament production, and resort to threats of war in international disputes were not only dangerous, but since they fundamentally

11 Ibid., 133.

12 Ibid., 159.

belonged to the "militant" type, they were also retrogressive. He believed "that military expansion must by all means be halted."[13] Japan had to embrace just laws, property rights, abolish tariffs, build harbors, allow free access to Japan for foreigners, and above all, international free trade and democracy. The work was thus also an appeal to the young, and a critique of conservatives and nostalgic thinkers, and the nation's current leadership.

What historians have noticed less is that Tokutomi's use of Spencer also meant, perhaps counter-intuitively, a rejection of the idea that economic and military strength went hand in hand, an idea that in Japan had been expressed in the influential slogan *fukoku kyōhei* ("rich nation, strong army"), and which Tokutomi rejected as a "sophism."[14] Again and again, Tokutomi argued that the "industrial," was incompatible with the "militant." In an industrial age, production of weapons takes up industrial capacity, requires huge tax increases, and thus debilitates normal economic activities. Hence military buildup in an industrial age undermines its own conditions of possibility by paralyzing industry and thus diminishing wealth. Despite Tokutomi's reminding the reader many times of the real dangers of Western colonial aggression, he argued that the "struggle for survival" in the nineteenth century would no longer be one of physical force, but of intellectual and economic competition. In this sense, Tokutomi's Spencerism was faithful to Spencer himself, who certainly did *not* believe that the "struggle for survival" between nations was a force in future evolution. This being decades before the coinage of terms such as "spin-off" and the "military-industrial complex," Tokutomi's argument might have been persuasive, and it is important to see that Spencer was mobilized here for a political message against aggressive foreign policy and military spending.

For Tokutomi, and not a small number of Japanese like him, Spencer's theory of evolution from militant to industrial had an immense explanatory power. It explained the country's past, the Meiji restoration, and the dazzling changes following it, in terms of a "scientific theory" of evolution, and pointed towards a possible bright future. More importantly, Tokutomi's argumentation shows how Spencer's general theory of social evolution could be appropriated by a Japanese intellectual and applied to a particular case – Japan's transitions during the Meiji period – to convey and serve a particular political purpose.

---

13 Ibid., 126.

14 Ibid., 175.

A Diverse Legacy

The "dual use" theory has its limitations for understanding Spencer in Japan. An important addition to this theory has been made by the historian Douglas Howland, who has argued that the introduction of Spencer's sociology in Meiji Japan, regardless of the strife between liberals and conservatives, made possible the envisioning of "society" (*shakai*, one of the neologisms of this period) as such. Envisioning "society" as a phenomenon, to be understood scientifically, and potentially to be manipulated, Howland argued, signaled a shift in Japanese intellectual history from an enlightenment model of progress to a scientific model.[15] There are other limitations to the dual use interpretation of Spencer in Japan. Firstly, as can be seen from Tokutomi, the idea that the liberals drew on Spencer's individualism while the conservatives drew on his organic conception of society does not hold up. Both liberals and conservatives appropriated Spencer's organic conception of society for their own purposes. Secondly, the dichotomy of liberals versus conservatives itself is somewhat simplifying for this period, with various figures straddling both sides and crossing over from one to the other. Thirdly, and more important, since most scholars who have studied the role of Spencer's ideas in Japan have limited themselves to these highly visible political debates during the 1870s and 1880s, we have come to understand Spencer's role in Meiji Japan almost exclusively through the lens of these political debates, thereby ignoring other fields where Spencer's theory was used, such as psychology, biology, the position of women, education, and religion.

It is notable that the translation of Spencer's *Social Statics* (and a later separate translation singled out as "The Rights of Women") was one of the first texts to appear on women's rights in Japan. Mori Arinori (1847–1889), an important Meiji leader who drew much on Spencer, also advocated the reform of marriage and called to end mistreatment of women. When commentators interpreted the Tokugawa period in terms of the "militant" society, they noted how, as Spencer's theory would predict, women's liberties were most restricted among the samurai, but less in other status groups in Tokugawa society.

Spencer's relevance for biology has been forgotten in Japan and elsewhere due to the later characterization of his thought as "social Darwinism." Contemporaries did take Spencer's theories on biology seriously, and his role in this field is an important aspect of the history of science in Japan that has not yet been researched. It is noteworthy that Charles Whitman (1842–1910),

15 Douglas Howland, "Society Reified: Herbert Spencer and Political Theory in Early Meiji Japan," *Comparative Studies in Society and History* 42, 1 (2000): 67–86.

who taught biology at the University of Tokyo after Edward S. Morse, lectured while using Spencer's *Principles of Biology*, as he thought this was the best book to understand biology in general.[16] Hence the first generation of Japanese biologists, Ishikawa Chiyomatsu (1861–1935), Yatabe Ryōichi (1851–1999) and others, learned at least one part of their biology education through Spencer. How this influenced biology in Japan is a matter for future research.

Another, not immediately political, aspect of the history of Spencer in Japan is his importance for religion. Firstly, Spencerian agnosticism seemed to have been popular among young intellectuals in the early Meiji period, much to the complaint of Christian missionaries. As a result, Spencer was sometimes seen as a danger to religion, despite his own call for reconciliation between religion and science. This fear did not disappear overnight. To give just one example, the Christian Kishimoto Nobuta, at a speech in Chicago in 1893, argued that Japan was a "battlefield of religious and non-religious secular forces as much as of Christianity and other religions," and that "Many Japanese intellectuals were indifferent or hostile to religion in general, and thus religious forces in Japan should first fight such agnostic, nihilistic, and rationalist, and materialistic intellectuals, influenced by figures like Spencer, Comte, and Schopenhauer."[17]

But Spencer also played a larger, positive, and more lasting, role in religious thought in Japan. Firstly, Spencer's ideas played a role in the general understanding of the phenomenon of "religion" itself. Spencer was introduced in Japan at a time when religion was undergoing important changes. While there had always arguably been something like religion in Japan, the Japanese term "religion" (*shūkyō*) was a neologism (the term had previously been used to describe the "doctrine" *kyō* of a particular Buddhist sect, *shū*). And just like Spencer's thought was important in the catalyzing of the *idea* of "society" itself, so was Spencer also important in the Japanese construction of other large modern categories and phenomena such as "religion," "science" (*kagaku*), and "philosophy" (*tetsugaku*), all Meiji neologisms.[18]

Part of the appeal of Spencer's *First Principles* was that it discussed the relation between religion, science, and philosophy, in a very comprehensive and systematic way, and (at least initially) helped many to make sense of these new

16 Ishikawa Chiyomatsu, *Ishikawa Chiyomatsu Zenshū*, ed. Ishikawa Chiyomatsu kankō kai, vol. 4 (Kōbunsha: 1935–1936), 80–81.

17 Nirei Yōsuke, "Toward a Modern Belief: Modernist Protestantism and Problems of National Religion in Meiji Japan," *Japanese Journal of Religious Studies*, 34:1 (2007), 168.

18 For a discussion of new categories in Meiji Japan, see among others Howland, *Translating the West*, and for religion and philosophy, see G. Clinton Godart, ""Philosophy" or "Religion"? The Confrontation with Foreign Categories in Late Nineteenth Century Japan," *Journal of the History of Ideas* 69, no. 1 (2008): 71–91.

large categories and how to organize a mass of new knowledge and theories that was being introduced (and produced) in Japan. Moreover, Spencer seemed to offer a scientific explanation for why there was a differentiation into these different branches, according to a universal pattern of evolution. At the same time, Spencer, especially in his *Principles of Sociology*, also provided an explanation for the origins of religion, its development, its relation to psychology, and gave religion a function in the evolution of human societies. Spencer's "Religion: A retrospect and prospect," was also separately translated in 1888 as *The Evolution of Religion*.[19] Hence Spencer's theories also provided a scientific approach to what was, at least in terms of discourse and categorization, the new phenomenon of "religion." Meiji Japanese philosophers often explained the distinction between philosophy and religion in Spencerian terms. To give just one example, Inoue Enryō (1858–1919) was one of the most important Meiji-era Buddhist philosophers. He spent much of his life advocating a modern, philosophically valid Buddhism that was in accord with science. When discussing the distinction between religion and philosophy, he echoed Spencer when he wrote: "Roughly speaking, the world – in the broadest possible meaning of the word – consists of two parts. Technically these are called the knowable world and the unknowable world. In plain language, they refer respectively to the world that can be known with the human intellect and the world that cannot. In other words, they point to the distinction between philosophy and religion."[20]

Secondly, Spencer also played a role in religious theories propagated by Japanese religious thinkers. Firstly, this was felt in attempts to harmonize religion and science. In *First Principles*, Spencer argued that religion and science would further differentiate, with science relinquishing religious ideas, and religion giving up superstition and unfounded knowledge claims, both would crystallize into their pure forms. A clash between religion and science, therefore, was only temporary. In the 1870s and 1880s, this was potent material for Buddhist and Christian thinkers alike, both groups in a position of feeling on the defensive in a rapidly modernizing society in which science claimed an authority on truth-claims.

Further, while Spencer's thought was of course seen largely as new, Western, and modern, in some respects it touched common ground with religious

---

19 Ariga Nagao, trans., *Shūkyo shinkaron* (Tokyo: Makino shobō, 1888).

20 *Bukkyō tetsugaku* (1893), in Inoue Enryō, *Inoue Enryō Senshū*, ed. Tōyōdaigaku sōritsu hyakunen kinen ronbunshū hensan iinkai, vol. 7 (Tokyo: Tōyōdaigaku, 1987–2004), 107–113. Spencer discusses the distinction between religion and science as "opposite sides of the same, religion the invisible, science the visible," in Herbert Spencer, *First Principles* (London: Williams and Norgate, 1867).

thought in Japan. The choice of translating Spencer's term "principles" as *genri* ("fundamental principle" or "pattern") echoed Confucian philosophy, the term *ri* being the key object for inquiry into nature. The idea that the whole of nature emerged from a single mysterious and undifferentiated origin was recognized as similar to the Confucian metaphysical idea of the "Great Ultimate." Not a few Buddhist thinkers saw a useful parallel between Spencer's "Unknowable," and Buddhist notions of the Absolute, expressed as "Thusness" (*Shinnyo*): both could be interpreted as an all-pervading cosmic absolute, beyond the grasp of reason. Spencer's epistemological conclusion from his theory about the Unknowable, that "in ultimate essence nothing can be known,"[21] that ultimate reality is beyond discursive knowledge, was a position that resonated with a long history in East-Asian religious thought of the ineffability of the highest truth. For example, Inoue Enryō found much to his liking in Spencer's philosophy. Inoue pointed his finger to past Buddhists, whom, he argued, had not really understood or clearly formulated the relation between the relative world and the absolute, and often discussed the Buddhist idea of the Absolute, "suchness," (*Shinnyo*) as having much in common with Spencer's "Unknowable."[22] Spencer's ideas about the Unknowable, it must be mentioned, were at the time criticized in Europe, and in Japan too it was by no means uncritically accepted. For example, Inoue Enryō argued that an epistemological problem with Spencer's theory was that he had simply *posited* an unknowable outside the knowable.[23]

Evolutionary conceptions of religion, often along Spencerian lines, were widely talked about. Inoue argued along Spencerian lines that Buddhism itself had evolved from a singly homogeneous source, growing in complexity to diverse heterogeneous schools. This, he argued, was a proof of Buddhism's strength and its modernity. For example, in 1903, Tokutomi Sohō, discussed above, who by this time had reversed on his pacifism and embrace of Spencerian liberalism, continued to discuss religion in Spencerian terms. He argued that the essence of religion comes from an "authority," which sets man's hearts at peace, and derives from the "mysterious" or "Unknowable" (using *fukashigi*, the same word used for Spencer's "Unknowable"). The essence of

21 Spencer, *First Principles*.

22 See for example, Inoue Enryō, *Bukkyō tetsugaku* (1893), in *Inoue Enryō Senshū*, edited by Tōyō Daigaku sōritsu hyakunen shūnen kinen ronbun shū hensan iinkai, vol. 7, 107–181 (Tokyo: Toyodaigaku 1990), 109. For an English translation, see G. Clinton Godart, transl., "Inoue Enryō," in *Japanese Philosophy: A Sourcebook*, Edited by James W. Heisig, Thomas P. Kasulis, and John C. Maraldo (Honolulu: Hawai'i University Press, 2011), 619–623.

23 Inoue Enryō, *Tetsugaku yōryō* (1888), in *Inoue Enryō senshū*, vol. 1, 147.

religion, this belief in a supra-rational absolute will not change, but, he argued, religious institutions have to adapt to society. While in *The Future Japan*, he had been unabashedly anti-Buddhist, he now defended a role for Buddhism in Japan.[24]

Another example of the remarkable variety of the ways in which Spencer was appropriated in Japan was the case of Mori Arinori.[25] One of the early Meiji leaders, Mori had read Spencer during his stay in the United States, and when he became ambassador to England from 1879 to 1884, met Spencer, and consulted with him. Mori also spent time in the prestigious Atheneum Club, and held his farewell party there, attended by Spencer. Mori eventually became minister of education, and in this position employed Spencer's philosophy in his own ideas and policies. Mori does not fall into a dualist scheme of "liberals" versus "conservatives." He advocated freedom of religion, and probably inspired by both Mill and Spencer, the equality of women and the reform of marriage (which at the time still allowed for official concubines), but also installed military led physical training in schools, a top-down centralized education system, and was lukewarm about a parliamentary system for Japan.

When he became minister of education, the central elements of Mori's educational reform were to boost physical and moral education, two elements he found lacking in Japan. Spencer formed the theoretical basis for this endeavor. Spencer had argued that education should prepare students to live independent and full lives, as this was a prerequisite for a free society in which the external controls of the militant society had been removed. Mori shared this belief. Opposing some of the conservatives, he wanted to install an ethics education that was not Confucian in outlook, and wrote an *Ethics Textbook* (*Rinrisho*), which was inspired by Spencer's *Principles of Ethics*. Mori adopted Spencer's idea that the key in ethics was a balance between egoism and altruism, and Mori modified it by calling it *jita heiritsu*, which he rendered himself in English as "the *cooperation* of self and other."[26] Mori's interest in physical education came from his own samurai schooling and his experience with muscular Christianity in the United States, but it was Spencer's *Education: Intellectual, Moral, and Physical*, that led him to place great emphasis on the balance between the three aspects of education. Hence from 1886, worried that the physical component had been neglected, he advocated the

24 Shinbukkyō dōshikai, eds., *Shōrai no shūkyō* (Tokyo: Shinbukkyō dōshikai, 1903).

25 This section on Mori is largely based on Ivan Parker Hall, *Mori Arinori* (Massachusetts: Harvard University Press, 1973), and Alistair Swale, *The Political Thought of Mori Arinori: A Study of Meiji Conservatism* (London: Routledge, 2000).

26 Swale, *The Political Thought of Mori Arinori*, 170, and Mori Arinori, *Rinrisho*, 424–429.

installation of military-style physical education in Japanese schools, run by the army, probably not what the anti-militarist Spencer had in mind. It also caused a great controversy. The military aspect of physical education was of course abandoned after 1945, but the basic orientation of a balance between the three elements of education has remained to this day in the form of the basic principle of the "three educations" (*san'ikushugi*).[27] These are some of the more longer lasting but less noticed effects of Spencer in Japan.

Mori also consulted with Spencer over Japan's political future, and presented him with a draft of a constitution. In what became a famous episode, Spencer gave what he himself called "conservative" advice, to not prematurely replacing the old with the new, but "grafting" the new (democracy) on the old (family system). As Alistair Swale has pointed out, Mori's own *On a System of Representative Government for Japan* of 1883, the draft of which he showed to Spencer, set out with the question "How to engraft a Representative System of Government on the political institutions of Japan?" – suggesting Mori was strongly inspired by Spencer who had put the problem in the same words. Mori argued that Japanese society was not ready yet for a parliamentary system.[28] Spencer later wrote to statesman Kaneko Kentarō (1853–1942):

> Probably you remember I told you that when Mr. Mori, the then Japanese Ambassador, submitted to me his draft for a Japanese Constitution, I gave him very conservative advice, contending that it was impossible that the Japanese, hitherto accustomed to despotic rule, should, all at once, become capable of constitutional government. My advice was not, I fear, duly regarded, and so far as I gather from the recent reports of Japanese affairs, you are experiencing the evils arising from too large an installment of freedom.[29]

Spencer, fearing Japan would fall prey to Western imperialism, also advised against allowing intermarriage. It is noteworthy that on this high level of government consultation with the great philosopher, Japan did not follow Spencer's advice, and allowed intermarriage, and adopted a constitution and a parliamentary system.

---

27 See Mark Elwood, *Principle, Praxis, and the Politics Of Educational Reform in Meiji Japan* (Honolulu: University of Hawai'i Press, 1995).

28 Swale, *The Political Thought of Mori Arinori*, 82–88.

29 David Duncan, *The Life and Letters of Herbert Spencer* (London : Williams & Norgate, 1911), 319.

The "Spencer Bust"

From the 1890s onwards, and especially after the turn of the century, Spencer's name appears less and less. Japanese scholars already began treating Spencer as a past chapter in the history of ideas. In 1927, when reflecting on the history of social thought in Japan, one commentator asked: "How did his [Spencer's] thought, that was for a time at the height of popularity, vanish without leaving a trace?"[30] It is a question not many scholars have asked. Amidst all the talk of the "Spencer boom," the "dual use" theory of Spencer in Japan did not tell us a lot about the limitations of Spencer's role in Japan, or why it declined so dramatically. The passing of time in itself is of course not a reason, since older figures like Marx, Goethe, and Hegel continued to be influential. But theories do sometimes also follow a pattern similar to fashion, and by the turn of the century, Spencer had a distinctive mid-nineteenth century aura around him, and his writing style did not help. Another commentator, Nishimura Tengai, who wrote in 1890 about Spencer, declared, "in a nutshell, he gets lost in long-windedness." Nishimura stated that Spencer gives so many examples that are out of place that "if one opens his book and just looks at the examples, one wonders if this is a ghost house (*bakemono yashiki*), or *The Journey to the West* (*Saiyūki*). In other words, his works are like a *manjū* [a bun with a bean-jam filling] with a really thick outer layer, using an abundance of weird examples, while talking only very little about the essence of his philosophy."[31]

To a large degree, the decline of Spencer's importance in Japan was part of a world-wide pattern, and can be attributed to the further specialization of disciplines with their own methodologies, such as in psychology and biology, a growing split between the sciences and the humanities, and the further rise of Marxist thought and the discipline of sociology to understand society (although Spencer's popularity started to decline in the 1890s, before Marxist thought made serious inroads in Japan). Philosophers criticized Spencer for his deductivist method and there was a rise in interest in epistemology and Neo-Kantianism, as well as German idealism.

However, just as there are local circumstances for the successful transmission of ideas, there are also local conditions for the decline of ideas. Firstly, Spencer's evolutionary pattern from "militant" to "industrial," from war and rigidity to liberty and peace, as he himself realized, was belied by later historical events. In Japan, it seemed to have been really the Sino-Japanese war of 1894–1895 that resulted in territorial gain, access to markets, a large indemnity

30 Shimoide Shunkichi, *Meiji shakai shisō kenkyū* (Tokyo: Asano shoten, 1932), 48.

31 Nishimura Tengai, *Wakan taisei kokon gakusha retsuden* (Tokyo: Kōbunkan, 1890), 299.

from China, a boost in national prestige and a wave of jingoism, which blasted away Spencer's optimistic model of modernity. When France, Germany, and Russia intervened to force Japan to give up its hold on the Liaodung peninsula, which most Japanese saw as legitimate war booty, the disappointment and anger was huge. Tokutomi Sohō experienced it as a turning point in his life, and he gave up his Spencerian liberalism and began to argue for a more aggressive foreign policy. The much larger and devastating, but nevertheless victorious, Russo-Japanese war of 1904–1905, did little to revive Spencerian pacifism. And the few who opposed the war had turned to socialism. Contrary to Tokutomi's predictions, Japan's military build-up, expansion, and war did not seem incompatible with industry and trade.

Secondly, concomitantly with these wars, Japanese intellectuals and ideologues began to revive (and re-invent) Tokugawa-era martial values, known as *Bushidō* (The Way of the Warrior) as the spiritual and timeless essence of the Japanese people. This was exactly the opposite of Spencer's theory that militant values, such as loyalty, would gradually decline. One of the most influential spokesmen for the revival of *Bushidō*, Nitobe Inazō (1862–1933), argued, in contrast to Tokutomi's earlier wholesale characterization of the Tokugawa society as "militant," that Spencer's characterization in fact applied only to the samurai class (conceding for example that women in the samurai class were the most restricted), which was a tiny minority, and even then only to a very limited degree. More importantly, he argued that the Tokugawa era and *Bushidō* had fostered in the Japanese the value of loyalty, an "ethical outcome" of the political theory of "the state as anteceding the individual – the latter being born into the former as part and parcel thereof, – he must live and die for it or for the incumbent of its legitimate authority."[32]

Thirdly, and most obviously, when capitalism and industry began to take off in Japan, as elsewhere, it did not seem to bring Spencer's promised moral improvement. On the contrary, the public attention was drawn to what came to be known as the "social problem," the slums, prostitution, alcohol abuse, violence, and crime.

Spencerism was not just replaced. Nor did it wither away. The intellectual credibility and appeal of Spencer was also actively demolished by influential intellectuals and academics. Let us look at Nakajima Rikizō (1858–1918), an important academic philosopher at the University of Tokyo, and a teacher of a new generation of important philosophers. Hence his criticism of Spencer, which he passed on to his students, is important and gives some insight into the decline of Spencer in the academy. Nakajima became one of the major

32 Inazo Nitobe, *Bushido* (1905, Tokyo: Charles R. Tuttle, 1969), 88–89.

proponents of the philosophy of Thomas Hill Green (1836–1882), himself a critic of Spencer, and whose Hegelian philosophy enjoyed a large popularity in the Japanese philosophical world from the 1890s. Nakajima's philosophical outlook was more religiously oriented than Spencer's (he later became involved with Theosophy), probably one reason for his interest in Green. Nakajima's treatment of Spencer in his 1898 *Short Biographic History of Western Philosophy* is revealing. Here he grouped Spencer in the "second period" of British philosophy (together with Reid, Hamilton, and Hartley), which was followed by a "third period" exemplified by T.H. Green.[33] Hence Nakajima, already in the 1890s, relegated Spencer to a past stage in the evolution of the history of philosophy.

Nakajima was very concerned about moral education in Japan and was personally involved in writing textbooks. Hence he had a stake in the evaluation of Spencer's theory of ethics. In 1909, Nakajima published *Spencer's Theory of Ethics*, one of the few book-length studies on his thought in Japan at the time.[34] In this work, probably intended for students and the general public, Nakajima presented a whole range of arguments against Spencer, not only his theory of ethics. We can probably assume that Nakajima had been teaching these criticisms of Spencer at the University of Tokyo, thus contributing to the demise of Spencerism in Japan.

Firstly, echoing a larger rejection of scientism around turn of the century Japan, Nakajima rejected Spencer's deductivist method as unscientific. Nakajima also doubted "that his [Spencer's] evolutionary philosophy, in other words the basis for his theory of ethics, is the complete truth is rather doubtful."[35] Moreover, Spencer was too selective in his base of evidence: "He [Spencer] tried to take the life of a certain kind of scholar [his own] and turn that into the standard of all men in the real world. I think that it is for this reason that what Spencer says will not survive in the real world."[36] Spencer's laws of ethics, Nakajima argued, were not eternal scientific laws on which basis human behavior could be prescribed.[37]

Secondly, Nakajima rejected Spencer's model of progress from "militant" to "industrial," and especially the idea of war as an inhibiting factor in the progress of morality. This criticism should be seen in the background of the Sino-Japanese and Russo-Japanese wars, both resulting in victories for Japan.

33 Nakajima Rikizō, *Rettai seiyō tetsugaku shōshi* (Tokyo: Fuzanbō, 1898).

34 Nakajima Rikizō, *Supensā shi no rinrigaku setsu* (Tokyo: Dōbunkan, 1909).

35 Nakajima, *Supensā shi no rinrigaku setsu*, 175.

36 Ibid., 193.

37 Ibid., 183.

War, Nakajima argued, often spurred the progress of morality, for example the virtues of patriotism, empathy, group cohesiveness, loyalty, all "beautiful virtues nurtured for war," and "such habits as valuing knowledge are often gained through military education."[38] Nakajima was no militarist, but was sympathetic to a revival of *Bushidō*, and in other writings lamented the decline of the samurai class as the loss of a moral example for the rest of the population. Japan, he argued, was therefore in need for a new "core class" in society that would place honor and service at the highest level, and would therefore be able to unite society.[39]

Nakajima also rejected Spencer's idea that a society becomes more moral as it becomes more "industrial." On the contrary, new rifts will appear within society, and people are not rewarded for what they are worth. "We can see that this kind of harm is already happening in today's society: war is [indeed] gradually declining, but in the struggle on the dimension of industry, people, according to their qualities, do not gain much reward, and therefore strife occurs. Therefore his theory that as society evolves from militant to industrial, the world will be one in which the principles of justice are sufficiently realized, is very doubtful."[40] As society evolves from warlike to industrial, the gap between rich and poor grows, and it is doubtful that this will be a world in which people will receive rewards commensurate with their abilities. Nakajima rejected Herbert Spencer's vision of society and justice, an extreme of "reaping what one sows" (*jigyōjitoku*) as ultimately immoral.[41]

Thirdly, Nakajima rejected Spencer's theory of justice, conceived as the maximizing of individual freedom as long as it does not infringe upon others' right to exercise their freedom. One problem, Nakajima argued, was that "People differ in their constitution and abilities due to their education, environment, and heredity [also according to Spencer's own theory of evolution]. Therefore I think it is impossible to establish justice on the basis of only standardizing freedom."[42] Furthermore, Spencer's idea of justice was "simply a principle of social morality, and exists only between adult individuals, while denying it exists in the family, between parent and child, or between husband and wife. This goes against common sense."[43] The right to marry, property rights, rights of inheritance, rights between parent and child, and between

38 Ibid., 188.

39 Nakajima Rikizō, *Tokuiku to rinri* (Tokyo: Meguro shoten, 1907), 178–179.

40 Nakajima, *Supensā shi no rinrigaku setsu*, 202–203.

41 Ibid., 202.

42 Ibid., 206.

43 Ibid., 196.

husband and wife, could not, Nakajima argued, be explained with Spencer's theory of justice. This criticism reflected a specific late Meiji-era renewed emphasis on conservative family values (as stated in the official vision of the Meiji state as a "family state"). Nakajima, and many other Japanese at the time, asked: why indeed should a theory of ethics and justice take an abstract individual, stripped of its qualities and place in the family, as the ultimate unit and measure of justice?

In conclusion, it is probably safe to say that those aspects of Spencer's thought that gained the most attention at the time of the introduction of Spencer's thought in Japan, especially the model of evolution from militant to industrial, have proven not to have had a lasting appeal in Japan, and came to be widely rejected, accelerating Spencer's demise in Japan. Spencer's agnosticism also did not seem to have a wide following in Japan. In the medium-term, it should be noted that Spencer's thought was an important way for the early liberal movement to find its voice. But other less noticed aspects, such as in biology, his contribution to Japan's envisioning and construction of the categories of "society" "religion" and "science," and evolutionary conceptions of these phenomena (even although Spencer's own theories on these have been forgotten) might have been more lasting. Spencer's ideas on education, through Mori Arinori's efforts, have left the most visible impact on Japanese society. It is probably no coincidence that Spencer's *Education*, first translated and published by the Ministry of Education in 1880, was translated again in 1941 and again in 1955, long after his popularity as a philosopher had waned.

Perhaps the rise and fall of Spencerism worldwide can to some degree be understood as a boom and bust; Spencer's "Synthetic Philosophy" was a giant systematic attempt to explain almost everything according to a couple of principles of evolution (such as from homogeneity to heterogeneity, and from militant to industrial) and adaptation. While it is often said that a good theory is simple in its formulation but able to gather many phenomena under its umbrella, unlike Darwin's theory of natural selection, Spencer's philosophy generated some inflated expectations that the theory eventually seemed not to be able to meet, and perhaps unfairly, at the first sign of trouble, phenomena that did not seem to fit the theory, and competition with rival ideas, most investors ran. In Japan's case, after an initially enthusiastic embrace amidst the changes of the 1870s and 1880s, the subsequent successes of the wars against China and Russia, which did not fit the prediction of a social evolution from militant to industrial, concomitant with an influx of new philosophies and a renewed interest in conservative and nationalist ideas, all ensured Spencerism dramatically lost value for Japanese intellectuals.

TABLE 3.1 *Major translations of Herbert Spencer's works into Japanese, 1870–1920 (place of publication is Tokyo, unless otherwise indicated)*

Years	Original	Title	Publisher	Translator	Comments
1877	*Social Statics* (1851)	*Kenri teikō*	Maruya Zenshichi (Maruzen), 2 vols, 162 and 103 pages	Ozaki Yukio	Ozaki Yukio (1858–1954) became one of the leading liberal politicians of Japan
1879	"Over-Legislation" (1853)	*Kanshōron*	Yamanaka Ichibei, 114 pages	Suzuki Yoshimune	
1880	*Education: Intellectual, Moral, and Physical* (1861)	*Kyōikuron*	Monbushō (Japanese Ministry of Education), 479 pages	Seki Shinpachi	Seki Shinpachi (1839–1886) was a scholar of English who had worked for both the Shogunate and the new Meiji government
1878	"Representative Government" (1857)	*Daigi seitai ron*	Suzuki Yoshimune, 86 pages	Suzuki Yoshimune	
1881	*Social Statics*, Chapter 16: "The Rights of Women"	*Joken shinron*	Shiseidō, 60 pages	Inoue Tsutomu	
1882	*First Principles* (1862)	*Tetsugaku genri*	Katō Shōshichi	Yamaguchi Matsugorō	
1882	*Social Organism* (1860)	*Shakai soshiki ron*	Matsunaga Yasutarō, 113 pages	Yamaguchi Matsugorō	
1882	*Prison Ethics* (1860)	*Keihō genri goku soku ronkō*	Yamaguchi Matsugorō, 117 pages	Yamaguchi Matsugorō	
1882	*The Morals of Trade* (1859)	*Shōgyō rigai ron*	Katō Shōshichi	Yamaguchi Matsugorō	
1883	*Principles of Ethics*, vol. 1, Part I: *The Data of Ethics* (1879)	*Dōtoku no genri*	Suhara Tetsuji, 683 pages	Yamaguchi Matsugorō	

TABLE 3.1 Cont.

Years	Original	Title	Publisher	Translator	Comments
1883	*Representative Government* (1857)	*Daigi Seitai ron fuku gi*	Ogasawara shobō	Miyazaki Masaaki	
1883	*The Study of Sociology*, Chapter 1–10 (1873)	*Shakaigaku*	Zega shobō, 5 vols, 775 pages	Ōishi Masami	Ōishi Masami (1855–1935) became the secretary of the Liberal Party of Japan and in 1898 Minister of Agriculture and Commerce
1883	*Political Institutions*, Part V of *Principles of Sociology*	*Seitai genron*	Nishimura Gendō, 118 pages	Ōichi Masami	
1884	*Social Statics* (1851)	*Shakai heiken ron*	Hōkokusha	Matsushima Kō	
1884	*First Principles* (1862)	*Tetsugaku genri*	Katō Shōshichi 3 vols.	Yamaguchi Matsugorō	
1884	(*Chapters 12–14 of First Principles*)	*Manbutsu shinkaron*	Mintokukan	Sugimoto Kiyohisa and Nishimura Gendō	
1884-1885	(*Partial translation of Principles of Sociology, Book V*)	*Seihō tetsugaku*	Ishikawa Hanjirō, 2 vols, vol 1 (390 pages), vol 2 (383 pages)	Hamano Sadashirō and Watanabe Osamu	Foreword by Fukuzawa Yukichi
1885	*Principles of Sociology* (1874–1875)	*Shakaigaku no genri*	Keizai zasshi sha	Noritake Kōtarō	Under supervision of Toyama Shōichi
1886	*Principles of Sociology, vol. III, Part VI, Chapter 16: "Religion: A Retrospect and Prospect"*	*Shūkyō shinka ron*	Matsuda Shūhei, 69 pages	Takahashi Tatsurō	Foreword by Itagaki Keisuke, leader of the Freedom and People's Rights Movement

Years	Original	Title	Publisher	Translator	Comments
1886	*Education*	*Kyōikuron*	Makino Zenbei, 2 vols., 571 pages	Ariga Nagao	
1888	*Representa-tive government*	*Daigi seitai tokushitsu ron*	Fuzanbō, 177 pages	Hiramatsu Kumatarō and Takada Sanae	
1894	*Individual Versus the State*	*Kojin tai kokka ron*	Hakubunkan, 161 pages	Nagai Kumaji	
1892	*Principles of Ethics*	*Su shi rinri gen ron*	Tsūshin kōgakkai, 564 pages	Tanaka Tōsaku	
1892		*Riseiron: ichimei kanpi minson ron*	Kobe: Shibamoto Seiji, 121 pages	Zenda Chō	
1893	*Over-legisla-tion*	*Kanshō ron*	Bunkōdō, 115 pages	Koshikawa Gennosuke	
1898	*First Principles*	*Sōgō tetsugaku genri*	Keizai zasshi sha, 833 pages	Fujii Uhei	
1923	*First Principles*	*Dai ichi genri*	Keizai zasshi sha, 826 pages	Sawada Ken	
1941	*Education*	*Kyōikuron*	Mikasa Shobō	Ishibashi Kinosuke	
1955	*Education*	*Kyōikuron*	Tamagawa daigaku shuppan bu, 268 pages	Okamoto Ninzaburō	

CHAPTER 4

Spencer and Science Education in China

Ke Zunke and Li Bin

In the 1980s Chinese philosopher Jin Yuelin recalled studying in Britain in the 1920s. He wrote, "Spencer, who was a British sociologist, wrote a long series of works, which was no more than a platitude, and made no real achievements. I remembered that his grave was near to Marx's, but the people, who coming to view Marx's grave today, had forgotten or even not know at all that there was a man like Spencer once existed."[1] Jin Yuelin's statement reflected the image and reputation of Spencer in China in the 20th century. His philosophy didn't attract detailed study, just some brief introductions.[2] He was regarded as a Social Darwinist, who applied Darwin's theory to sociology.[3] He and Huxley were viewed as the icons of thought of science education in 19th-century England.[4] Spencer's decline took place not only in China, but was almost a worldwide phenomenon. However, he still held considerable status among Chinese intellectuals in the early 1920s, which began with the translation of his essays and works.

Only a small part of Spencer's works was translated into Chinese, which mainly focused on education and sociology (see Table 4.1 at the end of this chapter). Through the circulation of these translations, Spencer's ideas of education and sociology made significant contributions to the modernization of Chinese culture and education. Previous Chinese scholarship on Spencer concentrated on the content and influence of his ideas, seldom paying attention to the process of communication and appropriation.[5] James Secord argues that a communication revolution took place in Britain from 1830 to 1850. This revolution allowed the publication of cheap science books for a larger reading

---

1 Jin Yuelin, *Jin Yuelin huiyilu*, ed. Liu Peiyu (Beijing: Beijing daxue chubanshe, 2011), 27.

2 Liu Fangtong et al., eds., *Xiandai xifang zhexue* (Beijing : Renmin chubanshe, 1990), 76–87.

3 Wang Yangchong, *Xifang jindai shehuixue sixiang de yanjin* (Shanghai: Huadong shifan daxue chubanshe, 1996), 47–76.

4 Wu Shiying and Ren Zhongying, eds., *Waiguo jiaoyu sixiang tongshi* (Changsha: Hunan jiaoyu chubanshe, 2002), 148–193.

5 Wang Xianming, "Yanyi mingzhu yu Zhongguo wenhua de xiandaihua: yi Yan Fu yi *Qunxue yiyan* weili de kaocha," *Fuzhou daxue xuebao(shehui kexue ban)*, no. 2 (2008): 24–30; Xiao Lang, "Kejiao xingguo de qiangyin: Sibinsai jiaoyu sixiang zai jindai Zhongguo," *Huadong shifan daxue xuebao(jiaoyu kexue ban)*, no. 2 (2008): 88–96.

 | DOI 10.1163/9789004264007_006

audience that included middle class and working class readers. And so, scientific ideas had more importance than ever. Simon Eliot has labeled this same process the "distribution revolution," and asserts that there also came the "mass-production revolution" in the fourth quarter of nineteenth century. This two-part industrial revolution in the production and sale of books had a profound effect on the dissemination of science.[6] After the Sino-Japanese War in 1894, a publishing revolution took place in China, a "distribution revolution" and a "mass-production revolution" came one after another from 1895 to the 1930s, and quickly changed the character of Chinese society and culture. Christopher A. Reed points out that from Late Qing Dynasty to the Republic of China woodblock printing was developed into mechanized printing by the introduction of western printing technology, and print capitalism was finally established in the 1930s.[7] Due to the great difference between China and Britain, the communications revolution in China had its own features. Chinese people chose printing technology according to the traditional culture and their own needs. At first lithography was actively adopted by publishers and spread quickly in the late nineteenth century. The communications revolution was not accompanied by a rise in literacy rates as in Britain. After the abolition of the imperial civil examination in 1905, which destroyed the traditional education system, the literacy rate decreased for more than two decades and increased again in the early 1930s.[8] In this chapter we will argue that the impact of Yen Yung Kiung's translation was initially restricted, and that the appropriation of Spencer in China was more important before the communications revolution began. In the explosion of journals from 1895 to 1912, Yan Fu's political and educational proposition "San Yu Bing Ju," which was based on Spencer's classification framework of education, was adopted by Chinese intellectuals. Yen Yung Kiung's translation was also valued and reprinted. In 1923 Zen Hung-chun took part in "the science versus metaphysics argument" in journals and periodicals. His translation was published that year and included in the Wan You Wen Ku Series (The Complete Library Series) three years later. It was a cheaply priced series for general readers and was collected in nearly all libraries in China. Chinese intellectuals appropriated Spencer's ideas of education

6 Bernard Lightman, *Victorian Popularizes of Science: Designing Nature for New Audiences* (Chicago & London: University of Chicago Press, 2007), 30–33.

7 Christopher A. Reed, *Gutenberg in Shanghai: Chinese Print Capitalism, 1876–1937* (Vancouver: The University of British Columbia Press, 2004).

8 Luo Zhitian, "Keju zhidui feichu zai xiangcun zhong de shehui houguo," *Zhongguo shehui kexue*, no. 1 (2006): 191–204.

for their own purposes, and it was the communications revolution that provided the necessary arrangements to assure their success.

Spencer's *Education* included four review-articles, which, he notes, I "originally wrote with a view to their republication in a united form."[9] He first argued that science is the most useful knowledge, and then discussed successively intellectual, moral, and physical education. These essays made up an integrated whole and were published initially in America in 1860. After the *North British Review* allowed Spencer to include "Intellectual Education," the four articles were republished in England one year later. It was his first bestseller and translated into French, German, Italian, Russian, Hungarian, Dutch, Danish, Spanish, Swedish, Bohemian, Greek, Sanskrit, Arabic, Japanese and Chinese from 1860 to 1882.[10] Spencer's *Education*, which consisted of four articles, had extraordinary mobility, and could apply to many different conditions. The first two Chinese translations were not the full text. Yen Yung Kiung, a minister of the American Episcopal Church in China, translated the first article of Spencer's *Education* in Late-Qing Dynasty. After the founding of the Republic of China, scientist Zen Hung-chun translated the first two articles. In 1962 educationalist Hu Yi translated the whole book. Chinese traditional knowledge hierarchy gradually disintegrated in less than a century, and scientific education was finally founded in the 1920s after reforms of the political and educational system. Spencer's educational ideas played a significant role in this process, and became the most important of his ideas transmitted to China. Yen Yung Kiung, Yan Fu, Liang Qichao and Zen Hung-chun were key players in this process. Translation is a distinct stage of appropriation. By choosing articles and emphasizing some aspects of Spencer's thought, or modifying his meanings, different translators could appropriate Spencer's *Education* for their own use. In his translation Yen Yung Kiung omitted paragraphs related to science and religion to cater to Christian readers. He considered science education as a means of proselytizing, and urged that it replace Chinese traditional education. By propagating "San Yu Bing Ju," Yan Fu, Liang Qichao and other journalists demanded that the imperial civil examination be abolished, and a more secular scientific education be built. Though Zen Hung-chun seized the chance to publish his translation, his ultimate goal was to develop scientific research, not just scientific education.

9 Herbert Spencer, *Education: Intellectual, Moral, and Physical* (New York: D. Appleton and Company, 1864), xvii.

10 Michael W. Taylor, *The Philosophy of Herbert Spencer* (London: Continuum, 2007), 21.

1 Church Education and *Yi Ye Yao Lan*

Yen Yung Kiung was one of the earliest Chinese ministers of the American Episcopal Church in China. He was born in a poor family in Shanghai in 1838. When he was ten years old, he went to a school founded by the American Episcopal Church that was near his home, where he was baptized the following year. In 1854 he was sent to America to study. In 1861, he graduated from Kenyon College in Gambier, Ohio, and received a bachelor's degree. While he studied in America, Archdeacon G.T. Bedell of Ohio Parish was his guardian. In 1862 he returned to China and prepared for preaching, but he was unable to do so because of financial difficulties. In 1868 Yen Yung Kiung was ordained by Bishop C.M. Williams, and then promoted to priest next year. He became the second Chinese priest of the American Episcopal Church in China. He began to preach at Wuchang and assisted in the founding of Bishop Boone Memorial School. In 1878, Bishop Samuel Issac Joseph Schereschewsky planned to build a church college in Shanghai, and asked Yen Yung Kiung to return to his hometown to help. When St. John's College started in 1879, Yen Yung Kiung was appointed as the proctor who managed the students and kept order in the college. He also was the first professor of natural sciences of St. John's College, and taught mathematics, physics, chemistry, astronomy and theology. He believed that there was no conflict between religion and science, as teaching natural sciences would contribute to missionary work, and eliminate superstition in China.[11]

To spread Protestantism in China, Western Christian Churches encouraged their members to missionize by developing education and spreading knowledge. Some Chinese scholar-officials were interested in church education too, because it could provide some scientific knowledge. During his early years in St. John's College, Yen Yung Kiung translated Spencer's article, "What Knowledge is of Most Worth." He loathed the Chinese traditional education, and believed that it was the reason that China remained stagnant and backward. If China wanted to progress and prosper, scientific knowledge was essential. In his introduction to *Yi Ye Yao Lan* in 1882, Yen Yung Kiung wrote "after the useful knowledge takes the place in education, our country will irresistibly become strong and prosperous."[12] Spencer demonstrated in his article "What Knowledge is of Most Worth" that science provided the most valuable

11 Xu Yihua, "Yen Yung Kiung yu Shenggonghui," *Jindai Zhongguo* 10 (2000): 193–215; Xu Yihua, "Xifanghua yu Chujinghua: Shenggonghui sanwei huaren xianqu mushi zhi yanjiu," *Meiguo wenti yanjiu* 2 (2002): 310–341.

12 Yen Yung Kiung, trans., *Yi Ye Yao Lan* (Shanghai: Shanghai gezhi shushiben, 1895), iii.

knowledge, and advocated that science should take the place of the classical languages. Yen Yung Kiung believed Chinese traditional education, which was based on the study of Confucian Classics, mainly The Four Books and Five Classics, and learning how to write Eight-part essays, was very similar to the classical education in England. So he believed it should be replaced by scientific education too. The purpose of his translation was clearly to give an example to Chinese scholars. He wrote,

> I have observed the education in England at some time in the past. The curriculum in the universities focused narrowly on the classics, Greek, Roman classics and old English works, which all was not utility. The problem of learning in China is not a mirror image of that of England, but in some ways. However, our problem could have been solved with the methods in this book.[13]

He also wrote "Translator's Preface" in English, which was more straight and frank than the Chinese introduction. He avowed that "this translation is a work of patriotism," and declared it "will aid in reforming the present narrow education in my country, and thus be the means of promoting her progress." He appealed to his English-speaking countrymen to "lend their influence in circulating this essay."[14]

During the Self-Strengthening Movement, which began in 1864 and was sponsored by powerful army generals to consolidate the Qing's military power, the government and Protestant missionaries were the major agents that introduced western learning to China. Schools, newspapers and publishing houses were founded to spread western knowledge. There were numerous missionary schools, Tung-wen Kuans in Peking and Shanghai, which were foreign language schools run by the Qing Government, and the Shanghai Polytechnic Institution. The major publishing houses included the American Presbyterian Mission Press, the School and Textbook Series Committee, the Christian Literature Society of China, and the Translation Department of the Jiangnan Arsenal. In the second half of the 19th century more than 555 western books were translated, including 123 on philosophy and social sciences, 162 on natural sciences, 225 on applied science, and 45 other categories.[15] Almost every newspaper in late-Qing Dynasty introduced some western knowledge. *The*

13 Yung Kiung, *Yi Ye Yao Lan*, iii.

14 Ibid., ii.

15 Xiong Yuezhi, *Xixue dongjian he wanqing shehui* (Beijing: Zhongguo renmin daxue chubanshe, 2011), 8–9.

Chinese Scientific and Industrial Magazine and *Church News* were the most important of the newspapers. At that time Chinese printing technology had made some progress. William Gamble, the director of American Presbyterian Mission Press, invented the Chinese type electroplating and the Chinese composing frame. The Chinese type electroplating saved time and labor, and the type produced by it had perfect shapes and clear characters. The types and typefaces were produced by the American Presbyterian Mission Press in quantity, and sold to newspaper offices and bookshops in various regions. The Chinese composing frame improved the typesetting speed at least three times, and was widely adopted by the publishing houses. The American Presbyterian Mission Press became one of the most important Christian publishing institutions in China; it not only printed most of the newspapers and books of churches, but also published a number of books on science and technology.[16] These technical innovations did not make it possible to publish cheap science books. But even so, the new printing technology allowed for greater circulation of western knowledge among the Chinese elite. In Late Qing Dynasty, science and technology books written in classical Chinese language were expensive. Common people could not afford them and only the scholars could understand them. The literacy rate stayed at nearly 20% at that time, and fewer people could read and understand classical Chinese language.[17] The Jiangnan Arsenal published 98 western books from 1871 to 1880, and only 83,454 copies were sold in total.[18] Western learning was circulated to some Chinese intellectuals, but the traditional education still kept its dominant position in Chinese society.

Yi Ye Yao Lan was one of the first translations produced independently by the Chinese people, not by the means of "Kouyi bishu." At that time "Kouyi bishu" was the main way of translating western books, which meant the foreign interpreter dictated sentence by sentence while the Chinese recorder recorded, and then revised and modified the final drafts. This approach required communication, consultation, and cooperation between the interpreter and the recorder. However the accuracy of translation was mainly dependent on the interpreter's knowledge and the style of writing was decided

16 Zhang Jinglu, ed., *Zhongguo jinxiandai chuban shiliao* (Shanghai: Shanghai shudian chubanshe, 2011), vol. 1, 259–261; Hu Yuanjie and Jing Zhiyu, "Zhongxi wenhua jiaoliu de qiaoliang: meihua shuguan," *Dangan yu shixue*, no. 3 (2003): 63–65.

17 Hao Jinhua, "20 shiji qianbanye xiangcun renkou de shizi shuiping," *Shanxi daxue xuebao(zhexue shehui kexue ban)*,no. 6 (2004): 116–118.

18 Lei Qili, "Wanqing minchu de yinshua jishu yu wenhua shengchan," *Huadong shifan daxue xuebao(zhexue shehui kexue ban)*, no. 5 (2008): 54–62.

by the recorder's ability. As translators often purposely changed the meaning of texts to suit their own aims, accuracy was always hard to achieve, or perhaps never achieved. The quality of translations was not good in general. "The interpreter can not do without losing the original meanings and the recorder can not do without choosing the wrong terms," Liang Qichao criticized. Most translations were "heterogeneous and miscellaneous, erroneous, slangy and vulgar."[19]

The article "What Knowledge is of Most Worth" was translated solely by Yen Yung Kiung. *Yi Ye Yao Lan* did not have the common faults of the "Kouyi bishu" translation process, and was considered to be an excellent translation by contemporary readers. However, Yen Yung Kiung was one of a few special cases, and actually the power of language was in the interpreters' hands. They often were missionaries or had a church background, and controlled the process of translation and dissemination. They inserted natural theology in their translations, which made the meaning more religious.[20] Translation was often a means of appropriation that was even more subtle than appropriations resulting from the use of Spencer's ideas from texts that had already been translated. It was the School and Textbook Series Committee that made efforts to translate and edit the textbooks, and standardize the Chinese terminology in the late Qing dynasty.[21] In his "Translator's Preface," Yen Yung Kiung declared that he didn't "sympathize" with Spencer's opinions about science and religion.[22] He omitted some passages in his translation, and thus transformed Spencer's discourse into an argument for natural theology, stressing the ethical implication of the laws of nature. In this process of appropriation, he consciously and actively changed the meaning of Spencer's work. Yen Yung Kiung became the first and only Chinese member of the School and Textbook Series Committee in 1886, and one of the two Chinese members of the Educational Association of China in 1896. He was one of the few Chinese people who were admitted into the highest echelons of church education. *Yi Ye Yao Lan* was the textbook authenticated by the Educational Association of China, and Yen's work was highly praised in the Catalogue of Books published by the Educational Association of China. "He knows Chinese and Western literature so well that he can grasp the original meanings and express them with pure Chinese," the

19 Liang Qichao, *Yinbingshi heji* (Beijing: Zhonghua shuju, 1989), vol. 1, no. 1, 67–68.

20 Benjamin A. Elman, *On Their Own Terms: Science in China, 1550–1990* (Cambridge, Mass.: Harvard University Press, 2005), 320–351.

21 Wang Yangzong, "Qingmo yizhi shuhui tongyi keji shuyu gongzuo shuping," *Zhongguo keji shiliao*, no. 2 (1991): 9–19.

22 Yung Kiung, *Yi Ye Yao Lan*, ii.

肄業要覽序文

嗟乎天壤間之風土人情不甚相遠也若是夫我中朝與西國通好以來凡中土之利弊西人靡不知之間有剴切指陳以代爲匡救者則幾疑西人之於利弊無瑕可指矣而不知其風土人情亦有固結而莫可解者昔於咸豐時有英國名士史本守先生目擊情形慨然有感思欲挽救於將來著之於册名曰肄業要覽夫乃嘆天壤間之風土人情固不甚相遠也今取而讀其書亦不激不隨其所論列又確中時弊凡歐洲法德意俄荷丹恒諸國亦既譯以本文若美國本係同文可不必贅其後肄業之間具有實用而風俗亦蒸蒸日上恨吾華人曾未之見也雖曰以之自鏡未必盡同然其所謂學問之弊固有不類而類者故於其剔弊諸法亦足餉我無窮予因暇時譯以成書竊効芻蕘之一得　恐文辭蕪陋不足達其所言惟望三復及之以返觀內証是亦當世得失之林也爰爲之序

光緒七年嘉平月　　上海顏永京書於約翰書院

嘗觀列國之人皆以粧飾爲事即如衣服亦用爲美觀而不以爲保護身體之用皆如此已有一種文身之俗俗以刀針劃其肌膚而加以彩色不顧其剝膚之痛祇爲其自他有耀甚至寒凍逼人不思所以蔽體當炎威酷熱炙烈日中而不知遮掩也昔有一人名亨博曾至南亞美利駕之烏林拿哥見彼處土人勤苦半月盡以力作之資買粉漆塗抹其身以博人之嘖嘖稱美而不念油漆之染身殊爲不適其婦女雖若不睬一縷而出則恬不知怪若非以油漆塗抹其身則深爲恥辱一若大失體面者反至杜門不出矣更有周行天下之客嘗謂五色之玻璃料珠及銅錫之首飾化外之人爭以奇價購之而絨布及哆呢等物反無過而問者有時或贈以袍褲則嵬嵬之衆曆其衣服之正用而顛倒置之若冠履然相詡爲章身之新具眞足令人噴飯然又有愚之甚於此者有船主名士杯香言嘗至亞非利加所屆之土人以羊皮製爲披肩在盛夏時服之而遨遊自鳴得意設遇風雨霜雪之変則反卷而納之懷中情願赤身寒顫而行以去也觀夫無教化國之風土人情則知衣服初爲粧飾身體之用若以衣服爲保護身體之用迨後而悟及之耳即今稱爲有教化之邦但在其遠

格致彙編　肄業要覽　十

FIGURE 4.1 *A page of* Yi Ye Yao Lan, *in* The Chinese Scientific and Industrial Magazine, *vol. IV, no. 12 (1882): 10.*

catalogue announced. “It is a book which should be studied carefully by the Chinese teachers in missionary schools and the people trained for the educational departments, at least they should read it carefully and thoroughly.”[23]

Yi Ye Yao Lan was first extracted and published in *The Chinese Scientific and Industrial Magazine* in January 1882. After that, *The Chinese Scientific and Industrial Magazine* stopped publication for eight years, and so *Yi Ye Yao Lan* had no chance to serialize the remaining seven chapters. But the book was printed and published by the American Presbyterian Mission Press in the winter of 1882. It was translated in classical Chinese language, and Spencer was referred to as Shi Benshou. It was vertically printed without punctuation.

The draft of the introduction and some parts of first three chapters were published in *The Chinese Scientific and Industrial Magazine*, which discussed the phenomena of how the ornamental precedes the useful, classified human activities into five kinds, and stressed that knowledge must be conducive to these activities. *The Chinese Scientific and Industrial Magazine*, founded by John Fryer in 1876, was the first scientific magazine in China. It mainly published translations and popularized science and technology to Chinese readers. Fryer was originally a Protestant missionary sent to China, and was employed by the Translation Department of the Jiangnan Arsenal in 1868. He was one of the main interpreters in that Translation Department, and served as editor in chief of the School and Textbook Series Committee for a long time. *The Chinese Scientific and Industrial Magazine* expanded significantly. By the end of 1880, it had almost 70 sale agencies in the main towns of China. It sold generally 3000 copies, and issues were reprinted many times before 1897.[24] Accordingly, it was estimated that the printed copies of each volume numbered more than 10,000. So, it was the earliest main way that *Yi Ye Yao Lan* circulated. In 1895 the Chinese Scientific Book Depot established by John Fryer reprinted *Yi Ye Yao Lan*. The long and important translations in *The Chinese Scientific and Industrial Magazine* were often republished in separate editions. The republication of *Yi Ye Yao Lan* indicated that it was not only valued by the editor, but also was very popular with readers.

In addition to the article “What Knowledge is of Most Worth,” Yen Yung Kiung also translated the first half of the book, *Mental Philosophy*, written by American churchman Joseph Haven. Yen Yung Kiung thought that Haven's book was better than Spencer's *The Principles of Psychology* for Christians. His

23 Wang Yangzong, ed., *Jindai kexue zai Zhongguo de chuanbo* (Jinan: Shandong jiaoyu chubanshe, 2007), 636.

24 Wang Yangzong, “*Gezhi Huibian* yu xifang jindai keji zhishi zai qingmo de chuanbo,” *Zhongguo keji shiliao*, no. 1 (1996): 36–47.

plan to translate Spencer's *The Principles of Psychology* was not realized. *Xin Ling Xue* was one of the earliest translations of psychology in China.[25] The Educational Association of China identified it as school textbook, and praised Yen Yung Kiung in the Catalogue of Books for he "overcame the difficult of making effective Chinese terminology."[26] While teaching at Shanghai Tongren Hospital, he also revised the Chinese translation of Huxley's *Physiology*. Though Yen Yung Kiung was dedicated to translation work, he thought that Chinese people should firstly learn English in order to grasp Western learning. Therefore, he advocated providing complete English curriculums to students at St. John's College and teaching in English. Yen Yung Kiung showed an interest in a variety of science activities. He had been invited to give a speech on his trip around the world at the Shanghai Polytechnic Institution, once introduced the theory of evolution in the church public meetings hosted by himself, and often helped his friends to buy telescopes, microscopes and cameras and other scientific instruments from America.[27] Interestingly, in 1889 Li Hongzhang, the governor-general of the capital province Zhili, selected the history of western science as the theme of the examinations in the Shanghai Polytechnic Institution. He viewed Darwin and Spencer as the masters of Western science after Bacon, and required candidates to answer questions on the contribution of Darwin and Spencer in science. The candidates were not limited to students of the Shanghai Polytechnic Institution, and ordinary students in various regions also had the right to take the examinations. Their answers were at different levels, but overall, they still knew little about Darwin's theory or Spencer's.[28] After all, of their works only the article "What Knowledge is of Most Worth" had been translated at that time. Benjamin Elman demonstrates that Fryer and other missionary translator kept silent about Darwin and avoided translating works of Darwin or Spencer. Despite its more informed account of Darwin and Spencer, Zhong Tianwei's essay only received fourth-place. Elman pointed out it indicated Wang Tao's and other contest judges'

25 Zhao Liru, "Youguan *Xinlingxue* yishu de yanjiu," *Xinli xuebao*, no. 4 (1983): 380–388; Yan Shuchang, "Yan Yongjing dui xifang xinlixue yinru ji qi hanyu xinlixue shuyu chuangzhi," *Nanjing shida xuebao(shehui kexue ban)*, no. 4(2012): 116–120.

26 Yangzong, *Jindai kexue*, 633.

27 Wu Jianyon et al., trans., *Yan Huiqing zizhuan: yiwei minguo yuanlao de lishi jiyi* (Beijing: Shangwu yinshuguan, 2003), 2, 4, 15.

28 Wang Tao, *Gezhi keyi huibian* (Shanghai: Shanghai shuju, 1897), vol. 4; For the discussion on the examination, see Elman, *Their Own Terms*, 345–350; Chen Qiwei, "Shei shi woguo jindai jieshao xifang zhexue de diyiren," *Dong Yue Tribune*, no. 4 (2000): 93–95; Wang Yangzong, "03.4 Kexue shuji de yijie," in *Zhongguo jinxiandai kexue jishu shi*, ed., Dong Guangbi (Changsha: Hunan jiaoyu chubanshe, 1997), 258–259; Yuezhi, *Xixue dongjian*, 286–302.

antagonism to Darwin's views. But he did not recognize that Wang Zuocai, whose essay received second-place, was Zhong Tianwei's pseudonym. The lower ranking of Zhong's essay was partly due to its writing style. Zhong's essay had too many details of knowledge, and lacked rhetorical effects and a coherent manner. Such an essay did not accord with Chinese classical prose style. Between the two essays by one person, Wang Zuocai's writing style was obviously better. By his trick in the examination, Zhong Tianwei satirized the way Chinese traditional education behaved. Considering the impact of writing style, the situation will be more complicated. Chinese people knew little about Darwin or Spencer, the most important reason of which was perhaps limited communication channels. Yang Haiyan points out that before 1895 the independent press and journalism were very weak in China, and the earliest introductions of Darwin and evolution provoked little reaction. When Spencer's essay "What Knowledge is of Most Worth" was translated into Chinese, "there are only brief, scattered references to Darwin and evolution in the press."[29]

It was Spencer's educational theory that stirred Chinese intellectuals' interest in him, not his relationship with Darwin and evolution. But for the next two decades Spencer's educational ideas and Darwin's evolution theory circulated through the same routes. Both of them depended on the explosion of journals and periodicals, and benefited greatly from Yan Fu's and Liang Qichao's propaganda. At that time there was no group of scientists in China who could have been both receptive to Spencer and who could have helped disseminate evolutionary ideas. It was journalists and others who had to take on this role.

2 The Imperial Civil Examination and "San Yu Bing Ju"

After the Sino-Japanese War of 1894, some thinkers believed that the Self-Strengthening Movement had failed, and attempted to introduce various reforms in China. Then Yan Fu played a key role in the diffusion of Spencer's ideas in China. During the spring of 1895, his articles, "Yuan Qiang," "Bi Han" and "Jiu Wang Jue Lun," successively appeared in *Zhi Bao*. In October of the following year Liang Qichao proposed to Yan Fu the idea of reprinting his articles in *Shi Wu Bao*. So Yan Fu revised and supplemented "Yuan Qiang," but it never appeared in *Shi Wu Bao*. He explained the original intention of the article to

29 Yang Haiyan, "Knowledge Across Borders: The Early Communication of Evolution in China," in *The Circulation of Knowledge Between Britain, India, and China*, ed. Bernard Lightman *et al* (Leiden: Brill, 2013), 181–208.

Liang Qichao. "I try to envisage a strong and prosperous country in accordance with new scientific principles," he wrote, "based on the development of people's intellect, morality and physical power."[30] In the article, he first extended Darwin's theory of struggle for existence to different races, societies and nations, and pointed out that nowadays there was the danger of the ruin of China and the extinction of the Chinese people. He introduced Spencer's ideas of sociology and education and proposed the way of survival for China. He emphasized that society consisted of individuals, whose qualities determined the strength of the society, and he asserted that the fundamental way forward was to transform Chinese people by intellectual, moral and physical education. He also realized that it would be a very long and hard road. "China is surrounded by hostile powers, and the time pass too quickly," Yan Fu wrote, "I am afraid we have no time to do anything, and then China will follow India and Poland. Before the philosophy of Spencer is put into practice, Darwin's theory has come true."[31] However, he insisted that without the development of people's intellect, morality and physical power, only imitating Western economy, military and politics would be in vain.

Yan Fu transformed Spencer's classification framework for education so it could fit into his overall reform plan, and then put forward "San Yu Bing Ju," which was the core of his political theory. He declared that three urgent things to do right now were to raise "Min Li," enlighten "Min Zhi," and rebuild "Min De."[32] ("Min Li" means the physical power of people; "Min Zhi" means the intellect of people; and "Min De" the morality of people.) He extended Spencer's discourse on children's education to Chinese people, who were the emperor's children in Chinese traditional culture. Yan Fu believed that intellectual education was the most urgent thing, especially to study Western science. Before the Ming Dynasty, the western knowledge hierarchy resembled that of the Chinese in some way, but now science became the most useful knowledge. Yan Fu and Yen Yung Kiung both constructed and reinforced their argument by pointing to the analogy between Chinese traditional education and classical education in England. Just as the English had to reform their classical education and include more science, the Chinese had to do the same with their traditional education. So they adapted Spencer's critique of classical education in England to help them provide a criticism of Chinese traditional education. Yan Fu declared that to study science the imperial civil examination must be abolished as it supported and maintained Chinese traditional

30 Wang Shi, ed., *Yan Fu ji* (Beijing: Zhonghua shuju, 1984), vol. 3, 514.

31 Ibid., vol. 1, 9.

32 Shi, *Yan Fu ji*, vol. 1, 27.

education. He accepted the aim of Spencer's moral education, "to produce a self-governing being, not to produce a being to be governed by others."[33] He also agreed that it was a gradual, progressive process. The process of domestic moral education should typify the history of political rule, which progresses from autocracy to constitutional monarchy, and then to abdication. Yan Fu thought that moral education for Chinese people was the most difficult of all, because servility was trained by the absolute monarchy since the Qin Dynasty. "The emperor treated the people like slaves, and the people regarded themselves as if they were slaves too," he declared.[34] The moral improvement of the Chinese people first needed political reform. But in the current Chinese social conditions, the next best thing was finding a heaven-sent sage to govern the country by means of supervising children. He would educate people and gradually reform the political system at the same time, and then he would abdicate and return the power to people when they can govern themselves. Yan Fu's idea of physical education was different from that of Spencer. He aimed at a military, not an industrial, society, and emphasized that physical education was to cultivate bodily vigor and courage demanded in war. Concerned with Chinese peoples' health, he declared that many habits should be abandoned. First of all, two vices, smoking opium and binding the feet of young girls, should be suppressed as early as possible.

Yan Fu and Liang Qichao were part of a group of reformers who arrived on the Chinese political scene after the Sino-Japanese War. They founded newspapers to propagate their ideas of reform. Liang Qichao's *Shi Wu Bao* was the most influential in south China, which sold usually more than ten thousand copies, and Yan Fu's *Guo Wen Bao* in north China.[35] Their primary objective was to abolish the imperial civil examination, establish modern schools, and study western knowledge. The imperial civil examination was the main channel for the government to select officers, and endorsed the dominant status of Confucian classics. Those men who studied western knowledge were despised and suspected by others, and did not have good prospects. Even though they entered into the bureaucracy, it was difficult for them to get appointment and promotion. After Yan Fu finished his study at the Naval Academy in England and returned to China, he felt unappreciated and depressed. He took the imperial civil examination four times and failed each time. It was very evident that most scholars actively took the imperial civil examination, and didn't want to study western knowledge. Liang Qichao believed that if the imperial civil

33 Spencer, *Education*, 212.

34 Shi, *Yan Fu ji*, vol. 1, 31.

35 Jinglu, *chuban shiliao*, 92–93.

examination was not reformed, all wise men would choose to learn writing the eight-legged essay for power and money. He declared that "to establish modern schools and cultivate talents to strengthen china, the imperial civil examination must be reformed first."[36] Any attempts to reform the imperial civil examination and develop modern education brought the reformers into conflict with the conservative officials and old scholars who adhered to the Chinese traditional knowledge hierarchy. In the Hundred Day's Reform which lasted from June 10 to September 20, 1898, Emperor Guangxu issued a series of decrees introducing drastic reforms in national administration, economy, education and military affairs. The most important one of them was to transform the imperial civil examination from Confucian classics to contemporary issues. But it was revoked after the reform failed in a palace coup.

Yan Fu and Liang Qichao, as well as many of their follow reformers, approached the issue of translation with great care. They realized clearly that to study western knowledge it was necessary to start with language. But learning a language took a lot of time; it was the direct way to read translations in the current urgent situation. Also undertaking a translation gave them the power and opportunity to transform the work they were translating, and adapt it to the Chinese context, while pushing for their own sense of how the nation should develop. Liang Qichao declared, "To find a way of self-development for the nation, more western books must be translated, and the scholars must read more translations to self-help."[37] In 1896 Liang Qichao published *Xi Xue Shu Mu Biao*, which was a catalogue of translations of western books, and the long commentary "Du Xi Xue Shu Fa (How to Read Western Books)" for the readers. He pointed out that there were three hundred readable translations of western books since the Self-Strengthening Movement, which were concentrated in the fields of military, medicine, technology and science. And there were few translations on politics, education, law and economic etc., classed together by Liang Qichao and named "Xi Zheng Zhu Shu." He valued them because the reforms aimed at making social and institutional changes, primarily in the education system and government structure. Producing translations, new editions or journal essays, were part of a strategy for reform. The reformers intended to achieve their goals by translating Spencer in a certain way and emphasizing certain aspects of his thought. *Yi Ye Yao Lan* was one of the educational translations in "Xi Zheng Zhu Shu," and highly recommended by Liang

36 Qichao, *Yinbingshi*, vol. 1, no. 1, 27.

37 Ibid., vol. 1, no. 1, 123.

Qichao. He wrote, "It has new methods and thoughts,"[38] "and many essential ideas. If parents want to educate their children successfully, they must read it."[39] In 1897 Liang Qichao edited the series of books "Xi Zheng Cong Shu," which included thirty-two books on different subjects, and the only educational book was *Yi Ye Yao Lan*. Due to Liang Qichao's influence, it was also included in the series of books "Zhi Xue Cong Shu" and reprinted after several months. In 1901 it was included and reprinted in the series "Fu Qiang Zhai Cong Shu Xu Quan Ji," and in the series "Xin Ji Ge Guo Zheng Zhi Yi Xue Quan Shu" next year. It was classified as the first book of "the introduction of Western learning" in the Catalogue of Books titled "Zhong Xi Pu Tong Shu Mu Biao." The catalogue commented that "No matter what kind of Chinese knowledge merely has facade and vain skeleton and doesn't care about the real reasons. You will be excited by this book."[40] "Zeng Ban Dong Xi Xue Shu Lu," a popular catalogue of translations, advised its readers that *Yi Ye Yao Lan* "cut into the current problems" and they must "read it as soon as possible."[41] In the late 19th century and early 20th century, *Yi Ye Yao Lan* had been widely circulated among scholars. It showed, by analogy, the probability of educational reform and setting up a new knowledge system, and the reform-minded men devoted much attention to it. When Cai Yuanpei was appointed to the headship of Shan Shan School and Er Dai School in 1900, he first required the students to read *Yi Ye Yao Lan*. He advised them to study useful knowledge, and not to prepare for the imperial civil examination.[42]

After Liang Qichao fled to Japan, he realized the importance of Yan Fu's "San Yu Bing Ju," and asserted that "Min De, Min Zhi and Min Li were the source and foundation of politics, knowledge and technology."[43] In 1902 he founded the journal *Xin Min Cong Bao*, which was aimed at advancing moral and intellectual education. His long article "Xin Min Shuo" was serialized in it, which recommended improving the moral and intellectual qualities of people, and investigated especially the difficult problem of cultivating morality. *Xin Min Cong Bao* sold usually 5,000 copies and its highest record was 14,000.[44] "Chinese people raced to read it, and even the Qing Government's prohibition couldn't

38 Liang Qichao, *Xixue shumubiao (fu du xixueshu fa)* (Wuchang : Wuchang zhixuehui, 1896), 18.
39 Liang Qichao, "Du Xi Xue Shu Fa," 18.
40 Huang Qingcheng, *Zhongxi putong shumubiao* (Shanghai: Suanxue baoguan, 1898).
41 Wang Tao et al., eds., *Jindai yishumu* (Beijing : Beijing tushuguan chubanshe, 2003), 79–80.
42 Gao Shuping, ed., *Cai Yuanpei quanji* (Beijing: Zhonghua Shuju, 1984), vol. 1, 93–97; Gao Shuping, ed., *Cai Yuanpei quanji* (Beijing: Zhonghua Shuju, 1984), vol. 7, 288.
43 Liang Qichao, *Yinbingshi heji* (Beijing: Zhonghua shuju, 1989), vol. 6, no. 4, 6.
44 Li Yansheng, *Zhongguo baokan tushi* (Wuhan: Hubei renmin chubanshe, 2005), 32.

hinder the current." Liang Qichao said, "Once a volume was published in Japan, there would be more than ten illegal copies in China."[45] From 1898 to 1912 the number of Chinese journals had increased from nearly 100 to more than 500. Including those discontinued during this time, the total of Chinese journals was between 700 and 800. In 1913 the postal system delivered 5,152,480 items of printed matter, most of which were journals.[46] Chinese journalism had undergone a rapid expansion during this time. By being propagated in journals, Spencer's educational theory, rather than his other ideas, had a profound impact on Chinese society. In 1898, two Spencer's sociology essays were translated into Chinese by interpreter Zeng Guangquan and recorder Zhang Taiyan. The quality of their translations was criticized by Yan Fu, and then Zhang Taiyan abandoned his plan to publish a series of translations of Spencer's sociology works in *Chang Yan Bao*. Wang Kangnian, the publisher of Chang Yan Bao, initially invited Yan Fu to translate Spencer's sociology works. Yan Fu was unhappy with the fee offered to do the translation, and Wang Kangnian had to turn to others.[47] In 1903, Yan Fu's *Qun Xue Yi Yan*, the translation of Spencer's *The Study of Sociology*, caused a great sensation in Chinese intellectuals. But his classical Chinese language in this translation was too elegant, and its readers were greatly limited. Since the beginning of twentieth century various sociology theories were transmitted from Japan and the West, Spencer's sociology thought lost its appeal quickly.

Yan Fu's "San Yu Bing Ju" was gradually filtering into people's minds, ultimately taken up by the government, and became the basic framework of constructing Chinese modern education. Although the Hundred Day's Reform failed, the abolition of the imperial civil examination was inevitable. Yan Fu's "San Yu Bing Ju" was also gradually filtering into people's minds, and became the basic framework of constructing Chinese modern education. In 1905 Qing Government ended the imperial civic examination, and promulgated the educational guidelines next year. It advocated loyalty to the emperor and the worship of Confucius, but essentially also included moral, physical and intellectual education. Yan Fu immediately made a speech, which declared that since the educational guidelines had been determined, then "education of physical science is the most pressing matter today." He mainly discussed the

45 Ding Wenjiang and Zhao Fengtian eds., *Liang Qichao nianpu changbian* (Shanghai: Shanghai renmin chubanshe, 1983), 273.

46 Sang Bing, "Qingmo minchu chuanboye de minjianhua yu shehui bianqian," *Jindaishi yanjiu*, no. 6 (1991): 53–76.

47 Wang Tiangen, "Yan Fu yu Zhang Taiyan shehuixue sixiang de duizhi yu jiaoliu," *Guangxi daxue xuebao(zhexue shehui kexue ban)*, no. 1(2003): 49–54.

importance of physical science for changing people's mind and building a prosperous, strong country, and referred to chemistry, zoology, botany, astronomy, geology, physiology and psychology. He also pointed out that art could help the cultivation of human emotion, "so moral education made considerable use of art."[48] In 1912, according to the advice of Cai Yuanpei, the Minister of Education, the government of the Republic of China promulgated new educational guidelines, "to stress moral education, assisting with utilitarian education and militaristic education, to achieve moral individuals by aesthetic education." Cai Yuanpei explained that "militaristic education was physical education; utilitarian education was intellectual education; civic virtue and aesthetic education adjoined moral education."[49] In 1916, some Chinese students studying in Japan founded the journal *Min Duo*, which aimed to "advance Min Zhi, cultivate Min De, and develop Min Li."[50] In 1917 Mao Zedong published his first article "Study on physical education" in the journal *Xin Qingnian*. Though he did not mention Spencer, he discussed the question: which one of intellectual education, moral education and physical education was the most important? He emphasized that physical education was the most important, and proposed that "no body, no intellect and morality."[51]

3 Scientific Education and Scientific Research

Zen Hung-chun's translation was published by the Commercial Press in June 1923, when he was president of the Science Society of China and editor of the Commercial Press. He used *Essays on Education and Kindred Subjects* (1911), and only translated three essays, "Introduction" by Charles W. Eliot, "What Knowledge is of Most worth" and "Intellectual Education." This book was one of his favorites and he probably bought it when he studied chemistry in America. He was once the student of Zhang Taiyan, the great master of Chinese Classics, and then translated it in a style that was a cross between the classical Chinese language and the vernacular Chinese. His motive for the translation was to advocate science and scientific education, which also was the constant object of him and his friends in the Science Society of China. Though he took time from his weighty duties to translate it, he worked carefully and referred to *Education* (1860) and *Education* (1861) and compared the small difference

48 Wang Shi, ed., *Yan Fu ji* (Beijing: Zhonghua shuju, 1984), vol. 2, 278–286.
49 Gao Shuping, ed., *Cai Yuanpei quanji* (Beijing: Zhonghua Shuju, 1984), vol. 2, 130–137.
50 Jinglu, *chuban shiliao*, 307–308.
51 Ershibahua Sheng, "Tiyu zhi yanjiu," *Xin qingnian*, no. 2 (1917), 2.

between them in his commentary.[52] Zen Hung-chun didn't have enough time to complete the translation of Spencer's *Education*, but "was compelled by someone to publish it in the halfway."[53] However, the three essays he translated have constituted an integrated text for those who advocated science and science education. Zen Hung-chun published his translation when "the science versus metaphysics argument" was reaching its climax. This argument began in February 1923 and ended in December. In no more than one year there were 29 articles in journals and periodicals. *Science and Outlook on Life*, which was composed of these articles and two new introductions, was published at the end of 1923. It had 4 editions by April 1926.[54] The number of editions suggests the impact of this argument cannot be underestimated.

Chang Chun-mai, one of the most important Chinese politicians of the day and a philosopher who had studied under German philosopher Rudolf Eucken, declared that science couldn't solve the problem of outlook on life when he gave a lecture titled "Outlook on Life" at Tsinghua College in February 1923. He argued that the core of outlook on life was what a person thought of his life. It contained nine aspects: "I and my family, I and my wife or husband, I and my property, my attitude to social system, my mind and the material world, I and my group, I and the others, my hope towards the world, and whether I believe in a creator of the world or not."[55] Chinese intellectuals immediately split up into two camps "Xuan Xue Pai" and "Ke Xue Pai," and debated about the relationship of science and outlook on life. "Xuan Xue Pai" urged that outlook on life was beyond science and science could never dominate it. "Ke Xue Pai" believed that science could solve the problem of outlook on life and proposed a "scientific outlook on life." Before that, because of World War I, Liang Qichao had asserted the collapse of European culture and science. Liang Shuming argued that there were three types of civilization, Western, Indian and Chinese. Now it was vitally necessary to restore the Chinese traditional culture. In April 1922, Zen Hung-chun gave a lecture titled "Science and Modern Civilization," which was the first of the Spring Lectures, organized by the Science Society of China, and refuted two of Liang's criticisms of science and western culture.[56] "Xuan Xue Pai" was inspired by two of Liang's thoughts, then intensified and

52 Zen Hung-chun, trans., *Jiaoyu lun* (Shanghai: Shangwu yinshuguan, 1923), 90.

53 Fan Hongye and Zhang Jiuchun, eds., *Kexue jiuguo zhi meng: Zen Hung-chun wencun* (Shanghai: Shanghai keji jiaoyu chubanshe, 2002), 688–689.

54 Zhang Junmai et al., *Kexue yu renshengguan* (Jinan: Shandong renmin chubanshe, 1997), 8.

55 Junmai et al., *renshengguan*, 34.

56 Hongye and Jiuchun, *Kexue jiuguo zhi meng*, 272–280.

expanded their criticism. "Ke Xue Pai" included chiefly Zen Hung-chun, Ding Wenjiang, Hu Shi and Tang Yue, who also were the members of the Science Society of China. They were students who had returned from America and Britain, and became the earliest Chinese scientists except Hu Shi. They founded the Science Society of China, and aimed to diffuse science, advance scientific research, and professionalize science. They believed that science could answer the questions of the meaning of life, and used the theories, methods and categories of science to interpret society and humanity. Hu Shi also put forward a "naturalistic outlook on life" that accepted ideas of time and space in physics, refused to explain nature with supernatural causes, and interpreted nature and society in accordance with Darwin's theory.[57] The battle between them and "Xuan Xue Pai" actually was about how to evaluate science and Chinese traditional culture, and which should be dominant in Chinese society.

During the science versus metaphysics argument, Chang Chun-mai attacked science education too. He argued that whether men aspire after truth or love truth depended on the personality and inner life, and had nothing to do with science education. After the introduction of science education in British and Germany from the 19th century, he remarked, there were five defects. Science education made people trust sensual experience and the concept of cause and effect too much; it led them to pursue material happiness and vanity; it constrained them by leading them to study narrow specialized subjects; and scientific education aimed at earning a living rather than developing a sound personality. He especially critiqued the importance of the concept of cause and effect in science education. "From the 19th century science teachers told students that every phenomenon has its cause and effect," he wrote, "and human evolution was dominated by natural law, so what man can do was only try to adapt to his environment." Science teachers "instructed people to believe the concept of cause and effect, adapt the environment and be salve of status quo," which retarded the social revolution.[58] Ke Xue Pai didn't directly respond to Chang Chun-mai's criticism of science education. However Zen Hung-chun published immediately his translation of Spencer's *Education*. He believed that Spencer's book published more than half-century ago was still relevant for science development and education reform in China, for "our nation's disease is just like that of the west half-century ago," and "so Spencer's book fitted to our nation, even though maybe it was outmoded for the west."[59] Zen Hung-chun acclaimed that Charles W. Eliot's life had been devoted to Harvard and

57 Junmai et al., *renshengguan*, 23–25.
58 Ibid., 102–108.
59 Hung-chun, *Jiaoyu*, 1–2.

America education, and he translated Eliot's introduction to illustrate that Spencer's idea of science education had guided educational reform in British and America in the past half-century. But Eliot also pointed out that "Spencer's doctrine of the supreme value of science" was very radical, and believed that "science is no more universally fitted to develop mental power than the classics or mathematics," and "if some science is to be learnt by everybody, the amount of such study should be limited."[60] So Zen Hung-chun emphasized in the commentary that "Spencer's idea was remedy for education above all, and it may have exceeded the proper limits in righting a wrong," but it was more applicable to the Chinese present situation than other ways.[61]

The science versus metaphysics argument ended quickly in Ke Xue Pai's victory. According to Zen Hung-chun and his friends, Xuan Xue Pai didn't really understand science. Zen Hung-chun wrote, "The argument only showed a small part of our country's thinkers' misunderstanding of science, so they couldn't help doubting the value of science."[62] He believed that it was caused by the failure of science education and lack of scientific research in China. A major disadvantage of Chinese science education was that "it focused too much on science teaching and didn't care about scientific research."[63] Science teachers just repeated the textbooks, yet had little experience of scientific practice. Without true scientists teaching, scientific education couldn't be successful. He attributed the misunderstanding of science to a lack of scientific research in China, and asserted that "further and deeper study of science was needed to be developed. Without that, people cannot approach the essentials of science. To cultivate science education, we have to develop scientific research first."[64] Chinese universities took teaching as their principal function and research the subordinate one. There was a direct way to advance research by establishing new institutions of research outside universities. Zen Hung-chun proposed that the key project of the Science Society of China should be to establish different kinds of institutions of research. The second remission of a portion of the Boxer Indemnity by the United States was announced in May 1924 and it was suggested that the remitted indemnity subsidize Chinese education and culture. Zen Hung-chun and the Science Society of China immediately claimed that there must be a foundation to manage the returned

60 Ibid., 3–4; Introduction by Charles W. Eliot, accessible online at: <http://oll.libertyfund.org/index.php?option=com_staticxt&staticfile=show.php&person=165>.

61 Hung-chun, *Jiaoyu*, 3–4.

62 Hongye and Jiuchun, *Kexue jiuguo zhi meng*, 309–310.

63 Ibid., 283.

64 Ibid., 310.

indemnity, which must "be invested in research of pure sciences and applied sciences, especially setting up institutions."[65] After plenty of propaganda work and behind-the-scenes activities, mainly executed by Zen Hung-chun, Ding Wenjiang and Hu Shi, who controlled the China Foundation for the Promotion of Education and Culture, most of the returned indemnity was used to subsidize scientific research.[66]

Zen Hung-chun's translation was included in the first Wan You Wen Ku Series and reissued in October 1929, and reprinted in July 1934. The Wan You Wen Ku Series was the product of the entrepreneurial vision of Wang Yunwu, who was the head of the Commercial Press. It was a unique experiment in Chinese publishing and aimed to provide a small but complete library for young students and the adults who could read. Two series had been planned in 1928. The first series composed of 800 introductory books on particular subjects, such as agriculture, industry, business, medicine, mathematics, teacher training, physical training, history and geography etc., including 100 Chinese classics, 100 translations of world classics and 10 reference books. It was a popular cyclopedia of human knowledge and published in five phases from 1929 to 1934.

To confront the competition of the publishing market, the entire first series was set at a very low price, only 360 yuan. Wang Yunwu expected orders from libraries in China. In 1928 the government of the Republic of China began to develop libraries, and required that every school must establish a library and allocate funds every year for purchasing books. And then, local governments, schools and institutions eagerly ordered the entire first series, which sold 8000 series and was republished in 1934. By purchasing the first series, more than 2000 new libraries, which constituted about half of the number of the libraries in China, were established. Most of middle schools in China had the first series by 1937.[67] Through the reading space of libraries, Zen Hung-chun's translations were spread to small cities and towns, even in the country. The Wan You Wen Ku Series brought Spencer's educational ideas to a large reading audience that included more and more young readers from low-income families.

The processes of translation, communication, and appropriation of Spencer's *Education* reflected dimly the changes of the Chinese knowledge hierarchy and educational system. Yen Yung Kiung, Yan Fu and Zen Hung-chun

---

65 Ibid., 313.

66 Ibid., 760–762.

67 Hui Ping, "Wang Yunwu wanyou wenku cehua jianlun," *Helan tushuguan xuekan*, no. 5 (2006): 124–127; Guo Taifeng, "Wei wanyou wenku danjingjielv de Wang Yunwu," *Shiji*, no. 5 (2001): 52–53.

FIGURE 4.2 *The book cover of* Jiao Yu Lun *included in the first Wan You Wen Ku Series* (*Zen Hung-chun, trans.,* Jiaoyu lun. [*Shanghai: Shangwu yinshuguan, 1929*])

were students studying abroad, but everyone had his own education background. They received respectively church education, military education and scientific education, which indicated the approaches of learning Western knowledge adopted by Chinese elites in different historical periods. For different purposes, Spencer's views on education were transmitted into China, but their interpretation and impact depended on different communication situations. *Yi Ye Yao Lan* was initially a church school textbook whose readers were Christians and some Chinese scholars interested in western knowledge. With the explosion of journalism in late Qing, "San Yu Bing Ju" propagated by Yan Fu, Liang Qichao and others was widespread, and finally accepted by the Chinese people and government. *Yi Ye Yao Lan* was also emphasized and got a larger audience. After the Imperial Civil Examination was abolished, "San Yu Bing Ju" became the basic framework of the Chinese educational system, and the importance of science education was emphasized constantly. As there was no group of scientists in China, it was journalists who disseminated Spencer's educational theory and advocated science education. With new developments in print technology, the competition of the publishing market and the expanding library network, Spencer's educational ideas have had a widespread and lasting influence in China. By emphasizing that science education was based on scientific research, Spencer's educational theory was appropriated by Zen Hung-chun to help to develop a cadre of scientists in China. Science had prevailed over Chinese traditional culture when the science versus metaphysics argument ended, and then Zen Hung-chun and his friends, who had been the first generation of Chinese scientists, could gather up their strength for the development of scientific research. Due to their consistent efforts in the next decades, the organization of science was basically formed in China, and the cultural hegemony of science was established firmly. Although Spencer's philosophy and sociology have probably been ignored, Spencer's *Education* was translated in 1962 and republished in 1997, because of its importance in science education.

With a communications revolution in China, Spencer's educational ideas were disseminated to more sites and more readers than before. Perhaps we can say a global communications revolution helped to form a global Spencerism. Even though we know little about readers, they were obviously not passive recipients. Under the Spencer's classification framework of education, they had their own thinking and different ideas. What was the relation between intellectual education, moral education and physical education? Which one was more important and how to arrange them in order? What was the content and goal of each of them? They often discussed those questions, and then demanded educational reform. Spencer's ideas were kept alive in the practices of translation, transmission and discussion, and were appropriated in different places and time. The influence of Spencer's educational ideas in Japan, as Godart's essay shows, was similar and connected in some way with their influence in China. In the process of communication, a diversified, but interactive, global Spencerism was formed.

TABLE 4.1 *Translations of Herbert Spencer's works into Chinese, 1880–2000*

Year	Language	Original	Title	Publisher	Translator	Comments
1882	the classical Chinese language	"What Knowledge is of Most Worth?"	*Yi Ye Yao Lan*	the American Presbyterian Mission Press	Yen Yung Kiung	Extract in *The Chinese Scientific and Industrial Magazine,* Vol. IV, No.12, Chinese 12nd moon,1881.
1897	the classical Chinese language	*The Study of Sociology,* Chapter One and Chapter Two.	"Bian Yu" and "Chang Xue", together entitled "Quan Xue Pian"		Yan Fu	"Bian Yu" published in *Guo Wen Hui Bian,* Volume 1, 3 and 4. "Chang Xue" never published.
1898	the classical Chinese language	"Progress: Its Law and Cause", "Manners and Fashion"	"Lun Jin Jing Zhi Li", "Lun Li Yi"		Interpreter Zeng Guangquan, Recorder Zhang Taiyan	In *Chang Yan Bao,* Volume 1,2, 3, 4, 5, 6, 8.
1902	between the classical Chinese language and the vernacular Chinese	"The Rights of Women"	"Si Bin Sai Nv Quan Pian"		Ma Junwu	"Da Er Wen Wu Jing Pian" and "Si Bin Sai Nv Quan Pian" printed in one book. "Da Er Wen Wu Jing Pian" translated from *The Origin of Species,* Chapter Three, Struggle for Existence.
1903	between the classical Chinese language and the vernacular Chinese	*The Principles of Sociology,* Vol. I. Part II. The Inductions of Sociology.	*Si Bin Sai She Hui Xue Yuan LI*	Kaiming, Wenming and Guangzhi shuju, Shanghai	Ma Junwu	

TABLE 4.1 Cont.

Year	Language	Original	Title	Publisher	Translator	Comments
1903	the classical Chinese language	The Study of Sociology	Qun Xue Yi Yan	Wenming shuju, Shanghai	Yan Fu	
1923	between the classical Chinese language and the vernacular Chinese	"What Knowledge is of Most Worth?", "Intellectual Education"	*Jiao Yu Lun*	the Commercial Press	Zen Hung-chun	
1962	the vernacular Chinese	Education: Intellectual, Moral, and Physical	Jiao Yu Lun: Zhi Yu, De Yu he Ti Yu	Renmin jiaoyu chubanshe, Beijing	Hu Yi	
1996	the vernacular Chinese	*Social Statics: Abridged and Revised*	She Hui Jing Li Xue	the Commercial Press	Zhang Xiongwu	
1997	the vernacular Chinese	*Herbert Spencer on Education* ed. Andreas M. Kazamias (Teachers College Press, Teachers College, Columbia University, 1966)	*Si Bin Sai Jiao Yu Lun Zhu Xuan*	Renmin jiaoyu chubanshe, Beijing	Hu Yi, Wang Chengxu	

CHAPTER 5

The Reforming Spencerians: William James, Josiah Royce and John Dewey

Mark Francis

The study of Spencer in late nineteenth-century America is primarily a philosophical task: he was regarded by his philosophical readers as the thinker who had expounded scientific method in such a way as to combine it with the progressive development of the psyche and of society. While Americans could, and did, read philosophers of science such as August Comte, J.S. Mill and William Whewell, it was only Spencer who offered a theoretical vision that stretched out to encompass both scientific theories about the brain and the mind. It was this last feature that made Spencer so enticing and dangerous because he threatened both conventional metaphysics and religion. A theory which linked the brain with the mind implied that the latter was not equivalent to the spirit or the soul. This, in turn, suggested that without the presence of a spirit or a soul it was difficult to speculate about theology or religion. Following the launch of his *System of Philosophy* in 1862, Spencer was feted in America. However, a reaction against Spencer began at the time he re-issued his *Principles of Psychology* (1870–72). The leader of this reaction was Chauncey Wright[1] (Fig. 5.1).

Shortly before he died, Wright caused a great deal of trouble in the embryonic discipline of philosophy in America. Everyone had been enamoured with both Herbert Spencer and science, but Wright forced a choice between them by claiming that Spencer's synthetic philosophy was not truly scientific, but a mere metaphysical synthesis which contained fundamental statements which were "anticipations of Nature," and therefore, were incompatible with the true spirit of Baconian inductivism.[2] Since Wright's contemporaries equated a scientific method with a Baconianism in which facts were collated without the prospect of an *anticipated* outcome, the game was up: there seemed to be no way of reconciling science with a philosophy such as Spencer's which began with a metaphysical *First Principles*. Once the authoritative Wright had

1 On Wright, see Elizabeth Flower and Murray G. Murphey, *A History of Philosophy in America*, New York: Capricorn Books, 1977, vol. II, 541.

2 John Fiske, "Chauncey Wright," *Darwinism and other Essays*, London, MacMillan, 1879, 85.

 DOI 10.1163/9789004264007_007

FIGURE 5.1
Wright and the destruction of Spencer's metaphysical science. Ralph Barton Perry, The Thought and Character of William James *(London: Oxford University Press, 1935), vol 1, facing p. 506.*

announced that inductivism did not allow for metaphysics unless it abandoned its basis in reality, the imperative philosophical task became the construction of a metaphysics that was linked with science and that would buttress the Spencerian synthesis. The correct way of rescuing, and of appropriating, Spencerianism split American philosophers. Spencer was too important to be abandoned so this division was not between the Spencerians and anti-Spencerians, but between two sides in an intellectual civil war in which both claimed to be the rightful heirs to the true appropriation of Spencer's synthesis of philosophy and science. The first group, led by John Fiske, claimed to be the orthodox one and had the virtue of being recognized by the great man himself. These took upon themselves the duty of defending Spencer against any and all criticism. The second group were the reformers who rewrote Spencer's philosophy and psychology so that these would be scientifically grounded and protected from the charges that Wright had levelled.

The chief Spencerian reformers were William James, Josiah Royce and John Dewey. What made them chiefs was not just their political acumen, but the popular quality of their writings. While critical of Spencer, they proclaimed their continued adherence to his doctrines: reformers such as James, Royce and Dewey were not anti-Spencerian, but, in their own eyes, revisionist philosophers who appropriated what was central to Spencerianism. Royce and James both used Spencer's texts when teaching at Harvard. Royce even elevated Spencer to the same level as Spinoza in making his work a text for his course in the philosophy of nature.[3] James used Spencer's *Principles of*

3 John Clendenning, *The Life and Thoughts of Josiah Royce*, Vanderbilt University Press, revised edition, 1999, 144.

Psychology as a stalking horse during the many years it took him to write his own *Principles of Psychology*. Dewey, though more critical of Spencer than Royce and James had been, also found Spencer a convenient whipping boy during his own attempts to reconcile science and philosophy. To Spencerian revisionists, Spencer was a great man who had done much, but was unjustly facing oblivion unless the orthodox Spencerians were defeated. If Spencer's reputation collapsed, it was argued, this would be "the result of having a school and of being worse injured by your disciples than by your enemies."[4] Reformed Spencerians were not a coherent philosophical group in the sense that that they preached the same doctrine, or that they avoided quarrelling with each other. They were held together only by the common belief that evolutionary psychology posed the chief challenge for American philosophy, and by their desire to appropriate Spencer's psychology in order to blunt its materialist qualities. For the rest, they went their own several ways: James alternately admired and despised Spencer's attempt to establish a half-way house between religion and science. Royce, who had become a Hegelian by the time he wrote on Spencer, devoted himself to re-focusing Spencerian philosophy so it would fit with traditional philosophy of nature. Dewey, who eventually shed his Hegelianism, ended by throwing aside Spencer as a man whose comforting philosophy falsely signalled to the existence of a God. When the reforming Spencerians were not revising Spencer's philosophy, or castigating Spencer for some intrinsic error that could not be corrected, they kept up hostilities with what they called the Spencerian "School," the orthodox Spencerians whose ideas are chronicled by Bernard Lightman in the next chapter of this volume. If one has to assign a chronology to the Spencerian reformers one would have to single out a time period that began in 1878 when James issued his "Remarks on Spencer's Definition of Mind as Correspondence"[5] and 1884 when John Dewey published the "New Psychology."[6] The high water mark of Spencerian reform came in 1890–91 with the publication of James' *Principles of Psychology*[7] and the final revised edition of Dewey's *Psychology*.[8] The end of Spencerian reform came in 1904 with Josiah Royce's elegiac memoir, *Herbert Spencer, an Estimate and Review*.[9] After that year, it is no longer useful to refer to James,

4 Letter from Josiah Royce to William James, September 19, 1880, *The Letters of Josiah Royce*, ed. John Clendenning, University of Chicago Press, 1970, 90.

5 William James, *Collected Essays and Reviews*, ed. Ralph Barton Perry, London: Longmans, 1920.

6 John Dewey, *The Early Works, 1882–1895*, vol. 1, Carbondale: Southern Illinois Press, 1969.

7 New York: Henry Holt.

8 New York: American Book Company.

9 New York: Fox Duffield.

Royce and Dewey as Spencerian reformers: their goals had shifted and it was no longer useful for them to re-cycle a variety of Spencer's psychology and philosophy of science.

Spencer's *Principles of Psychology* is often overlooked by scholars who concentrate their analytic energies upon his writings on education, sociology, politics and ethics. In a sense this neglect of psychology is reasonable because Spencer's public image for later generations often rested chiefly upon his accomplishments in these areas. Then too, Spencer's psychological perspectives are not easily consistent with the ideas that he generated when thinking about social, political and ethical evolution. However, scholarly neglect of Spencer's psychology makes no sense when examining late nineteenth-century American reactions to his philosophy. As Lester Ward, one of Spencer's followers, remarked, it was the *common* view to see Spencer's *Principles of Psychology* as the pinnacle of the great man's work.[10] Spencer's thoughts upon the human psyche preceded his *System of Philosophy*, and, while his *Principles of Psychology* was re-issued as part of the system, it continued to echo the voice of a younger, bolder and coarser man – a voice that was never entirely muffled by the more cautious and nuanced tones of his later works. The *Principles of Psychology* had appeared in the mid-1850s when Spencer was seriously thinking about evolution for the first time,[11] and when his ideas were closely allied to those of neurophysiology. In this period, Spencer's evolutionary theory, while rooted firmly in the development of the brain, stretched out to encompass a notion of intellectual culture that served as part of a theory of mind.[12] This included a number of seemingly diverse phenomena such as the inherited mental equipment of individuals, historically acquired knowledge and scientific laws. In the mid-1850s Spencer came to the conclusion that these phenomena had developed in harmony with each other. That is, Spencer's theory of mind, which appeared in his *Principles of Psychology*, included a notion of evolutionary culture in which individuals, together with the society in which they lived, progressed at the same rate. This cultural dimension of mind was often ignored by his Americans whose training in Germany, or affili-

10 Lester Ward, *Dynamic Sociology, or Applied Social Science*, New York, D. Appleton, 1923 (originally published in 1883 and 1892), vol. I, 168. Ward himself preferred Spencer's *Principles of Biology*.

11 Herbert Spencer, *The Principles of Psychology*, London, Longman, 1855 (Spencer recycled this work as Volume I of his two volume *Principles of Psychology* (1870, 1872)).

12 Spencer's 1855 psychology was a response to contemporary neurophysiolology as well as philosophy. See L.S. Jacyna, "The Physiology of Mind, the Unity of Nature, and the Moral Order in Victorian Thought," *The British Journal for the History of Science*, vol. 14, no. 2 (July 1981), especially 109–118.

ation with German psychology or 'philosophy,' had predisposed them to view psychology as *either* experimental neurophysiology *or* metaphysics. Then too, Wundt, the chief mentor of late nineteenth-century American psychologists, had excluded general theories of evolution from consideration. "That mental life itself is a problem of psychology," Wundt wrote, "this is mere dogma, handed down to us by past ages."[13] This meant that Spencer and others who spoke of the mind as if it were part of the evolution of life were materialists whose speculation was, at best, a heuristic hypothesis.

While German impatience with the evolutionary attempt to link the brain with the mind completely captured Americans like G. Stanley Hall, others, like Lester Ward, were still drawn to the materialist attempt to show that mental associations reflected the material reality of which humanity had acquired accurate perceptions over preceding generations. It would be convenient historically if everyone fell into two groups: the experimental psychologists, who only speculated about brain, and the evolutionary theorists who provided a scientific basis for mental life. However, some late nineteenth-century philosophers could not decide which side to support, and opted for both while feeling uneasy about their ambiguity. Such indecision particularly applied to William James.[14]

1 James: The Founder of American Philosophical Psychology

James, who thought that he had spent a dozen years ruminating upon Spencer before issuing his 1400 page *Principles of Psychology* in 1890[15] constantly attempted to reconcile philosophical differences between the major schools of his discipline (Fig. 5.2). Part of him stayed faithful to the objective tenets of association psychology while part yearned to subscribe to the beliefs of a more

13 Wilhelm Wundt, *Principles of Physiological Psychology*, trans. from the 5th German edition by E.B. Titchener, London: Swan Sonnenschien, 1910, vol. I, 9.

14 William James (1842–1910): twenty-first century readers chiefly remember James as the author of *Varieties of Religious Experience* (1902) which attempted to account for the way in which the effects of the spiritual world followed through to the phenomenal world. In the late nineteenth century, he was best known as a psychologist.

15 The gestation period was longer than this. Robert J. Richards notes that James had Spencer's *Principles of Biology* in his study plan for January 1870 (*William James, In the Maelstrom of American Modernism, A Biography*, Boston Houghton Mifflin, 2006, 110). R.B. Perry dated James' reading of Spencer to between 1860 and 1862 (Paul Jerome Croce, *Science and Religion in the Era of William James*, Chapel Hill: University of North Carolina Press, 1995, 270, note 81).

FIGURE 5.2
William James: the doyen of American psychologists. Henry James (ed)., The Letters of William James *(London: Longmans, 1920), vol. II, frontispiece.*

spiritually-inclined psychology. He described the former as a school – whose numbers included Johann Friedrich Herbart in Germany, David Hume in Scotland and James and John Stuart Mill in England – that professed "psychology without a soul."[16] The other form of psychology was the polar opposite, brimming with souls and spirits which operated on a different plane than the one composed of matter. Instead of adopting one of those alternate accounts of mental phenomena, James drew from both while, at the same time, complaining of their respective faults.

James' insistence on avoiding the dichotomous perils of associationism and spiritualism was what had originally driven him to adapt Herbert Spencer's evolutionary theory. In particular, he borrowed Spencer's emphasis upon the way in which the "inner" or physiological parts of the mind had adapted so as to be in correspondence with the "outer" world by reacting to environmental inputs.[17] That is, James relied upon Spencer's theory that mind/brain had

16 William James, *The Principles of Psychology*, London MacMillan, 1890, vol. I, 3.

17 Ibid., vol. I, 6 and William James, "Remarks on Spencer's Definition of Mind as Correspondence" (1878), *Collected Essays and Reviews*, ed. Ralph Barton Perry, London: Longmans, 1920, 44.

evolved to its present state by adjusting its operations in order to survive environmental change. James believed that Spencer's mental evolution was inferior to Darwin's theory of the natural selection of species as a scientific theory, but that, with sufficient modification, it could be used to negate the flawed 'cerebral' psychologies that were then holding centre stage. James' refashioning of Spencer's evolutionary theory was a good example of the way in which Spencer's work was appropriated by his American reformers.

Spencer's theory of evolution needed modification because, according to James, it was primarily a hypothesis about the necessity of a particular type of organic adaptation. He believed that Spencer's theory was an argument that all complex organic systems of sensory apparatus, including the human brain, must have undergone a process that would fit them to changing environments. This Spencerian argument seemed axiomatic. James' complaint was that there was no way of looking behind Spencer's hypothesis: essentially it was a speculation that minds must have evolved so as to adjust to nature or the organism of which they were part would have perished. James felt particularly aggrieved by the Spencerian account of the probable origin of the mind because it appeared to include the suggestion that our theoretical notions of space and time had changed as evolutionary responses to the environment.[18]

Spencer's evolutionary theory was modified by James who, surprisingly, invoked Darwin as an advocate of progression towards a goal of development. This strategy now seems inexplicable as we have become accustomed to regarding Darwinism as the enemy of such progressive theories. In the twenty-first century it seems obvious that a primitive organism did not intend, or will, change along a particular line of development, but that, retrospectively, we postulate that there is a genetic inheritance from an ancestral form in the past. However, this interpretation was not available to James.[19] Instead, he surprisingly emphasized, as Darwin's chief concepts, the notion of "spontaneous variation" as the producer of changed forms of life, and the idea that the environment acted as a randomizing agent that could either preserve or destroy the forms.[20] The element of randomness in conjunction with "spontaneous variation" appealed to James because this pairing of ideas supported his liberal

18 James (*Principles of Psychology*, vol. II, 620–624 (ff. 622) quotes Spencer at great length in order to make this point (ibid., II, 639).

19 Louis Menand (*The Metaphysical Club*, London: Flamingo, 2002, 141) paradoxically noticed the peculiar nature of James' views on Darwin's evolutionary theory by remarking that he "was Darwinian, but he was not a Darwinist."

20 James, "Great Men and Their Environment" (1880), *The Will to Believe and other Essays in Popular Philosophy*, New York: Longmans, 1897, 247. Also see James, "Remarks," 63.

belief that more evolved individuals and societies also possessed greater quantities of variation.

James' evolutionary language was spliced to a liberal view of science. The chief task of this sort of liberalism was to protect the highest minds so that when a variety of ideas was produced by the unrestrained activity of these minds, it would give an impetus to the progress of science and civilization.[21] For James, Darwinism supported liberal progressive science, while Spencer's evolutionary ideas did not – since they were without direction and did not value the random variation of ideas. Even worse from James' perspective, Spencer was hostile to the idea that genius played a decisive role in scientific progress.[22] Since he agreed with Spencer that scientific progress was part and parcel of evolution in general, James experienced a strong desire to appropriate Spencerism so that it would resemble a source of causal statements that would promote future intellectual and scientific progress. Unless it was modified, James believed that Spencerism would lack scientific respectability and would amount to nothing more than a series of hypothetical scenarios about how evolution might have occurred.

James' use of Darwin and Spencer seems paradoxical because the twenty-first century identifies Darwinism exclusively with natural selection and the struggle for survival, while Spencer is discussed as someone who was less enamoured with these ideas. However, in James' era, it was possible to assign quite different values to both Darwin's scientific work and Spencer's evolutionary theory. In James' case he claimed to be a Darwinian who was sceptical about natural selection and survival of the fittest because he believed that these had hidden teleological implications. Since he also considered Darwin to be without faults he blamed Spencer for this non-scientific and teleological blemish while claiming that, properly considered, James' 'Darwinism' was a theory of benign randomness and 'spontaneous variation' based upon the idea that individuals directed their own development. Since individuals were capable of comprehending causal laws once these were spontaneously produced, this would direct the growth of scientific knowledge because such laws were predictive. In other words, James' 'Darwinism' consisted of two strategies – one allowing for the understanding of the individual psyche, and the other for the explanations of the advancement of science in such a way that the life sciences would be explicated by the same sort of laws that applied in the physical sciences. Opposed to this, James glossed Spencerianism as an evolutionary

21 James, "Great Men," 249.

22 James ("Great Men," 232–3) thought that Spencer's dismissal of the importance of great men in *The Study of Sociology* was impudent.

theory that would erroneously group the development of the individual psyche with organic change in general, and as resistant to the notion that scientific progress was a matter of subjecting phenomena to causal laws. He believed that Spencer's evolution was a theory of undirected change within which some accidental processes were arbitrarily assigned a prepotency over others.[23] Since the processes were accidental, rather than necessary, they interacted with each other in capricious ways, which meant that the direction of change could shift at any time.[24]

When James re-focused Spencer's evolutionary theory as a teleological doctrine, his target was not simply Spencer, but all empirically-minded natural scientists who had employed induction as a method. That is, James' insistence that evolutionary arguments lacked the scientific rigour which one expected from science was a claim that all scientific explanation should be predictive, and, therefore, it needed to specify outcomes. Since English scientists such as Lyell and Darwin had explained evolution as a process of a development *from* a beginning rather than *towards* an end, James' remark that historically-based evolution was a particularly Spencerian flaw demonstrates his lack of serious interest in sciences outside psychology. However, despite its lack of substance, James' attempt to prove that Spencer's theory should have been teleological helped establish a twentieth-century solecism that Spencer's arguments were based upon a covert teleology.[25] Of course, the reality was that it was James, not Spencer, who desired an idea of human development which would direct him towards a spiritually-satisfying and ultimate goal.

2 Royce: The Hegelian Who Could Sympathise with Induction

At Harvard, James' most combustible colleague was Josiah Royce (Fig. 5.3).[26] He could be counted upon to disagree with almost any philosophical position

23 Ibid., vol. I, 593.

24 James, *Psychology*, vol. II, 619.

25 As examples of modern writers who follow James' teleological argument, see works by Valerie Haines and J.D.Y. Peel.

26 Josiah Royce (1855–1916): Royce was better known for his work on metaphysics and the philosophy of religion. His studies in German with Rudolf Hermann Lotze left him with more sympathy with science than one might have expected from a philosopher who is usually identified with American Hegelianism. Royce originally believed himself in complete argument with James' philosophy but split from him because of James' insistence on regarding consciousness as a product of individual human minds while Royce preferred to believe that consciousness was a universal principle.

FIGURE 5.3
Royce: Spencer's Hegelian sympathizer. Ralph Barton Perry, The Thought and Character of William James *(London: Oxford University Press, 1935), vol 1, facing p. 765.*

which James held. Royce did not start out as a Spencerian reformer, but as an opponent. His earliest philosophical essay was a complaint that Spencer and his disciples had denied the possibilities of any thought that was not derived from experience. That is, Royce criticized the evolutionary insistence that all thought was descriptive, and was either direct data or the result of conditioning.[27] On the subject of social evolution, the young Royce was also at odds with Spencer because he disagreed with the notion that development began with simple forms and progressed to complex ones. Royce's contrary view was that development moved towards simple forms.[28] Royce's first book, *The Religious Aspect of Philosophy* (1855), was also critical of Spencer's evolution but was much more non-specific. He treated Spencer's arguments as a sub-species of British hedonism – a tradition including Hobbes, Bentham and J.S. Mill.[29] By the time he wrote the work, Royce had adapted William James' general argument that Spencer's evolutionary theory was only cogent if it was refocused so that evolution was read as goal-directed. Royce believed that, through

27 Clendenning, *Life of Royce*, 85.

28 Ibid., 89.

29 Peter Fuss, *The Moral Philosophy of Josiah Royce* (Cambridge, Mass.: Harvard University Press, 1965), 11 and 93, note 22.

experience, human beings were able, though with difficulty, to determine the end towards which evolutionary processes had moved.[30]

Royce, who had more familiarity with the history of philosophy than anyone else in the United States, was able to discern that Spencer's philosophy was not based upon innate ideas or upon individual experience, but upon a comparison between these. That is, Spencer's ideas were somehow residual ones gleaned from the experiences of the human species as a whole and built in the individual nervous system. This meant that, for the species, ideas were caused by experience, but, for the individual, they were inherited.[31] For Royce, as for James, the juxtaposition of the species and the individual when considering the formation of ideas was a puzzle. Both preferred to analyze ideas as if they were solely products of individuals rather than of species, and they regarded the species' experience as a biological process, not a mental one. This, of course, made evolved ideas mysterious because they implied that the species would have to possess a long-lasting and perpetual biological ability to respond to experience.

As has already been suggested, Royce's early views on Spencer were close to those of James; he saw Spencer as a philosopher who did not understand that his ideas should have been expressed in a metaphysical form in which change was teleological – not in the sense of a Christian telos, but in the sense of being directed towards a goal. In his mature writings, Royce shed these teleological interpretations of Spencer because he now wanted to protect him from James and from English philosophers, such as Shadworth Hodgson, upon whom James was dependent. Royce now provided a sympathetic analysis of how Spencer had achieved his evolutionary insights.[32] In particular, he recounted the way in which Spencer had first employed a super-organic evolutionary theory that stressed the dyadic usage of evolution and dissolution as mechanical necessities to which all teleological interpretations must be subordinated.[33] By "mechanical," Royce meant a process free from direction. It was James, Hodgson and their German counterparts, not supernaturally-focused Christians, who were the real enemies of Spencer and evolutionists of his type.

30 Ibid., 11.

31 Ibid., 116.

32 Why Royce changed in this way is not clear. It seems likely that his late adherence to Hegelianism made him more sympathetic to evolutionist philosophy than he had been before.

33 Josiah Royce, *Herbert Spencer, An Estimate and Review* (New York: Fox Duffield, 1904), 69–70.

Spencer had overlooked the fact that his great historical enemy was not supernaturalism, but:

> *the tendency to conceive the universe as an eternal, and so, temporally viewed, as an essentially permanent order* whose laws may be studied, and where events often include what we call growth, but whose main outlines, classification, processes, forms are the same yesterday, today and forever: so that the story of the origins of things, even when true, is of secondary import.[34]

Royce saw Spencer as the inheritor, not of the mechanical legacy of seventeenth-century science with its conception of the persistence of force, but of the nineteenth-century fascination with the "historical" aspect of things.[35] By "historical" aspect, Royce was referring to the transitory feature of living organisms as distinct from things which had a more permanent nature. Royce saw Spencer's philosophical task as that of reconciling the development of living things with material ones; or "real" evolution with "real" mechanical nature. Spencer, he concluded, was one of those who shed light upon the synthesis of knowledge that was so needed by the new age.

Royce grouped Spencer with other philosophers who produced realist systems such as Kant, Herbart and Spinoza.[36] Royce stressed Spencer's realism partly because he wanted to hammer home the point that it was not idealists, but realists who produced these 'unknowable things in themselves' with their "substances" and their "Unknowns." It was the realists who were most likely to acknowledge "beings" that were far removed from experience and from common sense.[37] Part of Royce's analysis was reliant upon his insight that some of Spencer's ideas were original, and that while the origins of his philosophical interests could be found in the past, they differed from those of the present. Royce noted that Spencer was seemingly uninvolved with the debates that had convulsed the nineteenth century. When it came to popular controversies over how species had transformed over time, the descent of man, or whether the conscience or religion were purely natural products of evolution, Spencer was either above the debate or somehow off-side. Instead of acting like a Huxley, and finding answers to popular enquiries and opposing the supernatural,

34 Ibid., 35

35 Ibid., 39.

36 Josiah Royce, *The World and the Individual, First Series, The Four Historical Conceptions of Being* (New York: MacMillan, 1916), 109.

37 Ibid.

Spencer continued to operate as he had always done – attempting to erect inductively-based hypotheses for discrete evolutionary problems in psychology, biology, sociology and ethics. Unlike the many German philosophers whom Royce had studied, Spencer did not employ evolutionary theory in order to determine the relations of essential processes, did not define the inner unity of such processes, and did not empower us to conceive of a series of types of evolutionary processes in orderly relations to one another.[38] The sort of speculations that were of interest to German philosophers, and to Royce himself, did not attract Spencer's attention. It is to Royce's credit that, even as he perceived Spencer as philosophically dissident, he could still regard his philosophical enquiries as viable. It is also impressive that in Royce's hands, Spencer's inductively-based evolutionary theory retained its distinct character, while his contemporaries, James and Dewey, were more careless with Spencer's intentions. Anything in Spencer's philosophy with which they disagreed they caused to vanish by the re-configuring of Spencer's discourse so as to avoid inductivism and so that the ideas could be consistent with Absolute and phenomenological values. This appropriation of Spencer's philosophy did not so much extend its ideas to their logical conclusions, but transformed it into a variety of American idealism.

3 Dewey: The Jamesian Psychologist and Liberal Progressive

It was his firm adhesion to phenomelogical values that caused Dewey's philosophical stance to be more hostile to Spencer's evolutionary theory than that of other Spencerian reformers. While some familiarity with Hegelianism was the common occurrence of late nineteenth-century American philosophy, John Dewey's Hegelian ideas were more deeply rooted.[39] There is a minor controversy as to whether John Dewey's philosophical naturalism was a home-grown American product in its origins, or whether it originated in Germany as a result of the study of Hegel or from immersion in the writings of

38 Royce, *Herbert Spencer*, 116.

39 John Dewey (1859–1952). For the first half of the twentieth century, Dewey was regarded as the chief source of progressive ideas on education. In his many publications on education, he stressed the need for schooling to be adjusted to the industrial revolution and to democracy. He was also well known as a socialist and as a political commentator. However, in the period he was a Spencerian reformer much of his work was in psychology and in philosophical idealism of a Hegelian kind.

Fichte, Schelling and Schopenhauer.[40] In either case, there is little support for the commonly held belief that Dewey's ideas were *deeply* rooted in his early reading of Spencer or of other scientifically-minded writers who had explored the habitual and instinctive origins of the human mind, though, as has been remarked earlier, Dewey had read Spencer's *Principles of Psychology* before embarking on his own work on the human mind. Dewey's early immersion in Spencer's writings was probably not as important as his appraisal of how James had re-cast Spencer's psychology to suit Jamesian discourse. While Dewey's psychology was more anti-materialistic and whole-heartedly phenomelogical than James,' he never threw over the allegiance he owed James.[41] Insofar as any Spencerianism survived in Dewey's thought, it was in the reformed version that was preferred by James.

Dewey's philosophical naturalism did not allow for an alignment with the findings of evolutionary science: instead, it focussed him on experience in such a way that gave a minor role to any data that depended upon instincts, reflex action or any other biological features of the human brain. For example, when in his *Psychology* Dewey examined the kind of habitual actions that Spencer believed had promoted evolution, he concluded that they were unimportant mechanical processes: that is, these parts of cognitive life that adjusted themselves automatically to the permanent demands of their surroundings were not living features themselves. Dewey's argument here depended upon his belief that the conscious mind was in charge of apprehending variable inputs. This meant that, with the help of the will, the mind could order itself in a particular way when responding to any novel changes in its environment.[42] Dewey's psychology was a liberal or progressive one in that it emphasized freedom and choice, but it was essentially *not* a response to the findings of evolutionary science.

Dewey found Spencer frivolous because his treatment of the "Unknowable," which had appealed to the feeling that there must be something just beyond the edge of scientific knowledge, was infuriatingly casual. In the Hegelian terms preferred by Dewey, Spencer was guilty of abstracting just one element

40 Thomas C Dalton, *Becoming John Dewey, Dilemmas of a Philosopher and Naturalist* (Bloomington, Indiana: University Press, 2002), 9.

41 Dewey's respect for James was not due to his support for James' attempt to balance between the scientific and metaphysical realms. Rather it was owing to James' pre-eminence as an American psychologist at the time when Dewey too was writing in psychology.

42 John Dewey, *Psychology* (1887), *Early Works*, 1882–1898, vol. II (Carbondale, Ill.: Southern Illinois University Press, 1975), 102.

from the great "organism" of life, and then setting this up as the "unknowable."[43] Dewey was indignant that men and women should bow down and revere this thin abstraction as if it was the cause and basis of all reality and knowledge. This abstraction meant nothing when it was removed from the actual consciousness which people had of the role of life in spiritual freedom.[44] In brief, and in the Hegelian language of Dewey, Spencer was condemned for abstract thinking.

In his early writings Dewey had struggled to hold his Hegelian convictions and his grasp of scientific psychology at the same time: it was not until the final edition of his *Psychology* in 1891 that he gave up on attempting to update his textbook with recent psychological data. However, in the long term this did not matter as scientific data never seriously affected his views on mind. When Dewey dealt with human instincts, or with suggestions that behavioural patterns were not controlled by the will, he was careful to note that these *may* not exist. Similarly, if he found a remark to the effect that some behaviour patterns were inbuilt, and might be shared by related animal species, he would label such a phenomenon "unconscious cerebration" and warn that it might not extend to higher mental processes.[45]

Essentially, Dewey's objection to Spencer was a by-product of his own philosophy of science. Like James and Royce, he perceived the proper role of science as the explanation of phenomena by reference to physical forces. This is why Dewey analyzed Spencer's organic theory in terms of the application of physical forces.[46] For Dewey, there was no alternative to this; scientific procedure was a matter of applying deductive logic to the results of experiments, and he was unwilling to extend himself in order to countenance induction. The fact that Spencer, like Darwin, provided mountains of empirical evidence was of no interest to him since this was not given in an experimental form, nor was it relevant to an investigation of the mind (Dewey did cite a very large number of data-laden English and German neurophysiologists, but at no point did he see that their investigations established propositions about mental life). Further, Spencer's 'mental experiments' did not count since these were based upon an illegitimate mixing of subjective and objective matters.

It is well known that Dewey saw philosophy as having only two distinct phases – empirical and transcendental – which produced different kinds of

43 John Dewey, "Psychology as Philosophic Method" (1886), *Early Works*, vol. 1, 162.

44 Ibid.

45 Dewey, *Psychology*, 104.

46 Ibid.

knowledge.[47] This distinction between objective and subjective knowledge seems tame and conventional, but it acquires interest with Dewey's addendum that the latter kind, subjective knowledge, promoted the mind's self-development, and would function better if the subject possessed a notion of the Absolute. In terms of nineteenth-century psychological evolutionary development, these suggestions were awkward and seemed to offer little promise for scientific investigation. However, as Dewey's mentor, G. Stanley Hall, remarked, the appropriate way of reading Dewey here was as an ethicist, not as a scientist.[48] Dewey was not interested in the restrictions that neurophysiology placed on individual development, but only in how human beings developed control over their actions and their environment.

Dewey's emphasis upon how the mind responded to change by use of the will makes his theory of self-development resemble an eighteenth-century one – such as that of David Hume – that he professed to abhor.[49] Dewey seems to have avoided the rather ostentatious sense of being modern that is so conspicuous in the work of James and Royce. Even when Dewey was critiquing James Mill's work, *Analysis of the Human Mind* (1829)[50], he did not revert to modern evolutionary psychology. Instead, he simply complained that the work relied upon a mechanical image of the mind, and that it read all the aspects of the mind by reference to one principle. Instead of basing his critique on organic material from neuropsychology, Dewey referred to non-biological factors such as the life of society and education. Which factors came into play depended on which others had been suppressed by the consciousness. While Dewey's notion of 'mind' was not as carefully constructed as that of David Hume, it was still one in which development took place without reference to biological inputs such as automatic reflexes and instincts.

Dewey's theory of mind as a self-developing consciousness may have resembled Hume's in being free of instincts and infantile passions, but it carried extra metaphysical freight that could not be found in the work of an eighteenth-century Scot: for Dewey both scientific knowledge and perception in general required the presence of an ideal element. This idealization of science rendered explicit and definite any meaning that an idea already contained in

---

47 Shadworth Hodgson, "Illusory Psychology" (1886), in Dewey's *Early Works*, vol. 1, XLIX. This essay was a reply to Dewey's earliest essays in psychology.

48 Herbert W Schneider, "Introduction," Dewey's *Early Works*, vol. 2, XXV.

49 Significantly Dewey was quite gentle in his treatment of Hume (Dewey, "The New Psychology," 65).

50 James Mill (1829), *Analysis of the Phenomena of the Human Mind*, 2 vols., eds. J.S. Mill & Alexander Bain (London: Longmans, 1869).

perception.[51] Like William James, though less humorously, Dewey's mind was trapped in the concept of eternity. He believed that the very concept of existence had to be eternal because it could not be something that past ages had done without, and which some moment in the future would see springing into existence. Further, at any point consciousness must have included consideration of an end and a sense of unity that bound all events together.[52] For Dewey, these eternal qualities of consciousness interpreted and supplied meaning to all processes. From this perspective Spencer's analyses of evolutionary processes, which were based on data about psychology, biology and sociology, were essentially limited historical perspectives; at some point in the past they would not have existed. Therefore, the only way to make sense of Spencer's evolutionary processes was to modify them so that they would demonstrate the unfolding of an eternal idea.

There is no bridge between Dewey's earnest beliefs and Spencer's emotionally detached speculations about evolution because the former insisted upon the presence of teleology. Dewey knew that "physical evolutionists" argued as if evolution were not directed towards an end, but "facts show there *is* one."[53] Spencer, in Dewey's eyes, was a special case here because he displayed more awareness of the facts than other "physical evolutionists." For this reason Dewey followed William James in rewriting Spencer so he would resemble a teleological idealist. Since Spencer, Dewey argued, had claimed that the world had changed from an incoherent and indefinite homogeneity to a coherent and definite heterogeneity, he must also imply that there is a goal for this development and that the goal was an end of perfect harmony.[54] In his early and psychological writings, Dewey had modified Spencer rather than abandoning him; the *Synthetic Philosophy* was too prestigious and totemic to be overthrown. It was more suitable to refashion Spencerianism so as to be compatible with German philosophical ideas that portrayed the "organism" as something which was alive with spiritual and ethical qualities. This modified Spencer – who had been transformed from an inductive philosopher of science to a half-hearted and vague religious guru with an "Unknown" – needed an injection of philosophical idealism if he were to be plausible.

Later, when Dewey was no longer concerned with psychology or with the philosophy of nature, he became much more sympathetic to a secularly-minded evolutionary theory. This shift should also have made Dewey more

51 John Dewey, "Knowledge as Idealization" (1887), *Early Works*, vol. 1, 186.
52 John Dewey, "Ethics and Physical Science" (1887), *Early Works*, vol. 1, 223.
53 Ibid., 224.
54 Ibid.

sympathetic to Spencer, but his new enthusiasm for evolution was not thorough-going enough to cause him to re-examine the human mind in biological terms. Instead of freeing Spencer from the Hegelian language with which he had shackled him, Dewey now sneered that Spencer was a man who handled "used" metaphysical goods, and who had called his unknowable "God."[55] Spencer was abandoned in favour of a set of 'Darwinian' values which stressed the importance of competition in the natural selection of ideas. Instead of gesturing to genetics and the struggle of individuals to pass on their genes, Dewey's 'Darwinianism' was a fanciful way of advocating liberal values.

Dewey's late Darwinism was a liberal trope which dignified individual expression of opinion by giving it a role in the advancement of knowledge. If this had little to do with Darwin, it was at least a comforting echo of Darwin's contemporary, J.S. Mill, who gave individual freedom some standing in the production of scientific knowledge. Dewey's use of Darwin's name was not a homage to evolutionary biology or the development of the brain, but a response to a mode of thinking that had unsettled his grasp of traditional philosophy of a Hegelian kind. For the later Dewey, 'Darwinism' would prevent a recurrence of the philosophy of the Absolute and of the belief that philosophical knowledge differed from that of scientific knowledge. With the help of his 'Darwinism,' Dewey could now look to science to correct such errors.[56] Spencer was of no use here because Dewey still read him as an advocate of the God whom he had himself abandoned. There was no need to appropriate this Spencer in Dewey's new and progressive secular mission.

4 Conclusion

To speak about Spencer's late nineteenth-century critics – William James, Josiah Royce and John Dewey – is to plunge headlong in the maelstrom of American intellectual life at the fin-de-siècle. These three "public intellectuals," whom I have labelled Spencerian reformers, were the spokesmen for the ways in which philosophy and psychology balanced between the spiritual claims made on behalf of the mind or metaphysics and the scientific materialism that threatened to topple humanity off its pedestal, and reduce it to a historically-evolved set of biological responses. I call these men "public intellectuals," not to diminish their philosophical or academic standing, but to

55 John Dewey, *The Influence of Darwin on Philosophy* (1910) (New York: Peter Smith, 1951), 16.

56 Dewey, *The Influence of Darwin*, 18.

make the point that, like Spencer, they spoke to a general intellectual forum, not to restricted academic disciplines that resisted their members wandering too far into the preserves of others. While the Spencerian reformers held significant academic positions, this did not hinder them when addressing political and cultural questions about the purpose of the human mind, political progress or even about life in general. The resolution of these issues seemed vital to America's future, but in order to discuss this, the reformers had to rely on the widely read, but resolutely foreign, ideas of Herbert Spencer. To some extent there was a tension between Spencerian reformers and Spencer's works which was a conflict between domestic spiritual values and an imported philosophical materialism. The latter seemed a necessary accompaniment to the philosophical speculations of Spencer's *Principles of Psychology* which had its roots in neurophysiology. To the reformers, this materialism clashed badly with Spencer's metaphysical beliefs, and with what they perceived as his half-hearted gesture towards spiritualism in his idea of the "Unknown."

Part of the reformers' plans was to reconcile Spencer's psychology, which placed much weight on the evolved brain, with a spiritually-focused notion of the mind so that it could apprehend the object of religious feeling in a more reliable way. However, another part of the reformers' mission was to reject inductive philosophy of science – which they often credited to Spencer – in favour of an experimentally-based deductive science that only took cognizance of permanent forces in the way that physical science did. In the eyes of the reformers, science was not to be held hostage to historically transient data: a psychological phenomena on such as consciousness could not, in their eyes, exist later in an evolutionary process while being absent earlier. Spencerism had to be purged too so that it was free of misleading speculations based on induction. There were additional tasks before the reformers, and, curiously, one of these often caused them to reverse the usual (by that is meant ones we are likely to recognize) connotations of Spencerianism and Darwinism. In the reformer's world, these were reversed. Spencer, who detested violence and upheld pacifist views while praising freedom was credited with importing deterministic natural selection into science and political conflict, while Darwin, who restricted himself to biology and cultural evolution, was read as enshrining the values of choice and spontaneity in modern life.

Spencer's *Principles of Psychology* acted as a novelty mirror for American philosophers. This mirror had two facets which simultaneously reflected a portrait of the philosophy of mind that was both idealist and realist. The popularity of his evolutionary system rested on this twofold reflection. It could not be ignored even though his Germanized American readers would have been more comfortable with theories that were *either* transcendental *or* experimental. Since Spencerian reformers, such as James and Dewey, were dazzled by Spencerian reflections at the same time, while cursing Spencer for his smug assurance that seemed to side-step the angst which often accompanied thoughts about the "Absolute," they were discomforted. Then too, Spencer seemed to feel no metaphysical pain, nor any desire to experiment in order to test his ideas about the relationship between the psyche and the brain. Instead, he seemed content to employ introspection as a tool for understanding the brain when that kind of investigation should have been the exclusive preserve for experimental knowledge – following the fashion of Wilhelm Wundt and G. Stanley Hall. Royce, who was untroubled by the tension between the ideal world and the material one, was less dyspeptic about Spencer's failings than the other reformers. Unlike James and Dewey, he believed that Spencer's evolutionary theory was not a simplistic and progressive form of Darwinism, but a philosophy of change that focussed upon death and dissolution as much as it did upon life and evolution.

There are three concluding points that need to be made about American Spencerian reformers. First, their philosophical commentary upon Spencer was not an attempt to mediate between science and philosophy; instead, it was essentially a reaction against biological science. Other kinds of science were more acceptable as a basis for scientific reasoning, especially the physical sciences, because they seemed less disturbing to the philosophy of mind and less threatening to philosophical idealism. Second, Spencer's philosophy of science, which was inductive, was seen as illegitimate by Spencerian reformers. Spencer, who had theorized about the empirical world in much the same way as Darwin, Lyell and other Victorian natural scientists, was singled out as the sole exponent of a scientific method when his procedures were common to most natural scientists of his time. It could be said that Spencerian reformers had a bias against inductivism of a similar kind to that of Thomas S. Kuhn and Karl Popper in the twentieth century. Third, when, late in their lives, Spencerian reformers abandoned Spencer, they called themselves Darwinian. However, this re-labelling only came about because they were able to transform Darwinism into a liberal theory of freedom and choice. This was possible because, since Darwin did not engage in philosophical speculation, his ideas possessed a more plastic form than Spencer's had done. Paradoxically, "Darwinism" could be adapted as a term by Spencerian reformers while "Spencerianism" was eventually shunned because it seemed to contain a residue of biological speculation even after it was reformed.

CHAPTER 6

Spencer's American Disciples: Fiske, Youmans, and the Appropriation of the System

Bernard Lightman

In 1863 an article titled "Evolution of Language" appeared in the October issue of the *North American Review*. The anonymous author argued that Herbert Spencer's evolutionary theory could help to revolutionize the science of philology. Impressed by the article, Edward Youmans (1821–1887), science lecturer and advisor to the Appleton publishing house, was curious about the author's identity. Youmans was at the center of a Spencerian circle in New York and was constantly on the outlook for others who were like-minded. During the same month that the article on language was published, Youmans happened to be in Boston to raise money to underwrite a reprint of Spencer's *Essays* in America. He was shocked to discover that the *North American Review* article had been written by a young Harvard College student of twenty-one by the name of John Fiske (1842–1901). Anxious to meet him, Youmans tracked Fiske down at Cambridge.[1] Fiske later recalled that meeting Youmans that day changed his life. He declared, "I have always dated a new era in my life from the Sunday afternoon when Youmans ... came to my room in Cambridge."[2]

When he met Youmans, Fiske knew only four people who had any conception of the significance of Spencer's work.[3] Their enthusiastic admiration for Spencer was ridiculed. According to Fiske, when they insisted that Spencer's work as "as great as Newton's, and that his theory of evolution was going to remodel human thinking upon all subjects whatever, people used to stare at us and take us for idiots."[4] But through Youmans, Fiske was introduced to the Spencerian circle in New York,[5] and later, again through Youmans, the British scientists at the center of the development of evolutionary thought, including

1 Milton Berman, *John Fiske: The Evolution of a Popularizer* (Cambridge: Harvard University Press, 1961), 54.

2 John Fiske, *Edward Livingstone Youmans: Interpreter of Science for the People: A Sketch of His Life* (New York: D. Appleton & Company, 1894), 167.

3 Ibid., 167.

4 Ibid., 167.

5 Fiske met the New York Spencerian circle in March 1864. Youmans introduced him to Henry Ward Beecher, abolitionist and Congregationalist minister, the partners of the Appleton

 | DOI 10.1163/9789004264007_008

Thomas Huxley, John Tyndall, Charles Darwin, and Spencer himself. Being a part of this Anglo-American intellectual community enabled Fiske to become one of the foremost American popularizers of evolution.

Past scholarship has not neglected Fiske and Youmans's intellectual debts to Spencer.[6] Both Fiske and Youmans were important communicators of Spencerian evolution who were committed to demonstrating Spencer's distinctive originality and importance, especially to American audiences. There were certainly other American disciples of Spencer. However by focusing on Fiske and Youmans I can delineate how their Spencerism led them to take two different, but complementary, approaches to discipleship. Fiske was the popularizer, the productive writer and lecturer who tried to help Spencer complete his philosophical system. Youmans was primarily the organizer, working behind the scenes as Spencer's American agent. They both took advantage of the new publishing terrain after the American Civil War to bring Spencer's system to the forefront of the American intellectual scene, and even beyond. They both appropriated Spencer's thought in a way that resonated with the American reading public. If we examine the careers of Fiske and Youmans in light of the rich scholarship on nineteenth century print culture, a new picture of their discipleship emerges.

The Communications Revolution and the American Publishing Scene

Only two years after Fiske and Youmans met, Spencer's ambitious plan to write a multi-volumed synthesis of knowledge seemed to be on the verge of collapse. The initial volume, *First Principles*, which started to appear in parts in 1860, was published as a book in 1862. But in the autumn of 1865 Spencer announced that he would no longer be able to issue his series of philosophical works. Spencer had been publishing the books at his own expense, and sales had been disappointing, both in Britain and the United States.[7] Despite the establishment of a subscription scheme early on, where subscribers would each pay 10s yearly to receive installments, and the help of Youmans with the American market,

publishing house, Manton Marble, editor of the New York *World*, and John W. Draper, chemist and author of *History of the Intellectual Development of Europe*. See Berman, *John Fiske*, 54.

6 These are important themes in Berman's *John Fiske* and in discussions of Youmans in Richard Hofstader's *Social Darwinism in American Thought* (1944) and Robert C. Bannister's *Social Darwinism: Science and Myth in Anglo-American Social Thought* (1979).

7 Fiske, *Edward Livingstone Youmans*, 199.

Spencer could no longer afford the cost. Publishing popularizations of evolution was a risky business. In general, the market for books that popularized science was still unpredictable although the publication of cheap science books was on the rise. The System of Synthetic Philosophy was saved only through pure luck, when Spencer's father suddenly died and left him an inheritance. But only six years later Youmans had exciting news for Spencer about the sales of popularizations of evolution. On April 21st, 1871, he wrote to Spencer, "things are going here furiously. I have never known anything like it." Ten thousand copies of Darwin's *Descent of Man* were nearly all sold and five or six thousand copies of Huxley's *Lay Sermons* had been printed. "The progress of liberal thought," Youmans declared, "is remarkable."[8]

Fiske and Youmans became enamored with Spencer just as the worlds of British and American publishing were being transformed. In Britain, a communications revolution took place in the second quarter of the nineteenth century that, as James Secord has put it, represented the "greatest transformation in human communication since the Renaissance" by "opening the floodgates to a vastly increased reading public."[9] The communications revolution allowed publishers in Britain to eventually reach a mass reading audience in the second half of the century that included more and more members of the working class. As literacy rates increased, publishers found that the market for cheap books became viable. The introduction and development of the Fourdrinier machine, steam-driven presses, case binding, in addition to the reduction of the "taxes on knowledge" and the development of the railway system, were features of the revolution in this period.[10] The revolution was not limited to the production of more and more affordable books. During the middle of the nineteenth century the British periodical press grew by leaps and bounds due to the greater demand from increasingly literate and leisured reading audiences.[11] The British evolutionists took advantage of the public interest in science and the new publishing venues open to them. They produced books

8 Ibid., 266.

9 James Secord, *Victorian Sensation: The Extraordinary Publication, Reception, and Secret Authorship of Vestiges of the Natural History of Creation* (Chicago: University of Chicago Press, 2000), 2.

10 Bernard Lightman, *Victorian Popularizers of Science: Designing Nature for New Audiences* (Chicago and London: University of Chicago Press, 2007), 30–31.

11 Gowan Dawson, Richard Noakes, and Jonathan R. Topham, "Introduction," in *Science in the Nineteenth-Century Periodical: Reading the Magazine of Nature*, ed. Geoffrey Cantor, Gowan Dawson, Graeme Gooday, Richard Noakes, Sally Shuttleworth, and Jonathan R. Topham (Cambridge: Cambridge University Press, 2004), 16.

written for a popular audience and arranged for their articles to be published in the general periodical press.

The American publishing system was less well developed, but after the Civil War ended in 1865 the publishing industry experienced continuous ferment and growth as a result of the industrialization of the book. As in Britain, publishers benefited from technological developments in presses, stereotyping, and engraving, the growth of education, the rise of public libraries, and, most importantly, the rapid expansion of the reading audience. More books, magazines, and newspapers were published as part of the creation of a mass literature for a mass audience. New publishing houses were established and university press publishing began.[12]

However scientific books by American authors seldom sold well. There were fewer periodicals in which scientific articles could be published. But when it came to reprints by authors from outside the United States, American publishing houses had the option of practicing piracy. They could choose to benefit enormously from the results of the British communications revolution and reprint books by British scientists without having to pay them for the copyright. But in reality American publishers were often cautious when it came to reprinting new scientific books, fearing that the readership was not large enough. Initially, Spencer had planned to approach the Boston publishers Ticknor and Fields to undertake his System of Synthetic Philosophy. But when in 1860 he offered them his book on education, they declined to publish it. It was not obvious in the 1860s that scientific books could be profitable. Most American publishers, Fiske recalled, were "likely to view the matter in the same light as Messrs. Ticknor and Fields. The Americans were at that time excessively provincial."[13] Reprints of English books, Fiske asserted, and translations from French and German, "were sadly behind the times." Undertaking the reprinting of new scientific books, such as Spencer's, required "some boldness."[14] That is why Youmans was excited in 1871 when reprints of Darwin and Huxley sold so well. But prior to that, in the 1860s, when they first took on the task of disseminating Spencer's evolutionary system, Fiske and

12 John Tebbel, *Between the Covers: The Rise and Transformation of Book Publishing in America* (New York, Oxford: Oxford University Press, 1987), 80–81; John Tebbel and Mary Ellen Zuckerman, *The Magazine in America 1741–1990* (New York, Oxford: Oxford University Press, 1991), 57; Scott E. Casper, Jeffrey D. Goves, Stephen W. Nissenbaum, and Michael Winship (eds), *The Industrial Book 1840–1880, Volume 3 of A History of the Book in America* (Chapel Hill, North Carolina: University of North Carolina Press, 2007), 7–9.

13 Fiske, *Edward Livingstone Youmans*, 110–111.

14 Ibid., 111.

Youmans were facing an uphill battle due to the condition of American science publishing.

John Fiske and His Early Relationship with Spencer

Shortly after meeting Youmans, Fiske obtained a photograph of Spencer, and wrote to his mother about how impressive it was:

> Spencer's face is a magnificent one. There is something not quite perfect about the mouth; but the eyes are like those of a lynx, and the grandest I ever saw. Taken all together, the effect of the head and face is as imposing as Newton's; while at the same time the expression is gentle, humorous, and lovable, in the extreme.[15]

For Fiske, Spencer was as keen sighted as a lynx and looked as distinguished as Newton above the shoulders. Yet, Spencer's facial features attracted affection. Despite the imperfect mouth, Fiske felt irresistibly drawn to Spencer. Fiske not only worshipped Spencer, he also worshipped Spencer's Unknowable. Fiske subtly reshaped Spencer's system, making the religious dimensions far more central. While Spencer limited his discussion of the reconciliation of science and religion to the opening section of *First Principles*, Fiske considered the continuity between "cosmic theism" and Christianity to be the *raison d'être* of the synthetic philosophy. This is the message that American audiences needed to hear from a Spencerian disciple. But Spencer was also important for Fiske's attempt to establish himself as a popularizer of evolution. Fiske turned to Spencer for advice on how to navigate the difficult waters of the world of publishing.

John Fiske was born in Hartford, Connecticut, to Edmund Brewster Green, a lawyer and Whig journalist, and Mary Fisk Bound. Fiske was sent to live with his grandmother Polly Fisk Bound and great-grandfather John Fisk in Middletown, Connecticut, at the age of one after the financial failure of his father's political journalism. After the death of his father in 1855, he accepted his grandmother's request that he legally adopt his great-grandfather's name. The "e" was added in 1860. Raised in an orthodox family, Fiske formally joined the North Congregational Church in Middletown when he was fourteen. Shy and bookish, he lost his religious faith through his reading of contemporary

15 John Spencer Clark, *The Life and Letters of John Fiske*, 2 vols (Boston and New York: Houghton Mifflin Company, 1917), I, 265.

FIGURE 6.1
John Fiske, the cosmic philosopher. From John Spencer Clark, The Life and Letters of John Fiske, *2 vols (Boston and New York: Houghton Mifflin, 1917), I, frontispiece.*

philosophy, science, and literature. At the age of eighteen he announced his conversion to Comtean positivism, around the same time that he entered Harvard College at Cambridge. He found the Harvard curriculum uninspiring and created a self-directed program of reading in history, philosophy, and languages in preparation for a career as philologist (Fig. 6.1).[16]

Fiske became an evolutionist in 1859, but prior to the appearance of Darwin's *Origin of Species*. The "immediate cause" that "drove" him to the development theory was "the mental reaction experienced in reading Agassiz's arguments against that theory in his Essay on Classification, in 1859, shortly before Darwin's was published."[17] Fiske discovered Spencer in his first year at Harvard. He came across the original prospectus of Spencer's System of Synthetic Philosophy in a Boston bookstore. He wrote to a friend on June 24th, 1860, "my soul is on fire ... for Herbert Spencer is about to execute a gigantic series of Positive books on which he has been at work for years."[18] He told his mother that he considered it his "duty to mankind as a Positivist to subscribe."[19] At this

16 American National Biography Online.

17 Fiske, *Edward Livingstone Youmans*, 58.

18 Clark, *Life and Letters of John Fiske*, I, 139.

19 Ibid., I, 139.

point, he viewed Spencer as working within the Comtean tradition. After receiving the first installments of *First Principles*, Fiske started reading the earlier works of Spencer referred to in the footnotes to the prospectus. In July 1861 he was devouring a one-volume edition of the *Principles of Psychology* (1855). In August he started in on the *Social Statics*, first published in 1851. What excited Fiske was Spencer's aim of producing a comprehensive explanation of the universe that surpassed Newtonian physics and that was far more satisfying than Congregationalist theology.[20] But Fiske found that reading Spencer was a daunting, as well as an exhilarating, experience. In a letter to a friend from the early sixties he admitted that Spencer's "Laws of Organic Form" "was so *omnisciently* learned that I could barely understand it." Spencer brought up one hundred kinds of plants, about which Fiske knew nothing, as illustrations. "He brought them in with such perfect coolness," Fiske wrote, "and proceeded to argue from the way the leaves are cleft and the petals arranged in each kind, with such an apparent unconsciousness that other people didn't know all vegetables in creation that I began to think myself a block-head." Fiske was left with "a sense of awe after closing the book, as if I had been holding communion with Omniscience: and this I never felt when reading any one else." After reading "Laws of Organic Form," Fiske was even more convinced that Spencer was "the greatest thinker of this time." He went beyond Comte as he "not only printed out the methods of constructing a Positive philosophy, but has also constructed one."[21]

After they had met in 1863, Youmans encouraged Fiske to write to Spencer. His first letter to Spencer was written on February 20th, 1864, after he had graduated from Harvard the previous summer. Fiske wrote, "I have known you a long time through your writings and have felt a strong desire to become personally acquainted with you, but the fear of appearing presumptuous has hitherto restrained me from taking any steps to secure that end." He explained that it was Youmans who had suggested that he write to Spencer. Then Fiske recounted his spiritual journey, from being brought up "in the most repulsive form of Calvinism," to his attraction to Comte, Mill and Lewes, and then, finally, his discovery of Spencer's work. He emphasized the profound impact that Spencer had had upon him. "The influence of your writings is apparent alike in every line of my writings and every sentence of my conversation," Fiske declared, "so inextricably have they become intertwined with my own thinking, that frequently on making a new generalization, I scarcely know whether

20 Berman, *John Fiske*, 37.

21 Clark, *Life and Letters of John Fiske*, I, 230.

to credit myself with it or not."[22] Spencer responded on March 26th, 1864, commenting, "it is very refreshing to me to meet with so much sympathy as that expressed in your letter." He believed that Fiske's intellectual progress implied "an unbiased search after *truth*" that was "very unusual." Spencer noted that he had already seen Fiske's article in the *North American Review*, and that it showed "power of independent thought" and signaled the possibility that Fiske could "render more important service" by further elaborating the application of evolution to language.[23]

Fiske waited two years before writing his second letter to Spencer. On June 3rd, 1866, he wrote to Spencer that since they had last been in contact Fiske had graduated from the Harvard Law School, had been admitted to the Bar, had been married, had practiced law for a year, and had finally abandoned the profession in disgust. He had decided to devote himself to a life of study and writing, unless a philological professorship came his way. But his big news for Spencer was that he hoped to publish a volume of essays that would be "illustrative of your philosophy."[24] He asked Spencer for advice on finding an English periodical that would publish a piece on Positivism, for the editor of the *North American Review* was impossible to deal with. The rest of the periodicals in the United States were orthodox. Spencer replied on June 19th, 1866. He was worried that Fiske would find it difficult making a living by his pen. Even among the limited liberal periodicals there was reluctance to publish advanced opinions. "Periodicals established for the purpose of propagating liberal opinions," Spencer advised, "but presently having to struggle for existence from want of sufficient support, are prone to subordinate their original aims to the cultivation of a light literature that will bring more readers." Spencer promised to bring Fiske's work to the attention of the editor of the *Westminster Review*, and mentioned that the *Fortnightly Review* and *Macmillan's Magazine* were other possibilities. Then Spencer encouraged Fiske to continue with his book project. He wrote to Fiske that "the volume you name would I think help very much to popularize the general doctrine as well as strengthen it by further illustrative matter."[25]

When Charles William Eliot became president of Harvard he invited Fiske to give a series of lectures on positive philosophy in 1869–1870. It seemed possible that a position for Fiske might become available. But there was opposition to hiring a self-professed positivist on faculty, so Eliot appointed Fiske assistant

22 Ibid., I, 294.

23 Ibid., I, 296.

24 Ibid., I, 314.

25 Ibid., I, 316–318.

librarian in 1872. Since his library duties were not onerous, Fiske had the time to continue writing articles and to revise his Harvard lectures for publication.[26] In 1873 Fiske traveled to London, where, financed by his Boston admirers, he stayed to work on the Harvard lectures and met Spencer in person for the first time. On October 13, 1873, he wrote to his wife that Spencer was "exceedingly refined and elegant in manner, and appears like the great man he is, though he seems overworked."[27] But the more Fiske saw of Spencer, the more he realized that the man he had worshipped from afar had serious problems. Spencer's ill health led him to write his wife, "the more I see of the poor old fellow, the more I pity him from the bottom of my heart." Fiske wondered how Spencer was going to complete all of the volumes projected for his synthetic philosophy given his current health. Fiske also realized that Spencer suffered from living a hermit-like existence. "He is so lonely," he told Mrs. Fiske, "and so curtailed from want of human sympathy." Fiske became conscious of personal defects in his hero. Spencer was so cerebral that he "never seems to warm up to anything but *ideas*." Fiske protested that Spencer's powers of criticism had made it impossible for him to enjoy simple things, like "a perfect day." "So he picked flaws with the grey-blue sky," Fiske complained, "and the peculiar Turner-like light, and *everything*."[28]

Fiske's *Outlines of Cosmic Philosophy*

Fiske finished his revisions of the lectures in February 11th, 1874, while still in London. After commenting on how it had been a "profitable four months in London," he remarked that getting "that everlasting big book into shape has been no fool of a job; and it has been well done, too – O, sing Hallelujah!"[29] Fiske's massive two volume *Outlines of Cosmic Philosophy* was published later that year. In the preface Fiske described his work as a "complete outline of the new philosophy based on the Doctrine of Evolution."[30] Then he discussed the relation of his work to Spencer's. Fiske explained that he had originally intended to write a series of essays that illustrated Spencer's philosophy, but

26 American National Biography Online.

27 Clark, *Life and Letters of John Fiske*, I, 471.

28 Ibid., I, 473–474.

29 Ibid., I, 466.

30 John Fiske, *Outlines of Cosmic Philosophy Based on the Doctrine of Evolution, With Criticisms on the Positive Philosophy*, 2 vols (Boston and New York: Houghton, Mifflin & Company, 1874), I, vii.

that the work had grown to become more than that. "For while it does not, as a whole lay any claim to the character of an original work," Fiske declared, "it has nevertheless come to contain so much new matter, both critical and constructive, that it can no longer be regarded as a mere reproduction of Mr. Spencer's thoughts."[31] Fiske's project had become a more accessible summary of Spencer's synthetic system in two volumes. However, Spencer had not yet written projected volumes on the *Principles of Sociology* and *Principles of Ethics*. In effect, Fiske was attempting to complete Spencer's system. Although he could draw on Spencer's previous work to fill in some of the gaps, Fiske had to provide what he called "new constructive matter." Fiske's treatment of themes not covered in Spencer's publications would be more "thoroughly and satisfactorily presented" by the master in his forthcoming works.[32] Even though Fiske had consulted at length with Spencer about the revisions of the lectures through their correspondence, and then later during their meetings in London, there was still potential for significant disagreement between master and disciple (Fig. 6.2).

They differed on the most appropriate title for the evolutionary system that Fiske wanted to outline in his book. Whereas Spencer favored the term "Synthetic Philosophy," Fiske wanted to use "Cosmic Philosophy." Spencer objected that all philosophies could be termed "Cosmic" since their subject was an explanation of the universe or the cosmos. Spencer wanted to be sure that his system was presented as being distinct from all others, especially Comte's. Fiske countered that Spencer's alternative term was open to "a parallel objection." Fiske preferred "cosmic" because it conveyed what was meant by "Synthetic" but it also "hits the precise point by which Mr. Spencer's philosophy is fundamentally distinguished alike from Positivism and from all ontological systems." Spencer's philosophy was not merely a synthesis; it was a cosmic synthesis that made no appeal to theological or ontological agencies that were extra-cosmic.[33] Fiske emphasized that the disagreement on title was "purely a question of nomenclature" and not a difference of opinion on the system to which the nomenclature applied. The system that Fiske expounded was "essentially his [Spencer's]" and the supplementary illustrations that had been added were "quite in harmony with the fundamental principles which he has laid down."[34]

31 Ibid., I, vii.
32 Ibid., I, viii.
33 Ibid., I, ix-x.
34 Ibid., I, xi.

FIGURE 6.2 *Fiske cartoon depicting him as flying the "evolution kite" in America. Darwin is helping Fiske with the kite in the background, but the animal to his right is reading Spencer's* First Principles. *The cartoon appeared originally in the September 12th, 1874 issue of the New York* Daily Graphic. *From John Spencer Clark,* The Life and Letters of Jon Fiske, *2 vols (Boston and New York: Houghton Mifflin Company, 1917), vol. 2, opposite p. 56.*

Outlines is divided into three parts, "Prolegomena," "Synthesis," and "Corollaries." Spencer would have had few disagreements with Fiske on what was contained in "Prolegomena" and the first sixteen chapters of "Synthesis." In "Prolegomena," Fiske defines and illustrates the scope and methods of the Cosmic Philosophy before constructing a theory of the universe. Here he draws heavily on Spencer's *First Principles* and his *Principles of Psychology* – providing lengthy quotes from these works–as he discusses the doctrine of the relativity of knowledge and the law of evolution as the first generalization concerning the cosmos framed in conformity to the requirements of proper scientific method. Throughout "Prolegomena" Fiske distinguishes Comte's positivism from Spencer's evolutionary synthesis. He rejected his earlier tendency to group Spencer with Comte and the positivists. Although Comte was important for espousing the scientific method, his philosophy ultimately diverged from Spencer's. Not only would Comte "unquestionably repudiate us as disciples," Fiske insisted, "so do we unhesitatingly repudiate him as a master."[35]

The first sixteen chapters of "Synthesis" continues in this same vein. Fiske covers the formulation of the law of evolution, with the characteristic Spencerian emphasis on the movement from an indefinite, incoherent homogeneity to a definite, coherent heterogeneity, and how it is manifested in the evolution of the planets, the earth, the origins of life, and the development of mind. Fiske rarely deviates from Spencer. He constantly refers to *First Principles* and to the *Principles of Psychology*. His account of the evolutionary law is not intended "as a substitute for the study of Mr. Spencer's treatise, but rather as a commentary upon it."[36] Discussing the evolution of mind at the beginning of volume two he tells his readers that he is condensing Spencer's "exposition of one hundred pages" into half a dozen pages.[37] The critique of positivism also continues, as Fiske condemns Comte's rejection of psychology as a valid science. In offering phrenology as a substitute, Fiske argues, Comte committed a serious "scientific blunder."[38]

However, when Fiske begins the section on the evolution of society in chapter seventeen of "Synthesis," he no longer has a big book by Spencer to depend on as the basis for his discussion. Although he refers to some of Spencer's essays and *Social Statics*, he draws far more on Walter Bagehot's *Physics and Politics* (1872), though not uncritically. Fiske makes three main points when he deals with social evolution. First, social evolution consists in the integration of human families or tribal communities into larger and larger groups which became ever more heterogeneous and interdependent. Second, that civilization consists in the ever-increasing definiteness and complexity of the correspondence between the community and the environment. Third, and finally, Fiske asserts that in exploring these themes he is carrying on "the

35 Ibid., I, 144–145.
36 Ibid., I, 355.
37 Ibid., II, 89.
38 Ibid., II, 74.

inquiry to a point somewhat in advance of Mr. Spencer's exposition, as it now stands."[39] Not that Fiske sees himself as diverging from Spencer. He demonstrates his loyalty by making Comte his main target. Fiske argues that the claims of Comte to be regarded as the founder of Sociology "cannot be sustained."[40] This cleared the way for Spencer to be seen as the real creator of a scientific study of society.

Part III, "Corollaries," also contains new critical matter, especially the chapters on religion. Here Fiske affirms that the Cosmic Philosophy is "based upon the affirmation of God's existence, and not upon the denial of it, like irreligious Atheism, or upon the ignoring of it, like non-religious Positivism."[41] Fiske then goes on to explain the main difference between the religions of Cosmic Philosophy and of traditional Christianity. Christian anthropomorphic conceptions of God are not only ruled out of order because they are unscientific, evolutionary theory is used to explain why humans anthropomorphize.[42] Nevertheless, Fiske maintains that the Cosmic Philosophy is "eminently conservative" in regards to "the fundamental truths of Christianity," and that it has "no fellowship either with the radical Infidelity of the eighteenth century or with the world-mending schemes of Positivism."[43] Cosmic Philosophy is not a new religion that will replace Christianity. Rather it describes the less anthropomorphic phase of Christianity to come. "In the scientific philosopher," Fiske states, "there may be as much of the real essence of Christianity as there was in the cloistered monk who preceded him; but he thinks in the language of a man and not in the language of a child."[44]

The new insights into religion provided by the Cosmic Philosophy are the climax of the *Outlines*. Fiske emphasized the religious implications of Spencer's thought. He shared this tendency with many of Spencer's American admirers and with British dissident secularists who in the 1880s saw Spencer's Unknowable as the foundation for building a religious agnosticism.[45] Spencer did not object to the religious interpretation that Fiske had given to his thought. On December 11th, 1874, he wrote to Fiske to say that he had read sections of

39 Ibid., II, 162–163.

40 Ibid., II, 253.

41 Ibid., II, 377.

42 Ibid., II, 410.

43 Ibid., II, 472.

44 Ibid., II, 505.

45 Berman, *John Fiske*, 38; Bernard Lightman, "Ideology, Evolution and Late-Victorian Agnostic Popularizers," in *History, Humanity and Evolution: Essays for John C. Greene*, Ed. James R. Moore (Cambridge: Cambridge University Press, 1989), 285–309.

the *Outlines*. "What I have read, however," he told Fiske, "which has been chiefly in the new parts, has pleased me greatly."[46]

Although *Outlines* helped to establish Fiske's reputation as a philosopher it did not advance his position at Harvard. He resigned his Harvard library position in 1879 and devoted himself to lecturing and writing. He began working more and more on American history, applying evolutionary theory to this topic. Spencer's philosophy led him to foresee a future of peaceful evolution, and the expansion of American democracy. Fiske found evidence in Spencer's writings for the belief that America was divinely destined to lead the world.[47] Fiske and Spencer maintained their friendship over the years. They socialized in London when Fiske came to give a series of lectures at the Royal Institution in 1880 on evolution and American history and politics. Huxley declared that they were "the very best lectures he has ever heard at the Royal Institution."[48] When Spencer came to the United States in 1882, he visited Fiske and his family in Boston. Fiske showed him the room where in 1860 as a Harvard student he first became acquainted with Spencer's ideas by reading *Social Statics*.[49] At the dinner in Spencer's honor on November 9th, 1882, at Delmonico's, a New York restaurant, Fiske spoke about the religious and moral dimensions of his master's philosophy. Fiske asserted that in demonstrating the existence of an "infinite and eternal Power" in nature as a "scientific truth," Spencer had rendered as great a service to religion as to science.[50] After Spencer had returned to England, he wrote to Fiske on November 24th 1882, "I wanted to say how successful and how important I thought was your presentation of the dual aspect, theological and ethical, of the Evolution doctrine."[51] Fiske's appropriation of Spencer for the American context was done, in this case, with Spencer's approval. Fiske remained a loyal disciple of Spencer's until his death in 1901.

46 Clark, *Life and Letters of John Fiske*, II, 58.

47 Barry Werth, *Banquet at Delmonico's: Great Minds, the Gilded Age, and the Triumph of Evolution in America* (New York: Random House, 2009), xiii, 208.

48 Clark, *Life and Letters of John* Fiske, II, 178.

49 Ibid., II, 253.

50 John Fiske, "Evolution and Religion," in his *Excursions of an Evolutionist*, 8th edn (Boston: Houghton, Mifflin and Company, 1887), 301, 305. For more on Fiske as well as the banquet at Delmonico's see James R. Moore, "Herbert Spencer's Henchmen: The Evolution of Protestant Liberals in Late Nineteenth-Century America," *Darwinism and Divinity*, Ed. John Durant (Oxford: Basil Blackwell, 1985), 76–100.

51 Clark, *Life and Letters of John Fiske*, II, 264.

Edward Youmans and the Popularization of Evolution

Fiske outlived Youmans, who died in 1887. Youmans told Fiske that if his biography were to be written, he wanted Fiske to be the author.[52] Fiske's *Edward Livingston Youmans* was published in 1894. The volume was dedicated to Spencer in a letter to him from Fiske placed after the title page. Fiske asked Spencer to accept the dedication "in token of the long friendship we have shared with each other and with him who has gone from us."[53] Spencer was delighted with the dedication. In a letter written to Fiske on March 27th, 1894, he thanked Fiske "for the sympathetic expression of your dedication which took me by surprise," though he modestly suggested that it would have been more appropriate to give Youmans's devoted sister the honor. Spencer also remarked that Youmans deserved to be remembered as a self-sacrificing "servant of humanity" that "is rarely met with."[54] Fiske described Youmans somewhat differently in the opening pages of the book, as the herald for Spencer's evolutionary philosophy in the United States. "He did more than any one else to prepare the way in America for the great scientific awakening," Fiske declared, "which first became visible after the publication of *The Origin of Species*. In Youmans the approaching better era found its John the Baptist."[55] According to Fiske, Youmans's discipleship led to something akin to a religious revival.

But Youmans's methods for disseminating Spencer's philosophy went far beyond writing popularizations of his synthetic system. Youmans became Spencer's agent in the United States. Youmans's success in marketing Spencer's work was due to his sophisticated understanding of the transformation of the publishing world and the growth of new audiences for cheap science books. Like Fiske, he realized that Spencer's work had to be made more accessible to readers. He persuaded Appleton to publish Spencer's works in the U.S. During extensive tours he brought Spencer's supporters together to form a network of loyal readers. He ensured that subscribers to the System of Synthetic Philosophy continued their support for the project. As a result of his efforts, Spencer's American publications were a financial success. But Youmans's activities went beyond the U.S. He was the driving force behind the ambitious International Scientific Series, which he intended to be a vehicle for disseminating Spencer's

52 Fiske, *Edward Livingston Youmans*, [iii].

53 Ibid., [iii].

54 David Duncan, *The Life and Letters of Herbert Spencer* (London: Methuen & Co., 1908), 357–358,

55 Fiske, *Edward Livingston Youmans*, 4.

evolutionary system worldwide. No wonder that Spencer felt indebted far more to Youmans than to Fiske. In his massive *Autobiography* (1904), Spencer mentioned Fiske only twice, referring to him as "one who had done so much as an expositor of the Synthetic Philosophy."[56] Youmans receives more than nineteen mentions and the coverage of their correspondence is extensive. Spencer asserted that Youmans "was of all Americans I have known or heard of, the one most able and most willing to help me." Youmans had "devoted his life mainly to spreading throughout the United States the doctrine of evolution."[57]

Youmans was born in Coeymans, New York, to his father Vincent, a farmer and mechanic, and his mother Catherine Scofield, a teacher. The Youmans children were encouraged to read and explore new ideas. Edward, a voracious reader, became interested in science at a young age. Due to an attack of ophthalmia in 1835, his eyesight was damaged. In 1839 he moved to New York to be treated for bouts of blindness that almost completely devastated him. One of Youmans's first encounters with evolutionary theory came when the *Vestiges of the Natural History of Creation* (1844) was read to him while he battled blindness. Eventually, when his medical difficulties were partially under control, he began a career as a writer and lecturer on scientific topics. Among his early works were *A Class-book of Chemistry* (1851), a successful textbook that went through three editions; *Alcohol and the Constitution of Man* (1853); and *The Handbook of Household Science* (1857), which dealt with the application of science to such domestic topics as heating, ventilation, food preparation and cleaning. He also edited two collections of essays focusing on the impact of science on education and modern culture. Youmans lectured on the lyceum circuit from 1851 to 1868, but health problems made it impossible for him to continue.[58] Fiske claims that "the hospitable reception of Herbert Spencer's ideas" was "largely due" to Youmans's lectures. Youmans had prepared American audiences for evolution.[59] But it is more likely that Youmans's early efforts to popularize science were important for providing him with experience from which to draw when he made the dissemination of Spencer's philosophy the major goal of his activities. Through his lecturing and science book writing Youmans had become familiar with American audiences interested in science. He knew what appealed to them (Fig. 6.3).

Youmans first came across Spencer in 1856, while he was working on the *Handbook of Household Science*, when he read an article on the *Principles of*

56 Herbert Spencer, *Autobiography*, 2 vols (London: Williams and Norgate, 1904), II, 405.

57 Ibid., II, 53.

58 Fiske, *Edward Livingston Youmans*, 42; American National Biography Online.

59 Fiske, *Edward Livingston Youmans*, 78.

FIGURE 6.3
Edward Livingston Youmans, Spencer's American agent. From John Fiske, Edward Livingston Youmans: Interpreter of Science for the People *(New York: D. Appleton and Company, 1894), frontispiece.*

Psychology (1855). He ordered *Principles* from London – there was no American edition – and after reading it judged that it was an important step towards the construction of a general theory of evolution. He then read *Social Statics* and realized that Spencer was the author of several anonymous articles on evolution in various quarterlies that he had previously read. In 1860 a friend showed him the prospectus announcing Spencer's System of Synthetic Philosophy. The following day, on February 23rd, 1860, Youmans wrote to Spencer, offering his help in lining up American subscriptions. This was the beginning of Youmans's extensive correspondence with Spencer and their enduring friendship.

Youmans believed that the dissemination of Spencer's evolutionary philosophy was absolutely crucial in light of the current situation in the United States. On April 12th, 1864, he told Spencer that his work promoted a "mental emancipation" needed in the United States, when the bloody Civil War was nearing its end. Youmans was a passionate abolitionist. "The great slave system," Youmans wrote to Spencer, "intrenched in conservatism, and the natural ally of everything old, superstitious, arbitrary, and barbarous – the sworn foe of all liberty of thought and expression, all reform and progress,...this system had well-nigh paralyzed the mind of the nation." But Youmans believed that the war had "broken the spell." He observed, "I have never before known such boldness of inquiry and demand for first principles."[60] Spencer, whose *First Principles* was almost ready for publication by Appleton thanks to Youmans, was just what was needed in America at this point in time.

Youmans produced few popular science books after he became involved with Spencer, but he did write a number of essays popularizing evolution. His "Herbert Spencer and the Doctrine of Evolution" (1874) was the text from a lecture that he gave before the New York Liberal Club on June 5th, 1874. Spencer read the published version, rejoicing at the "complete success of your address and exposition." He complimented Youmans on the "clearness and power" of the lecture.[61] The lecture was meant to draw attention to Spencer's important contribution to the acceptance of evolutionary theory. Spencer was not merely a popularizer of the views of Darwin and other men of science. Only he understood that the laws of the changes in the past leading up to the present had to be determined and that this would lead to an "extensive revolution of ideas."[62] Spencer had done "more than any other man" to make the evolutionary view of nature the "starting point of a new era of knowledge."[63] Spencer had arrived at a law of evolution as a universal principle of nature, and announced it "two years before the appearance of Mr. Darwin's *Origin of Species*."[64] Then Spencer had begun to construct an entire system, and, by 1874, was halfway through that daunting project.[65] Youmans went into detail about how Spencer struggled with adversity in order to get that far. Ill health, little encouragement, and "very shabby support" had not stopped him from taking on "the most formidable intellectual project that was ever undertaken by any single mind." Spencer "was not wanting in heroism."[66] Like Fiske, Youmans was careful to delineate Spencer's distinctiveness. Spencer was "no follower of Comte, Darwin, or any other man." Darwin had established a theory of organic development, but "he has made no attempt to elucidate the general law of Evolution." It was due to "popular confusion" that "Darwinism" and Evolution were seen as the same thing. In being the first to elucidate and establish the "law of Universal Evolution" Spencer was comparable in stature to Copernicus, Newton, Harvey, Priestly, Young, and Darwin.[67]

60 Ibid., 178.

61 Ibid., 322.

62 Ibid., 502–504, 508.

63 Ibid., 509.

64 Ibid., 526.

65 Ibid., 532.

66 Ibid., 538.

67 Ibid., 541–543. Nine years later Youmans sounded similar themes in the publication covering the banquet in Spencer's honor at the conclusion of his trip to the United States in

In addition to writing essays that popularized Spencerian evolution, Youmans also defended his hero from critics. In 1867 he rejected the charge that Spencer was a materialist, and by implication, an atheist. Youmans declared that the doctrine of evolution stood apart "from all theological or ontological or metaphysical questions."[68] In 1884 he ridiculed the author of an article in the *North American Review* who had attempted to discredit Spencer's work. "There is a good deal of this sort of 'crushing' of Herbert Spencer by aspiring literary adventurers," Youmans wrote, "which is not worth serious refutation; and the best rule that can be given for judging of such efforts is to ask what the writer has ever done to vindicate his claims to be heard, when he proposes to undermine and overthrow the careful and life-long work of a thinker who has made a profound and lasting impression upon the mind of his age."[69]

Disciple as Agent

But Youmans's role as disciple encompassed far more than writing popularizations of Spencer's ideas and defending him from his critics. As Spencer's American agent, he relied on his understanding of the book trade to ensure that Spencer's books and articles were published in the U.S., that his readership grew, and that he was suitably compensated for his intellectual work. Having published with Appleton previously, Youmans was a trusted advisor to the publishing house. Appleton was the second most powerful publishing firm in the U.S. before the Civil War. After the war Appleton began a period of expansion. Since New York would become the center of American after 1870,

---

1882. Stressing Spencer's originality, Youmans credited him with establishing a "new organon of philosophy, founded upon science and embodying throughout the theory of evolution." This is why, according to Youmans, Spencer was more important than Darwin. Whereas Darwin's theory of natural selection was only a "subordinate part" of evolution, Spencer had remapped "whole branches of knowledge" in light of "an all-comprehensive and reconstructive principle." See E.L. Youmans, "What Mr. Youmans Did Not Say," in *Herbert Spencer on the Americans and the Americans on Herbert Spencer* (New York: D. Appleton and Company, 1883), 73, 75.

68 E.L. Youmans, "Herbert Spencer and His Reviewers," *Christian Examiner* 82, no. 2 (March 1867), 206, 209.

69 E.L. Youmans, "Herbert Spencer's Latest Critic," *North American Review* 139 (November 1884), 479.

Youmans was well placed to help Spencer find an audience throughout the 1860s, 1870s, and 1880s.[70]

On Youman's recommendation Appleton reprinted Spencer's book *Education: Intellectual, Moral and Physical* (1861) after Ticknor and Fields declined to publish it. Youmans told Spencer that he wanted to "open the campaign with it" rather than wait for *First Principles*, the first volume of the System of Synthetic Philosophy.[71] At Youmans's urging, Appleton published a reprint of *Social Statics* and then two selections of Spencer's essays titled *Illustrations of Universal Progress* and *Essays, Political, Moral, and Aesthetic*. When they were published Youmans wrote positive reviews and notices that were placed in magazines and newspapers and, in addition, made numerous pithy extracts for the reading columns of various daily and weekly papers.[72] Youmans wrote to Spencer on April 12th, 1864, that he had put a notice of Spencer's essays into the *Tribune*, a daily that had a circulation of 400,000. He told Spencer, "considering that the Tribune circulates mainly among that class which it is important to reach, and is moreover of great influence with others newspapers, this gain is a telling one."[73] Youmans also oversaw Appleton's publication of reprints of Spencer's volumes in the System of Synthetic Philosophy for as long as he was still alive.

In 1872 Youmans hit upon another plan for disseminating Spencer's thought.[74] He started a new periodical produced by Appleton, the *Popular Science Monthly*, and published articles by Spencer on sociology.[75] He wrote to Spencer on April 3rd, 1872, that he was depending on a second article by Spencer to open the second number of the journal. Youmans also used the *Popular Science Monthly* to conduct a running campaign against Spencer's critics.[76] Early issues consisted primarily of selections from English periodicals.[77]

---

70 Tebbel, *Between Covers*, 81, 100.

71 Fiske, *Edward Livingston Youmans*, 113.

72 Ibid., 115.

73 Ibid., 177.

74 J.D.Y. Peel, "Galton Lecture 2003: Spencer in History: The Second Century," *Herbert Spencer: The Intellectual Legacy*, Ed. Greta Jones and Robert A. Peel (London: Galton Institute, 2004), 132.

75 Fiske, *Edward Livingston Youmans*, 301.

76 Werth, *Banquet at Delmonico's*, 130.

77 Alan Nourie and Barbara Nourie, *American Mass-Market Magazines* (New York; Westport, Connecticut; London: Greenwood Press, 1990), 386. The magazine became an outlet for the writings and ideas of Charles Darwin, T.H. Huxley, Louis Pasteur, Ernest Rutherford, Henry Ward Beecher, William James, Thomas Edison, John Dewey, and future editor J. McKeen Cattell. See ibid., 386.

Youmans and Spencer were surprised by the success of the *Popular Science Monthly*. Sales increased steadily in the first year. After the new magazine had been running for about a year and a half, Youmans told Spencer that they were printing 12,000 copies of each issue, and that the demand seemed to be steady.[78] But early in 1874 Youmans began to worry that the novelty of the journal had worn off and he asked Spencer if he should made it less intellectually demanding. Spencer wrote that Youmans should be satisfied if circulation remained at its current level, which was "far beyond what you originally expected to reach." Then he expressed his disappointment that Youmans was considering "making it more popular." "This is just the tendency which all things set up with a high standard have to guard against," Spencer declared, "– the tendency to make material success the object, to the overlooking of the original object."[79] Youmans's journal, which he edited until his death, became one of the most successful popular science journals ever founded, and is still published today under the title *Popular Science*. Appleton was forced to sell the journal in 1900 for economic reasons.

Although Youmans ensured that Spencer's books and articles were published in America, he conceived of his role as agent in much broader terms. To him, it included editing responsibilities, which were intended to make Spencer's work more accessible to the American audience. In a collection of Spencer's articles titled *Essays on Progress*, Youmans had put headings to the pages and divided paragraphs to make Spencer's dense prose more reader-friendly. Spencer was grateful. He wrote to Youmans on May 18, 1864, that the *Progress* was "evidently more fitted to be popular than it was in its original form." Spencer was astounded by the time and effort Youmans had put into producing this book. "In this, as all along," he told Youmans, "you amaze me by the amount of labour you expend in furthering my undertaking."[80]

Youmans's efforts on behalf of Spencer did not end there. Youmans toured the United States in order to gauge the interest in Spencer's works before Appleton had started to publish his books. On November 4th, 1863, Youmans wrote to Spencer that he had spent the last three weeks "in a tour of observation and inquiry through New England to make the acquaintance of some of those who have taken interest in your writings." He reported, "there was a uniform and strong desire that your books should be issued here in a form suited to the American market."[81] Later that month, on November 23rd, Youmans

78 Fiske, *Edward Livingston Youmans*, 313.

79 Ibid., 314–315.

80 Ibid., 179.

81 Ibid., 156.

gave an account of his recent trip to Maine to meet those interested in Spencer's work. Stopping at Bangor, and then Portland, he visited several clergymen, who were students of Spencer's writings. Youmans described one of them as "a confessed and ultra-enthusiastic disciple." Then Youmans traveled to Boston, where he met with Asa Gray. It was on this trip that he sought out Fiske.[82] Drawing on his experience meeting enthusiastic Spencerians, Youmans could persuade Appleton that they should go ahead and begin publishing Spencer's works. But through these tours Youmans also began to build a network of American Spencerians, who offered strategic advice on which books to publish first.

Youmans also handled the financial side of Spencer's publications in the United States. He told Spencer in a letter of November 4th, 1863, that his purpose in becoming Spencer's agent was "twofold – to circulate your writings as extensively as possible, and to do it in such a manner that you might share the pecuniary results." Youmans stated that it was easy to do the first, but the condition of American publishing during the Civil War made the second objective more difficult.[83] It was only by drawing on the support of the network he had established that he was able to help Spencer reap the rewards of his intellectual labor. Fearing that Spencer's books would be pirated by another American press, who would not pay him any royalties, Youmans took steps to ensure that Appleton became the publisher. Appleton was hesitant to publish anything by Spencer at first. But Youmans persuaded Appleton to pay Spencer double the usual copyright, or twenty per cent on the sales, for the first book if the stereotype plates were supplied free of charge. Youmans told Spencer on November 23rd, 1863, that Spencer's American friends were conscious of the "injustice of our copyright law" and they resolved that Spencer would not be the "loser." They pledged to raise the funds to pay for the stereotype plates.[84] By April 12th, 1864, Youmans was able to report to Spencer that the first volume of essays was done, *First Principles* was nearly ready to be published, and the remaining essays in press would be published as soon as *First Principles* was done. Then *Social Statics* would be published, thus completing "the American series of your works, a large copyright being allowed on the whole."[85]

Later, in 1866, when it looked as though Spencer might have to suspend his work on the System of Synthetic Philosophy, Youmans raised seven thousand dollars from the network, and had it invested in Spencer's name in American

82 Ibid., 162–163.

83 Ibid., 154.

84 Ibid., 162.

85 Ibid., 176.

securities. This allowed Spencer to continue his work, and to employ an amanuensis and assistant.[86] The following year Spencer expressed surprise at the size of the royalty cheque that Youmans had sent him. Anticipating much less, Spencer wrote, "What a wonderful steward you are! I never dreamt a few years ago of any such results arising, and had it not been for you it is clear that no such results would ever have arisen."[87] Youmans's clout with Appleton, his understanding of the American reading audience, and his clever plan for setting up a favorable royalty scheme with the help of the network he had established led to the financial success of Spencer's American publications, despite the weakened state of the book trade due to the Civil War. Instead of receiving no royalties, the fate of many British authors whose works were pirated by American presses, Spencer enjoyed a handsome cheque for his work.

Youmans relied on his knowledge of the book trade to popularize Spencer's evolutionary worldview on an almost global scale through his creation of the International Scientific Series. In 1870 Youmans proposed to Appleton the idea of launching a new "International Scientific Library" for the general reader composed of the finest work of the most important, current scientific thinkers. To attract eminent scientists, Youmans dreamed of recruiting publishers in several different countries to participate in a scheme that would counter the absence of legislated international copyright agreements that had made piracy prevalent.[88] All authors would receive a fixed percentage royalty on their books and would receive that royalty in every country in which their books were sold. Youmans sailed to England in June 1871 determined to put this ambitious plan in place. After Youmans talked to Spencer about the project, the latter wrote on July 4th, "I am anxious to do all that is possible to extend and establish the arrangement you are making with English authors – arrangements which practically amount to international copyright."[89] Huxley, Tyndall, and Darwin also wrote letters of support.[90] In August, Youmans went to the British Association for the Advancement of Science meeting in Edinburgh to enlist potential contributors to the series. Spencer accompanied Youmans, and introduced him to important men of science.[91] A meeting

86 Ibid., 217–218.

87 Ibid., 232.

88 Adrian Johns, *Piracy: The Intellectual Property Wars from Gutenberg to Gates* (Chicago and London: University of Chicago Press, 2009), 295.

89 Fiske, *Edward Livingston Youmans*, 273.

90 Ibid., 274.

91 Ibid., 278.

organized to discuss the project was attended by Spencer, Huxley, Joseph Dalton Hooker, W.B. Carpenter, W.K. Clifford, Balfour Stewart, and Alexander Bain. Here it was decided that Huxley, Tyndall and Spencer would form an advisory committee that would consult with the London publisher.

After the Edinburgh meeting, Youmans went to London where he began negotiations with Henry S. King to undertake the role of publisher for the British component of the project. King, formerly a banker and East India agent, had recently set up a publishing firm and was developing an impressive list of books, both fiction and non-fiction.[92] By October an agreement had been reached, and Youmans was now confident that his dream would become a reality. He estimated that in five years there would be seventy-five to one hundred volumes and that the ISS would be "the world's popular cyclopaedia of reading science." Spencer, he wrote to his sister, was delighted but "half bewildered; every once in a while he breaks out, 'Who would have thought such a result would arrive from your first beginnings with me?'"[93] Youmans' next task was to find a French and a German publisher to join Appleton and King. In November, Spencer and Youmans went to Paris, and persuaded Henri Paul Baillière to participate in the scheme. Youmans went to Germany alone as Spencer had no connections there. During Youmans's trip he arranged a scheme that involved publishers from New York, London, Paris, and Leipzig. Before he returned home he had also opened up negotiations that would extend his plan to Milan (Dumolard Fratelli) and St. Petersburg (Znanie). Each publisher operated in a semi-autonomous fashion, keeping independent accounts and deciding whether or not to translate a book accepted into the series by another participating publishing house. But the provision that made international distribution possible, and that practically amounted to international copyright, was unique. From Youmans' point of view, the main purpose of the novel arrangement was to increase the circulation of evolutionary ideas from one country to another. For a time it did just that. The British committee dominated the entire series for the first ten years' of its existence, and Spencer provided the most editorial guidance.[94]

Early on in their relationship, Spencer was grateful for Youmans's assistance. On November 18, 1863, he told Youmans, "the energy and self-sacrifice you continue to show in the advancement of my scheme quite astonishes me; and

92 Leslie Howsam, "An Experiment with Science for the Nineteenth-Century Book Trade: The International Scientific Series," *British Journal for the History of Science* 33 (2000), 196.

93 Fiske, *Edward Livingston Youmans*, 285.

94 Bernard Lightman, "The International Scientific Series and the Communication of Darwinism," *Journal of Cambridge Studies* 5 (December 2010), 27–38.

while in one respect it is very gratifying to me, yet in another it gives me a certain uncomfortable sense of obligation more weighty than I like to be under." But for Spencer's recognition that Youmans's work on his behalf was prompted by a "love of truth" and the desire to "aid the spread of it," his feelings about Youmans's contributions "would be really oppressive."[95] Youmans assured him on December 14th, 1863, this is "my kind of work." He enjoyed promoting the "diffusion of valuable thought." He saw himself as "an ultra and thoroughgoing American." In Youmans's opinion, there was "great work to be done here for civilization" and what was needed was "ideas – large, organizing ideas – and I believe there is no other man whose thoughts are so valuable for our needs as yours are."[96]

The Disciples and Spencer's Decline

In 1864 a visitor at the Appleton publishing house delivered a long tirade against Spencer, who he viewed as a mere "upstart." Spencer's system would soon die, just like Comte's and Mill's. But W.H. Appleton responded, "I can tell you one thing – Spencer won't die as long as Youmans lives!"[97] Youmans, and Fiske, kept Spencer's system before the eyes of the American public from the early 1860s up until the end of the century. By the 1880s they had moved Spencer's ideas and his name from the periphery to the center of American intellectual life.[98] Fiske's popularizations of Spencer's ideas made them more accessible. By completing the system of his master, Fiske gave Spencer's work even more coherence, and by stressing the religious themes in Spencer's writings he made them more attractive to Americans living in a country where religion continued to be an essential element in public life. Youmans, who not only popularized Spencer's evolutionary philosophy in his writings, also ensured that Appleton published his books and articles, and that there was a receptive audience to read them, both in the United States and throughout Europe. The mastermind behind both the *Popular Science Monthly* and the International Scientific Series, Youmans's understanding of the publishing scene allowed him to organize successful large-scale projects that contributed significantly to the dissemination of Spencer's evolutionary ideas. As Spencer's

95 Fiske, *Edward Livingston Youmans*, 159.
96 Ibid., 169.
97 Ibid., 184.
98 Werth, *Banquet at Delmonico's*, 227. Werth makes this point about Youmans, but it is also applicable to Fiske.

American agent, he astutely handled the philosopher of evolution's business affairs. By the end of the century Appleton had published over fifteen of Spencer's books, most in multiple editions.[99] Although Youmans could not prevent Spencer's works from being pirated by other American presses, he ensured that Spencer was paid considerable royalty fees for those books that were published by Appleton.

Both Fiske and Youmans understood that emphasizing the religious dimension of Spencer's thought and adapting his ideas to the American context was the best strategy for gaining him a large American readership. They recognized that Spencer was particularly appealing to liberal American Protestants.[100] Without Youmans and Fiske, Spencer would have struggled to find an audience. Their understanding of the American politics, religion, and publishing made all the difference. But their appropriation of Spencer's ideas specificially for American audiences of the 1860s, 1870s, and 1880s in the aftermath of the Civil War was bound to lose its appeal when the context inevitably changed. By the time Youmans died in 1887 Spencer's reputation had already begun to decline. Spencer also outlived Fiske, who died in 1901. Two years later Spencer succumbed. By then, Spencer's reputation had sunk so low in Britain and the United States that the Dean of Westminster could doubt that Spencer was sufficiently eminent to merit the national tribute of a memorial in Westminster Abbey, an honor bestowed upon Darwin.[101]

---

99 *The National Union Catalog. Pre-1956 Imprints*. London: Mansell, 1978; Chicago: The American Library Association, 1978), vol. 561, 319–359.

100 Moore, "Herbert Spencer's Henchmen," 76–100.

101 Owen Chadwick, *The Victorian Church*, 2 pts (London: Adam and Charles Black, 1970), pt. 2, 114.

CHAPTER 7

The Ideology of the "Survival of the Fittest" during the *Porfiriato* in Mexico

Rosaura Ruiz Gutiérrez, Ricardo Noguera Solano and Juan Manuel Rodríguez Caso

"Herbert Spencer is an exception among the great philosophers. His name is celebrated in England, his birthplace, and throughout the civilized world. There is no language into which his works have not been translated, nor book devoted to the development of a moral or political science that does not constantly cite his name. Newspapers, magazines and publications of all types mention him repeatedly." The preceding paragraph is taken from a newspaper that supported the current Mexican President, Porfirio Díaz, in 1887.[1] It stands out in a long article on Spencer, indicating the importance that his ideas had taken on in Mexico towards the end of the nineteenth century, highlighting above all the scope of his proposal in different disciplines, as well as the impact that it had in different parts of the world (Fig. 7.1).

Spencer's ideas, promoted predominantly in the press, were of great importance in the establishment and strengthening of Mexican positivism, and at the same time, in a national scientific culture that along with the evolutionary ideas of Charles Darwin, formed a political, social, and educational background that, among other things, set the stage for the foundation of the country's most important educational and cultural project, the National University of Mexico. The way in which Mexican intellectuals appropriated and used Spencer's ideas stands out more clearly in this project. In particular, education and social evolution at that time represented a response to Mexico's political, cultural and social situation, an alternative via peaceful change, in opposition to the imminent revolution, which sought a profound transformation of the country through violence.

This chapter is divided into three sections. In the first, we briefly outline the context in which positivism was established in Mexico, initially within the Comtean tradition, until the consolidation of Spencerism, and particularly among political groups close to then President Porfirio Díaz. Secondly, we focus on the presence of Spencer in the Mexican context, beginning with the

1 *El Diario del Hogar*, January 29, 1887, 1.

 | DOI 10.1163/9789004264007_009

AÑO VI. Ciudad de México Sábado 29 de Enero de 1887. Núm. 115

Diario del Hogar

PERIODICO DE LAS FAMILIAS

Asuntos del día.

LA OCUPACIÓN DE QUERÉTARO

Variedades.

HERBERT SPENCER

FIGURE 7.1 *Cover of* El Diario del Hogar. *Includes an article emphasizing Spencer as a philosopher.* CREDITS: HEMEROTECA NACIONAL DE MÉXICO <http://www.hndm.unam.mx>.

translations of his work, which were few in number compared with other countries and which focused directly on issues related to Mexico, as well as pointing out the extracts that were taken up and translated in various newspapers where Spencerian thought was well received. Finally, we concentrate on Justo Sierra, the most important proponent of Spencerian ideas in the late nineteenth century, who from his position as a politician within the Porfiriato, and based on Spencerism, laid the groundwork for the foundation of the National University of Mexico.

1 The Background to Mexican Positivism[2]

The nineteenth century was a period of particular importance in Mexico's history, as it was when the foundations of the country's future politics, economics and education were laid. After the first quarter century, the period when the country gained its independence from Spain, the new nation adopted the name the United States of Mexico, established the Federal Republic as its system of government, and, in parallel, created a series of symbols that strengthened its emerging identity.

Halfway through the century, as a result of the mismanagement of vulnerable parts of the country, the country's territory was reduced to slightly less than half its original area. Economic problems emerged in conjunction with independence, making it necessary to resort to loans that burdened the country with crippling debt, leading to successive foreign invasions that made the country's path to stability far more difficult.

In addition to the fact that Spain had never created a scientific tradition either at home or in its colonies,[3] the state of the sciences in Mexico was particularly precarious because of social instability. It was common for members of the educated Mexican elites to study such careers as law, medicine, architecture, and engineering, because, even though they may have been attracted to disciplines such as mathematics, physics or chemistry, the sciences in Mexico

---

2 For this section, we have taken as our reference Leopoldo Zea's classic work on the introduction of positivism into Mexico. We should like to point out that we seek to present our history from a different standpoint, unlike other recent works such as that by Luz Fernanda Azuela, who proposes a vision based on the relationship between positivism and the scientific research that has developed in Mexico since the Enlightenment, in what she calls a 'protopositivism.' See Luz Fernanda Azuela, "La ciencia positivista en el siglo XIX mexicano," in *Otras armas para la Independencia y la Revolución. Ciencias y humanidades*, edited by Rosaura Ruiz, Arturo Argueta and Graciela Zamudio (Mexico: UNAM, UAS, UMSNH, HCH, FCE, 2010), 172–188. See Leopoldo Zea, *El positivismo en México: nacimiento, apogeo y decadencia* (Mexico: FCE, 1968). For another account that considers the influence of positivism in different fields, such as science and culture, see Eli de Gortári, *La ciencia en la historia de México* (Mexico: Grijalbo, 1980), 298–337. For a study of the place of positivism in the history of philosophy in Mexico, see Elizabeth Flower, "The Mexican revolt against positivism," *Journal of the History of Ideas* 10, no. 1 (1949): 115–129.

3 Eli de Gortári, *La ciencia en la historia de México* (Mexico: Grijalbo, 1980), 201–205.

in the middle of the nineteenth century were barely emerging. Despite the existence of museums and scientific societies such as the Mexican Society of Geography and Statistics (the first scientific society in the Americas, founded in 1833) or the *Sociedad Científica Antonio Alzate* (founded in 1884) which were disseminators of scientific knowledge, there was no state or national science project. The development of the sciences in Mexico throughout the nineteenth century owed less to institutional interest, and was more indebted to individual efforts, applying methodologies and knowledge brought from Europe.

This social instability also affected the publishing industry. After the Reform, in addition to government censorship, excessively stringent regulations for setting up a printing house made publishing anything nearly impossible. During the Porfiriato, it was only possible to obtain authorization for publications through the government department responsible for encouraging foreign and domestic investment in Mexico (*Secretaría de Fomento*), subject to criteria set by the government.

In the final quarter of the century, the military leader, Porfirio Díaz (1830–1915) rose to power. He ruled Mexico for thirty-one years from 1876 to 1911, with the exception of a brief interim period when government was nominally in the hands of his ally Manuel González (1833–1893).[4] During this period, the capitalist system was implemented, and, with modifications, has continued until the present.[5]

This period of the country's history is known as the 'Porfiriato.' It is a period that has particularly attracted the attention of historians, specifically because of the advance of positivist thought, which arose in the late 1860s as part of the Liberal Reform of then President Benito Juárez (1806–1872). After a bitter struggle between the Liberals and the Conservatives (whose main stronghold was the Catholic clergy), the Liberal movement was able to take power and then begin a process of readjustment that would make the transition between Liberal theory and what the actual conditions of the country at that time required. An element in this process can be seen in the Reform Laws that enacted the nationalization of the Catholic Church's property and capital. The advance of the Liberals, who pressed strongly forwards with their agenda, found little support among the Conservatives, who backed French intervention and the coming of Maximilian of Hapsburg (1832–1867) to Mexico in 1864. With the outbreak of war with France, the Juárez government was forced to resell lands to any who could make immediate payment to cover war expenses.

4 Rosendo Bolívar Meza, *Historia de México contemporáneo II* (Mexico: IPN, 2008), 17–28.

5 Arnaldo Córdova, *La ideología de la Revolución Mexicana* (Mexico: ERA, 1975), 17–42.

This meant, in effect, the return of the old landholding oligarchy, and thus the advance of Liberal social ideals was curtailed.[6]

In addition, the war led Juárez to seek the support of the Liberals to maintain stability once the war was won in 1861. Although it had been the Liberal model that prevailed, putting it into practice was far more difficult. In fact, Liberal thinking was modified in such a way that it was counterproductive to keep the Liberal conception of freedom, in the sense of *laissez faire*.[7] It became necessary to limit and subordinate it to the country's real needs. These modifications required an appropriate philosophical *corpus* that found its clearest expression in the Comtean positivist theoretical ideas of Gabino Barreda (1818–1881). [8]

As with any process where an idea is introduced and established, Mexican positivism acquired unique and particular characteristics. Beyond adopting a theoretical proposal, the various positivist ideas were put into practice in terms of political activism, and this is explicable when one considers that positivism was introduced into Mexico as part of a specific political plan, Liberalism, providing a sound basis for carrying out a range of different proposals.[9] Barreda's 'Civic Prayer' was one of the first indications of this new ideology. He pronounced it on September 16, 1867 in Guanajuato, the birthplace of Mexican patriotism.[10] Barreda had been appointed chairman of the Education Reform Commission by President Juárez, and the main objective of the Commission was to standardize the National Preparatory School curriculum around Auguste Comte's positivism.[11]

---

6 Rosaura Ruiz Gutiérrez, *Positivismo y evolución: introducción del darwinismo en México* (Mexico: Limusa, 1991), 144.

7 Rosaura Ruiz Gutiérrez, *Positivismo y evolución: introducción del darwinismo en México* (Mexico: Limusa, 1991), 145.

8 On Barreda's positivism, Leopoldo Zea's scholarly work must be mentioned. His work is among the few in the twentieth century that detailed which were the branches of knowledge through which positivism, specifically Comtism, was introduced into Mexico, and the influence of these currents of thought in certain political and educational circles.

9 Leopoldo Zea, *El positivismo en México: nacimiento, apogeo y decadencia* (Mexico: FCE, 1968), 28–29.

10 Gabino Barreda, "Oración Cívica," in *Opúsculos, discusiones y discursos* (Mexico: Ed. Imprenta del Comercio, 1877); Rosaura Ruiz Gutiérrez, *Positivismo y evolución: introducción del darwinismo en México* (Mexico: Limusa, 1991), 145.

11 As Ruiz notes, Barreda's positivism was the most suitable instrument for establishing liberal order in the country after the bloody French invasion and the Reform War, albeit with differences from Comte's original proposal. While Comte perceived the forces of disorder in the liberal sectors that drove the French Revolution, Barreda thought that the Mexican liberals were the forces that guided society towards a new order. From this, Barreda

Although considered the most notable advocate of this movement, strictly speaking it was not Barreda who introduced positivism to Mexico. According to the American historian Charles A. Hale, the first Mexican positivist was actually Pedro Contreras Elizalde (1823–1875), a doctor who was also a member of the Education Reform Commission. Contreras was born in Spain to a Yucatecan mother and was able to study medicine in Paris with two of Comte's followers, even getting to know Comte personally and becoming a founding member of the *Société Positiviste de París* in 1848. He returned to Mexico in 1855 and established a close relationship with Juárez, marrying one of his daughters, and later serving as Minister of Justice and Public Instruction (1861–1863, 1867–1872).[12] As had Barreda, Contreras had studied in France and was a member of the Juárez government, which helped him to circulate his positivist convictions freely, especially among the political class.[13]

Barreda was born in Puebla in 1818 and was a doctor, philosopher and politician. He initially studied law, but his penchant for the natural sciences led him to study chemistry at the College of Mining, and later on, in 1843, to study medicine at the National School of Medicine. After the Mexican-American War, he moved to Paris to continue his medical studies. Before returning to Mexico in 1851, he attended various lectures given by Comte on the evolution of mankind, which prompted an intense interest in Comtean philosophy.[14] From that

amended the Comtean principles of Love, Order and Progress in favour of Freedom, Order and Progress, based on considering freedom as being compatible with the laws that determine both organic and inorganic phenomena, and in that sense, subject to the requirements and needs of progress. These ideas were to form the basis for the creation of the National Preparatory School and Benito Juárez' liberal government's policies of promoting education. See Rosaura Ruiz Gutiérrez, *Positivism and evolution: the introduction of Darwinism into Mexico* (Mexico: Limusa, 1991), 145–147. Another clear example of the relationship between positivism and ideology and education in Latin America occurred in Cuba with José Martí. See Pedro Pruna Goodgall, *Darwinismo y sociedad en Cuba. Siglo XIX* (Cuba: Ed. Científico-Técnica, 2010), 95–106.

12 Charles Adams Hale, *La transformación del liberalismo en México a fines del siglo XIX* (Mexico: FCE, 2002), 223–224.

13 Unfortunately, sources on Contreras are scarce. No in-depth work has been done so far into his particular contributions to the introduction and dissemination of positivism in Mexico. As regards the relationship between Barreda and Contreras, see Moisés González Navarro, "Los positivistas mexicanos en Francia," *Historia Mexicana* 9 (1959): 119–129; Leopoldo Zea, "El sentido de la historia en Gabino Barreda," *Aztlán: A Journal of Chicano Studies* 14:2 (1983): 221–233.

14 Leopoldo Zea, *El positivismo en México: nacimiento, apogeo y decadencia* (Mexico: FCE, 1968), 55; Charles Adams Hale, *La transformación del liberalismo en México a fines del siglo XIX* (Mexico: FCE, 2002), 224.

moment on, Barreda engaged fully with positivist philosophy and became a leading exponent in Mexico.

He did, in fact, write the first positivist essay in Mexico. This was "On moral education," a speech that he gave at the Humboldt Society in Mexico City, published on May 3, 1863 in *El Siglo Diez y Nueve*, a newspaper of a decidedly Liberal bent. From the Comtean interpretation of the history of mankind, and in line with the provisions of the Reform Laws, he set out the difference between the various religious persuasions and what is correct moral education. In addition to the speech being based on the philosophy of Nicolas de Condorcet,[15] it is also based on the phrenological assumptions of the German physiologist Franz Joseph Gall, providing his proposal with an empirical basis. In general terms, he believed that the moral instincts of man are rooted in his own physiology and not in what he calls "the disastrous influence of a wholly selfish education." In the final analysis, Barreda's purpose was to promote educational reform based wholly on Comtean principles. Beginning with the Barreda proposal, positivism managed to penetrate the official circle of President Juárez and the philosophy became the guide for the reorganization of higher education. These developments faced strong opposition, especially from the church, though this situation did not preclude its becoming firmly established within the field of education.

Although the main positivist tradition was Comtean, several historians have pointed out the importance of Herbert Spencer's ideas in the Mexican setting.[16] It is worth remembering here that Spencer himself sought to distinguish between his proposal and Comte's, a situation that did not stop his generally being considered a positivist. Positivism also advanced in Mexico in parallel with evolutionary ideas of Charles Darwin, and was subsequently referred to as *social Darwinism*. Some features of positivism that became known in Mexico were both Comte's and Spencer's. Although Spencer used Comte's term "positivism," his version differed significantly from the Comtean proposal, as it suggested that a universal law could account for the progressive development of the society, in short, for evolution. Spencer's proposal provided Mexican intellectuals like Justo Sierra with a theoretical basis that was much more suit-

---

15 Leopoldo Zea, *El positivismo en México: nacimiento, apogeo y decadencia* (Mexico: FCE, 1968), 109; Charles Adams Hale, *La transformación del liberalismo en México a fines del siglo XIX* (Mexico: FCE, 2002), 263–264.

16 Leopoldo Zea, *El positivismo en México: nacimiento, apogeo y decadencia* (Mexico: FCE, 1968): 303–304; Eli de Gortári, *La ciencia en la historia de México* (Mexico: Grijalbo, 1980); Álvaro Matute, "Notas sobre la historiografía positivista mexicana," *Secuencia. Revista de Historia y Ciencias Sociales* 21 (1992): 49–64; Charles Adams Hale, *La transformación del liberalismo en México a fines del siglo XIX* (Mexico: FCE, 2002), 320–334, 339–342.

able to their political ends, since Spencer's position lent support to Sierra's ideas of gradual, non-violent change, the inverse of Comte's non-evolutionary positivism. Positivism in Mexico derived from the Comtism that Barreda proposed, but it was his students particularly who took to Spencerism as an improved vision of the positivist method, and which could be incorporated into a scheme for the progressive evolution of Mexican society.

Their ideas laid the groundwork for the general ideas about humanity and society that were discussed in Mexico. Of these, the idea of society as a natural organism stands out, and that, like any organism in nature, it is subject to evolution, that is, to gradual change over time. Humans are a fundamental part of this changing organism, and their ideas, beliefs, and behaviors can only be understood through science, in other words, through the observation of social phenomena via the medium of experimentation, and the search for social laws. Another fundamental idea was that progress was the highest social law. In Comte's interpretation at least, this outlined an optimistic message of progress and even the regeneration of the human species. It must be emphasized here that one of the main elements of positivism in Mexico was its relationship with Darwin's ideas, which were introduced into Mexico in the mid-1870s. This introduction came separately, but the relationship between the two sets of ideas was consolidated jointly and it is, therefore, not surprising to find ideas and speeches in which we can identify characteristic elements of both positivism (whether Comtean or Spencerian) and Darwinism. At the time, Darwin's original ideas were revealed through other authors' interpretations, as in the case of Ernst Haeckel or of Spencer himself, a situation that frequently meant that when evolution was mentioned, it could be related to either Darwin or Spencer.[17]

2 Spencer in Mexico

Positivism in Mexico, then, developed from two sets of ideas, those of Comte and those of Spencer. As mentioned above, Comtism is first documented in 1863 in a text written by Barreda, while Spencerism principally emerged in the press from the beginning of the 1870s. The case that concerns us here, that of Spencer, has in fact been studied relatively little by historians, even though the importance and particularly the influence of Spencerism on both politics and

17 On Haeckel's Influence on the interpretation of the evolutionary ideas in Mexico, see Rosaura Ruiz Gutiérrez, *Positivismo y evolución: introducción del darwinismo en México* (Mexico: Limusa, 1991), 235–250.

science in Mexico during the Porfiriato have been recognized. Among the authors who have analyzed the work of Spencer in Mexico, we have already mentioned Leopoldo Zea. Within his broad explanation of positivism, he highlights Spencer's contributions and influence on national politics and culture. Charles A. Hale, meanwhile, also highlights the role played by Spencer's ideas in the consolidation of liberal thought towards the end of the nineteenth century.[18] Other authors have pointed out the influence of Spencerism in politics, not just in Mexico but throughout Latin America during the nineteenth century. A case study of the direct influence of Spencer's ideas on culture is to be found in the work that Carmen Ramos Escandón has done on Genaro García, one of the translators and disseminators of Spencer in Mexico. She highlights his influence in the case of the history of feminism in Mexico, where she analyses Genaro García's intellectual development.[19] Unlike the authors who have analyzed Spencer's ideas in Mexico, what we should like to emphasize is that Spencer was an evolutionist who believed in ideas such as the survival of the fittest; Sierra argued that education was the most important way of creating the fittest individuals. Spencer's vision of Lamarckian inheritance explains how education contributes to individual change and from this to the transformation of society.

Despite the repeated references to his importance, the presence of Spencer – in terms of publications or references – was limited until the mid-1870s. One of the first references to Spencer in the Mexican press was in *La Iberia*, a newspaper published by the Spanish writer Anselmo de la Portilla and, in his own words, a "newspaper of culture, literature, science, arts, agriculture, trade, industry and material improvement." The first reference to Spencer was made on June 15, 1875, and rather than highlighting any of his proposals or ideas, he is merely mentioned as an example, being considered one of England's great

18 Leopoldo Zea, *El positivismo en México: nacimiento, apogeo y decadencia* (Mexico: FCE, 1968); Álvaro Matute, "Notas sobre la historiografía positivista mexicana," *Secuencia. Revista de Historia y Ciencias Sociales* 21 (1992): 49–64; Laura A. Moya López, "México: su evolución social. 1900–1902. Aspectos teóricos fundamentales," *Sociológica* 14, no. 41 (1999): 127–156; Charles Adams Hale, *La transformación del liberalismo en México a fines del siglo XIX* (Mexico: FCE, 2002); Jussi Pakkasvirta, *¿Un continente, una nación?: Intelectuales latinoamericanos como unidad política y las revistas culturales en Costa Rica y el Perú (1919–1930)* (San José: Editorial de la Universidad de Costa Rica, 2005); Chester Urbina Gaitán, "La influencia de Herbert Spencer en el tiempo (1899–1900). Un examen de la nación, la sociedad y la cultura costarricense," *Revista de Ciencias Sociales* 133, no. 134 (2011): 91–97.

19 Carmen Ramos Escandón, "Genaro García, historiador feminista de fin de siglo," *Signos Históricos* 5 (2001): 87–107.

theorists seeking solutions to national problems. There is another reference on April 19, 1876, to the publication in England of *Mind*, the quarterly journal of psychology and philosophy, stating that it was edited by George Henry Lewes, George Croom Robertson and Spencer, with no further details as to the content.[20]

As regards Spencer's various publications, it should be noted, as Hale points out, that reading Spencer in Mexico meant reading him in French translation.[21] The works in French were distributed by such publishers as Garnier Frères, Felix Alcan, Hachette, Scheleicher Frères, and Guillaumin, whose head offices were all in Paris. Very few of Spencer's works in English arrived in Mexico and, as in the case of Darwin, it was thanks to French influences that his work became known.[22] Apart from three particular exceptions, there have never been large-scale publications of Spencer's work: in 1894 *Resumen sintético de los principios de moral* (The Principles of Ethics), translated and edited by Ezequiel A. Chavez, was first published.[23] It was reprinted three times (1905, 1910 and 1922), and consisted of a summary of Spencer's works on issues of ethics and morality, particularly his *Principles of Ethics*. What Chavez did, rather than transcribe the four volumes, was to pick specific topics, such as the development of social morality.

The other two were: "The Ancient Mexicans" (1896) and "Ancient Yucatan" (1898), both translated into Spanish by the García brothers, David and Genaro (7.1). In fact, in both cases the García brothers not only translated the work of Spencer, but also made several corrections to dates, data and complementary information that they believed added to and enhanced Spencer's work (Fig. 7.2). This case serves to distinguish between two types of appropriation that took place in Mexico: one via the translations and modifications applied to Spencer's work; and the other via the particular interpretations made by different readers through newspaper articles or books, such as Sierra himself.

20 The first editor of the magazine was actually Robertson. Lewes and Spencer contributed articles to the first issue, respectively: "What is Sensation?" and "The Comparative Psychology of Man."

21 Charles Adams Hale, *La transformación del liberalismo en México a fines del siglo XIX* (Mexico: FCE, 2002), 322.

22 It should be remembered that the translations into French by French writer Clémence Royer (1830–1902) were highly controversial because of Royer's particular interpretations of Darwin's ideas, particularly his alteration of the meaning of 'natural selection.' See Thierry Hoquet, *Darwin contre Darwin: Comment lire L'Origine des espèces?* (Paris: Editions du Seuil, 2009).

23 Nayelli Castro and Clara Foz, "La circulación de las ideas positivistas en Argentina y en México: editores y traductores (1850–1950)," *MonTI* 5 (2013): 365–388.

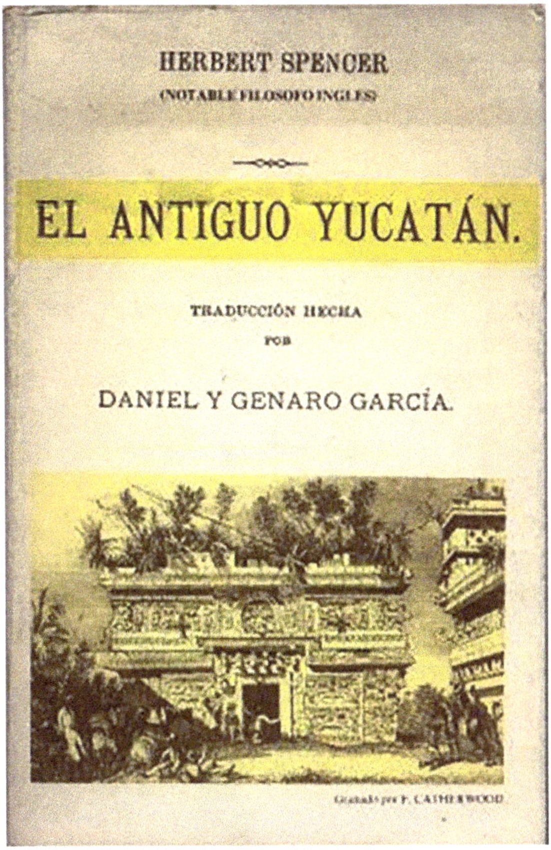

FIGURE 7.2

Portrait of El Antiguo Yucatán *("Ancient Yucatán"), a translation of the García brothers of Spencer's ethnographical and geographical descriptions of Yucatan. Facsimile edition of the original 1898, printed by José Diaz-Bolio, in Merida in 1980.*

In any event, it should be noted that possibly many of Spencer's mistakes were understandable given that he could not read Spanish. Although the variety of Spencer's writings was remarkable, it is striking that it was books focusing on Mexican topics that were considered for translation into Spanish, rather than other, far better-known works, especially in light of the importance and influence of Spencer's ideas in Mexico, mentioned above.[24] The three works were published by the Printing Office of the Ministry of Development, a government ministry in charge of economic issues in the country. On the topic of Spencer and the nineteenth century Mexican press, there were many newspapers that reported and discussed his ideas and work (some examples are given in Table 7.2 at the end of this chapter).

24 The originals of the works translated by the García brothers were a series of essays included in seventeen volumes of ethnographic and historical data by Spencer, called *Descriptive Sociology*.

The fact that great authors such as Darwin, Spencer and John Stuart Mill were read in French and not in English or via translations into Spanish, suggests that the influence of these authors was limited to a rather small intellectual community, and that the impact of their ideas in Mexican society was really through the occasional and sometimes brief mentions made in the press. To the present day, most of the works of Spencer that are preserved in Mexican libraries are original copies published in France, a few English twentieth-century originals and recent translations into Spanish, produced in other Spanish-speaking countries such as Spain and Argentina.[25] Despite Spencerism's powerful influence in the development of science, politics and education in Mexico, serious interest in the dissemination of his work has been minimal.

An example of Spencer's political appropriation can be seen in the press. One element contributing to the thrust of positivism as a political movement was that many of its leading exponents held important positions that helped the spread of Spencer's ideas. This occurred through the creation of mass media newspapers whose vision was based on Spencerism, a situation that helped to spread these ideas across a broad social spectrum that had no access to literature which was specialized and not in Spanish. *La Libertad* was one such newspaper, published between January 5, 1878 and December 2, 1884. Founded by a group of young positivists led by Justo Sierra, the newspaper's aim was to publicize a political point of view rather than a philosophical vision, but a political standpoint that was invariably based on positivism.[26] It was originally built on Barreda's outlook, but quickly moved towards Spencerism, because of the country's social needs, as these positivists argued for a new social order based on scientific ideas, because the position that Barreda took, although liberal, was perceived as conservative by students. In other words, they extolled the need to know the laws of nature and the relationships between them, in order to establish well-ordered conditions that would be suitable for the development of society. In addition, many of those who actively

25 We should note that the information does exist in the catalogues, but unfortunately access to the physical copies is complicated and in some cases, due to cataloguing problems, impossible. The exception is the works translated in Mexico, which are also available in electronic format through sites such as the UNAM's. Of the works translated into Spanish that do appear in the catalogues, there are three outstanding translations by the Spanish philosopher Miguel de Unamuno (1864–1936) made in the late nineteenth century.

26 This policy, as mentioned above, was maintained by supporting the Díaz dictatorship. See Carmen Sáez, "'La Libertad,' periódico de la dictadura porfirista," *Revista Mexicana de Sociología* (1986): 217–236.

participated in *La Libertad* were also prominent as politicians, for example, Justo Sierra, Pablo Macedo, Rosendo Pineda and Francisco Bulnes, all members of the Chamber of Deputies in 1880. The debates that were disseminated in *La Libertad* were the bases of many of the arguments that Sierra used in his books, particularly his concern for Mexico to be a well-educated nation, where science was the cornerstone of progress and one of the principal elements of education.[27]

In general, this group, which was known as the 'Scientists,'[28] thought of progress as a gradual process, and came back to the idea of natural selection as a framework to then speak of a struggle for existence between two different states, the primitive and civilized. These ideas consisted of a clear combination of both Spencer and Darwin's fundamental ideas, a situation that occurred continuously among various authors who saw in them a perfect blend of ideas for justifying the application of social Darwinism, by joining Darwinian natural selection to the Spencerian struggle for existence. Above all, the editorials of *La Libertad* sought to highlight the need to know the laws of nature, and especially the relationships between them, thus to establish, clearly and irrefutably, the correct order for the development of Mexican society.

For example, in the January 6, 1878 editorial, in his discussion of the importance of school as a factor in individual development and social transformation, Justo Sierra quotes *The Social Organism* to indicate that the state should only administer justice. All will then understand that a government "being a correlative function of the immorality of the governed," a phrase from Spencer's *The Social Organism*, is the result of a social state and will seek to improve it, improving society; the only way to obtain it.[29]

27 Eli de Gortári, *La ciencia en la historia de México* (Mexico: Grijalbo, 1980), 304.

28 The group was made up of lawyers, politicians and doctors, none of them truly being a scientist. As regards the role of the *Científicos* ("Scientists") in the Porfirian government, see Eli de Gortári, *La ciencia en la historia de México* (Mexico: Grijalbo, 1980), 314–316. For an alternative view of the role played by the group, in this case as firm opponents of the regime, see Walter N. Breymann, "The Científicos: Critics of the Díaz Regime, 1892–1903," *Proceedings of the Arkansas Academy of Science* 7 (1954): 91–97. On the social origins of the members of the group, see Alfonso de María y Campos, "Porfirianos prominentes: orígenes y años de juventud de ocho integrantes del grupo de los Científicos, 1846–1876," *Historia Mexicana* (1985): 610–661.

29 Justo Sierra, "La Escuela Preparatoria," *La Libertad*, January 6, 1878. The quotation from Spencer that Sierra mentions in his speech is not a literal quotation from "The Social Organism," *The Westminster Review*, January 1860. It is rather Sierra's own interpretation based on this and other texts where Spencer underlined the role of education and

Other examples that we might mention as regards the spreading of Spencer's ideas is *Revista Positiva*, a journal edited by a group of positivist enthusiasts headed by Justo Sierra and published between January 1, 1901 and December 3, 1914. It is noteworthy, however, that use of the concept of "Spencerism" was limited, not only in this journal, but in general. It is only mentioned once in *Revista Positiva* – on December 3, 1907 – in a brief article where the editor refers to his contact with Spencer's ideas during his youth, using the term 'Spencerism.' There were two other references to the concept, one in the December 1, 1909, issue of *Revista Moderna de México* and one in *El Abogado Cristiano Ilustrado* (The Illustrated Christian Lawyer) of March 7, 1918. In more substantial works, references are similarly few and far between. Highlights include "The Great *Maderista* Mystification: Jesuits and Pseudoscientists in the presence of Morals and Science" (1911) by Jesús Ceballos Dosamantes.[30] In this work, Ceballos makes a reference to the positivist school, using it as an example, as he saw it, of the corruption of science. There is also the case of "Philosophical Problems" (1915) by Mexican philosopher Antonio Caso, which mentions the concept when referring to the various philosophical schools, by way of a reaction against Spencer, which arose in many places after 1890.[31]

The case of Genaro García is also relevant. He was another politician, who, like Justo Sierra, was greatly interested in both Spencerian positivism and education. García was a member of the House of Representatives from 1892 to 1898, when he was appointed Secretary of the Chamber, holding the position until 1912, in addition to being, in his later years, director of the National Preparatory School. García acknowledged the influence of Spencer's thinking, especially in his translations, but what we want to emphasize here are his feminist convictions. As Ramos Escandón points out, García was introduced to Spencer's work through French translations.[32] His Spencerian vision can be seen in such works as "The status of women according to Herbert Spencer,"

knowledge as drivers of social change, among these being *Social Statics* (London: John Chapman, 1851) and *Essays on Education and Kindred Subjects* (1861).

30 Jesús Ceballos Dosamantes, *La Gran Mistificación Maderista: Jesuitas y pseudo-científicos, ante la moral y la ciencia* (Mexico: Imp. de A. Carranza e Hijos, 1911).

31 Alfonso Caso, *Los problemas filosóficos* (Mexico: Editorial Porrúa, 1915).

32 Carmen Ramos Escandón, "Genaro García, historiador feminista de fin de siglo," *Signos Históricos* 5 (2001): 87–107. In the University of Texas, there are located Spencer's books which belonged to García, such as *Essais de politique*, Paris 1890; *Essais sur le progress*, Paris 1891; *Essais scientifiques*, Paris, 1889, among others.

which emphasized ideas such as the rights and freedom that every woman should have, all within a vision of nineteenth-century liberalism.[33]

As discussed below in the case of Sierra, French spiritualism had a huge influence on Antipositivist circles, but it turned out that many of those involved in the discussions, both for or against positivism, displayed an eclectic attitude by equally accepting different sets of ideas, which in other contexts could have been incompatible. Another author who contributed to making Spencer's ideas, and positivism in general, more widely known was the journalist, writer and politician José María Vigil, although it was consistently his intention to strongly criticize the positivist model that the *Científicos* ("Scientists") hoped to establish in education policy. As a member of the National Preparatory School, he was involved in a bitter dispute with the physician and philosopher Porfirio Parra on which textbooks were to be used in courses, for example, of Logic or Ethics, where authors such as Spencer and Mill were the most widely used.[34] Instead, Vigil promoted texts based on spiritualist ideas or those of Krause, which were based on French spiritualism, and despite the criticism that they were merely favoring a fashion, really meant an extension of French educational discussions. Vigil's critique on Spencer rested on two arguments: that positivism was a philosophy that was tantamount to empiricism, and, as a philosophy, that it was defective because of the basic conflicts between those who defended it. For example, in an article published in *Revista Filosófica* on February 1st, 1882 and in *La Libertad* on the May 1st in the same year, he assumes responsibility for dissecting "positivist anarchy" and to do this, selects the works of the three most important positivists: *Classification des sciences*, Spencer (Paris, 1881), *Auguste Comte et le positivism*, Mill (Paris, 1879), and *La philosophie positive: M. Auguste Comte et M.J. Stuart Mill*, published in *Revue des deux mondes*, on August 15, 1866, by Émile Littré, demonstrating the disagreement on the question of cause between the three pillars of positivism.[35] To a large extent, Vigil was criticizing Spencer's attempt in *First Principles* to unite science, philosophy and religion, as, in his opinion, they should be linked to positivism only by empiricism.[36]

---

33 Carmen Ramos Escandón, "Genaro García, historiador feminista de fin de siglo," *Signos Históricos* 5 (2001): 97.

34 Charles Adams Hale, *Justo Sierra: un liberal del porfiriato* (Mexico: FCE, 1997), 266–268. See also Charles Adams Hale, "El gran debate de libros de texto en 1880 y el krausismo en México," *Historia Mexicana* (1985): 275–298.

35 Charles Adams Hale, *Justo Sierra: un liberal del porfiriato* (Mexico: FCE, 1997), 266–319.

36 Ibid., 273.

3 Justo Sierra: Spencerian Positivism as the Basis of Education

Moving on from this description of the general atmosphere of positivism, we will now focus on Justo Sierra, a man who stood out among all the positivists of the last quarter century. He was the main driving force behind Spencerian positivism,[37] and as we saw in general in the previous section, the battlefield between the various positivist camps was politics and education. As part of the reconstruction of the country after the War of Reform, President Juárez's plan to reform education led to the prominence of Gabino Barreda and his Comtean positivism, but over the years this influence declined, giving way to a Spencerian vision, and as already noted at the beginning of this section, by this time a position that was sufficiently distinguished from its Comtean predecessor. This was in turn deeply influenced by Darwin's ideas, which had been introduced into Mexico in the mid-1870s. In this regard, it should be emphasized that Sierra thought that gradual changes, such as those that occur in a Lamarckian-Darwinian evolutionary process were desirable for a country. Such changes should be based primarily on the education of the people, and would over time lead to progress. In Sierra's case, these are ideas that undoubtedly have a Spencerian origin.

Sierra can be described as an evolutionist, a public official, an active member of the group of the 'Scientists' and a supporter of Porfirio Díaz, all of which leads to his being a historically crucial character, if one is to understand the revolutionary phenomenon by observing the processes that occurred during the years of the Porfiriato. One of Sierra's main interests was the advancement of higher education. From 1881 onwards Justo Sierra was actively involved in the quest to establish a free and secular, public university, a huge achievement that was finally reached on September 22, 1910, when the National University of Mexico was inaugurated, coincidentally the same year when the Porfiriato came to its end at the outbreak of the Revolution.[38] Sierra reworked ideas like integration and differentiation in his desire that education in Mexico reach the level of Western countries.[39] Such ideas had their origin in Spencer's work on

37 In Mexico, Comtean positivism was understood as a system that took as its primary instrument of knowledge and transformation science and its method; while Spencerian positivism was similar to Comtism but with the addition of a general evolutionary view, going from the physical to the organic, from homogeneous to heterogeneous, from organic to the biological, from biological to social, and from social to moral.

38 Then in 1929 autonomy was granted to the University, in the sense of having full freedom to organize and provide education. Since then it is known as National Autonomous University of Mexico (UNAM for its acronym in Spanish).

39 Josu Landa, *La Idea de Universidad de Justo Sierra* (Mexico: UNAM, 2005), 42–43.

sociology, where their implications and impact on education are broadly explained.

At the opening ceremony, Sierra revisited Spencerian positivism, highlighting the importance of science and establishing the commitments that must be honored by future university students:

> When a young man becomes an adult, it is correct for the University to either throw him into the struggle for existence in a higher social field, or raise him to the sublime circles of scientific research, never forgetting, however, that all contemplation should be the foreword to action, that it is not right for a student to only think for himself, and that, while you may forget spirit and matter at the door of the laboratory, as Claude Bernard said, morally we can never forget either humanity or our country.[40]

Another example, from the same speech in which Sierra incorporates Spencerian elements on education from an evolutionary perspective, is when he refers to the process of change that the individuals in a society undergo:

> Because for individuals, as we have already said, to be strong is to encapsulate their whole development – physical, intellectual, ethical and aesthetic – in the determination of a character. It is clear that the essential element of a character is in the will; forcing it to evolve intensely, by cultivating it physically, intellectually, morally, growing from child to man: this is the sovereign role of the primary school.[41]

As we have already stressed, Sierra regarded education as a fundamental element in the transformation of the individual, and for this, inspired by Spencer,[42] he proposed and encouraged this thought as a slow and gradual process, thinking it would be the only way that Mexican society could evolve. He clarified education's plural and secular character specifying the tasks and commitments of those who give life to the university, necessary elements for proclaiming its autonomy:

---

40 Justo Sierra, "Discurso en el acto de la inauguración de la Universidad Nacional de México," in Justo Sierra, *Prosa* (Mexico: UNAM, 1990), 118.

41 Ibid., 112.

42 Landa mentions that in the "National University Project" proposed in 1881 by Sierra, he considered that our country could now enter that phase of evolution almost simultaneous with what Spencer calls integration [... and differentiation]. Josu Landa, *La Idea de Universidad de Justo Sierra* (Mexico: UNAM, 2005), 43.

> The University that is born today cannot, therefore, have anything in common with the other [referring to the Royal and Pontifical University]; both derive from the desire of representatives of the State to entrust men of high science with the mission to use the country's resources in education and scientific research, because they are the most appropriate body for these functions, because the state knows of no more important function, nor considers itself the best able to perform them. The founders of the University of long ago said, "The truth is defined, teach it." We say to the students of today, "The truth is being defined, seek it." They said, "You are a select group charged with imposing a religious ideal and a policy summarized in these words: God and King." We say, "You are a group perpetually selected from within the popular substance, and have been entrusted with the realization of a political and social ideal summarized thus: democracy and freedom."[43]

The principles that characterized Justo Sierra's thought were compulsory education and the non-intervention of the state in the struggle between individuals as the condition of the possibility of social progress, society as the result of natural laws that govern every living thing, and history as the result of the natural movement of society and independent of the will of men. These principles were taking form in the social reality of the Mexican people, thanks in large measure to Sierra's political role.

Inspired by Barreda's teachings, Sierra was a positivist who, in Spencer particularly, found scientific methodology to be a crucial element for supporting his political standpoint, which since the mid-1870s had merged with evolutionary thinking. In 1881, as a member of Congress, he proposed the creation of the National University; in 1894, he was made a Minister of the Supreme Court, where he later became president; and in 1910, as Secretary of Public Education and a founding member of group of the 'Scientists,' he finally realized the founding of the National University of Mexico.

As mentioned in the section above, Sierra's concern for education can be seen from the debates that appeared in *La Libertad*.[44] One of the fundamental disciplines, in his view, was history and, in fact, his major works took history as the starting point of the new education. Standing out among these works are his "General History," "Juárez, his Works and his Time" and "The Political Evolution of the Mexican People." This last was an edition of two essays he

43 Justo Sierra, "Discurso en el acto de la inauguración de la Universidad Nacional de México," in Justo Sierra, *Prosa* (Mexico: UNAM, 1990).

44 Leopoldo Zea, *El positivismo y la circunstancia mexicana* (Mexico: FCE, 1985) 398.

wrote for *Mexico: social evolution*, the work which he directed between 1900 and 1902: "Political History" and "The Current Era."

The positivism that Sierra displayed over the years was distinctly Spencerian,[45] with hints of the Comtism that dominated politics and education in Mexico from Barreda onwards, for Sierra had been a pupil of the latter. Yet it is clear that by 1874 it was Spencer's work that dominated his thinking.[46] Additionally, as Hale notes, Sierra's philosophical thought was equally dominated by French spiritualism,[47] which makes it clear how arbitrary it can be to delimit Sierra's conceptual thinking to a single sphere. Despite this mix of ideas, his discourse clearly shows the influence of Spencer.

Closely following Spencer's ideas, Sierra regarded society as an organism that was subject, as such, to natural laws that, in turn, were not under any circumstances to be questioned and were not therefore subject to change. The main work in which Sierra reveals these ideas is in his *Political Evolution of the Mexican People*, a work considered by the philosopher Antonio Caso to be the best example of the kind of discourse resulting from the Juárez Reform. In the same way that his teacher Barreda had outlined Mexico's history according to the Comtean vision,[48] in his own work, Sierra outlined a tour through the various stages of Mexican history: from pre-Hispanic civilizations, through the colonial period and independence, up to the Reform.[49] From an evolutionary and progressive standpoint, Sierra describes the steps by which the Mexican people has progressed, stressing at the end the final stage: how this period of change will lead Mexico towards a better state, based on, clearly, foundations derived from positivism. We can thus point out that Sierra specifically used Spencer's evolutionary theory to comprehend the social history of Mexico, something that Spencer himself never did.

As director of this work, we may observe how he reconciled historians of great stature with common politicians and, in the same way, observing the

45 We strongly emphasize that Sierra adopted the idea of survival of the fittest to justify his stance on the permanent quest of individuals for freedom under a scheme of evolutionary thought.

46 Charles Adams Hale, *Justo Sierra: un liberal del porfiriato* (Mexico: FCE, 1997), 12.

47 Charles Adams Hale, *La transformación del liberalismo en México a fines del siglo XIX* (Mexico: FCE, 2002), 281–282. As regards French Spiritualism as an alternative to positivism in Mexico, see *El positivismo mexicano*, edited by Water Beller, Bernardo Méndez and Santiago Ramírez (Mexico: UAM, 1985), 142–145.

48 Gabino Barreda, "Oración Cívica," in *Opúsculos, discusiones y discursos* (Mexico: Ed. Imprenta del Comercio, 1877).

49 Justo Sierra, "Discurso en el acto de la inauguración de la Universidad Nacional de México," in Justo Sierra, *Prosa* (Mexico: UNAM, 1990).

complex nature of history, made pre-revolutionary and revolutionary historical processes fit together. He translated the idea of biological evolution to social and political evolution. Sierra took his concern for educating Mexicans into the political arena, knowing that the problems the country faced were no longer external. To Sierra's mind, Mexico's biggest problem was to be found in the internal conflicts that disarticulated the social body and did not allow it to function properly. The evolution of the country should be marked by the passage of social evolution, and society should be seen as a body that works in an articulated way, in continuous dialogue with the problems that derive from a scarce scientific culture.

Sierra, in this sense, was an influential character who, from the political sphere, influenced the evolution of Mexican society and the institutional consolidation of a cultural space, that being the University. Sierra's importance does not lie in conceptual revolutions or major changes but in the importance that his thought and political practice had in discussions and in building a strong educational policy, which came to fruition in its institutionalization.

4 Conclusions

At the same time that positivism was progressing, during the final quarter of the nineteenth century, evolutionary ideas were being widely disseminated within groups of intellectuals, while education as a means of progress and freedom was becoming established as a fundamental element for an independent life. The followers of this program, José Ives Limantour, Francisco G. Cosmes, Eduardo Garay, Telésforo García, Francisco Bulnes, Justo and Santiago Sierra, among others, such as Michael S. Macedo, Joaquin D. Casasús, formed a political group known as the 'Scientists.'[50] This group saw Díaz as someone who could

50 This group was a small consolidated nucleus of members of the bourgeoisie holding political power, a situation that led to weighty discussions of the position of the various groups in society, such as the case of indigenous groups, the subject of a discussion of a notoriously racist nature, and which was strongly influenced by the evolutionary interpretations that began to be discussed in the last quarter of the century. On the specific case of positivist attitudes to indigenous peoples, see William Dirk Raat, "Los intelectuales, el positivismo y la cuestión indígena," *Historia Mexicana* 20, no. 3 (1971): 412–427. For a broader study of the subject of indigeneity during the Reform and the Porfiriato, see Martin S. Stabb, "Indigenism and Racism in Mexican Thought: 1857–1911," *Journal of Inter-American Studies* 1, no. 4 (1959): 405–423. For a different view, on how Spencerian positivism was the basis of Mexican criminology, see Gerardo González Ascencio, "Positivismo y organicismo en México a fines del siglo XIX," *Alegatos* (2010): 693–724.

and should be useful to their interests, hidden behind their apparent acceptance of the government (in other words, they wanted the dictatorship). Their real purpose, however, was to ally themselves with the government to promote their ideal of social progress, an idea inherited from Spencer.[51] The strategy they followed was a defense of the Constitution,[52] seeking to convert it into an impersonal authority of order, which would serve as a law to guide Mexican society towards progress.

The discourse of these intellectuals was constructed from positivist and evolutionary elements. A paradigmatic example is the case of Justo Sierra, who has been an important symbol in the structuring of public education. He was a key player in the discussions that took place in the newspapers and formal discussions in Congress, a character who allows us to go back into the past, touching on various turning points in the social construction of Mexico, a man of interdisciplinary style, committed to teaching and education, whose intellectual basis was Spencerism.

In general, historians have emphasized the importance of Spencer's ideas in both Mexican science and politics, despite the lack of in-depth studies. There is no doubt that positivism, firstly in its Comtean expression, and subsequently the Spencerian version, marked the evolution of politics and education in Mexico in a definitive way. The most outstanding propagator of Spencer's ideas was Justo Sierra, who, from taking key elements such as the idea of evolution and progress, and the conception of society as an organism, suggested a number of vital changes for the country, one of the most important being the foundation of the National University of Mexico, a bastion of science and culture to the present day, not only in Mexico but in Latin America.

In the late nineteenth century, Spencerism blended with other ideas, especially with Darwinism, based on Haeckel's interpretation, thereby producing an evolutionary ideology, which primarily permeated education and politics, to the detriment of the natural sciences. In the text, we have noted the influence that Spencerism had on the educational and political arena. It has not been possible to establish a direct relationship between Spencer's ideas and scientific practice in Mexico: firstly, because of the incipient nature of Mexican science, and secondly, because of the outbreak of the Mexican Revolution,

51 Leopoldo Zea, *El positivismo en México: nacimiento, apogeo y decadencia* (Mexico: FCE, 1968), 30.

52 For a more in-depth analysis of the interests of the "Scientists" in defence of the Constitution, see Charles Adams Hale, *La transformación del liberalismo en México a fines del siglo XIX* (Mexico: FCE, 2002), 166–220.

which changed social, political, economic and educational relationships in the country between 1910 and 1920.

Apart from certain individual efforts and some institutions whose function was the dissemination of mainly European scientific advances rather than research, science in Mexico in the late nineteenth century was virtually non-existent. To a large extent, this lack of development of Mexican science was influenced by the turbulent state of society, which governments had attempted to deal with since the mid-nineteenth century, and for which science was not an answer. The "Scientists," for example, give an idea of how science was used as a means to achieve political ends.

The publishing industry in Mexico was dominated by government control, and apart from agreements with French and Spanish publishers, book publication was infrequent and invariably controlled by the government. Newspapers became the principal means of freedom of expression, but interest focused mainly on political discussions, which as time went by led to the field of education.

Justo Sierra's thought was thus a clear example of these political and educational discussions looking to Spencerism as their philosophical basis (just as, in his day, Barreda had found justification for his proposals in Comtism). This was the case of the National University of Mexico, to the present day the country's greatest social and cultural project. Sierra's appropriation of Spencer's ideas can be seen in his description of the supremacy of certain races in Mexico's history, where he considers the important role to be played by some of them: "… that the new population, the mestizos, were those who were fit to form a social group destined to be increasingly important."[53]

53 Justo Sierra, *Evolución Política del Pueblo Mexicano* (Caracas: Biblioteca Ayacucho, 1977), 79.

TABLE 7.1 *Mexican translations of Herbert Spencer's works into Spanish*

Year	Language	Original	Title	Publisher	Translator	Comments
1894	Spanish	*The Principles of Ethics*, all volumes	*Síntesis de los Principios de la Moral*	Oficina Tipográfica de la Secretaría de Fomento, 159 pages + iv pages (Index)	Ezequiel A. Chávez	This is a summary of Spencer's works on morality.
1896	Spanish	*Descriptive Sociology*, volume II, 1874	*Los Antiguos Mexicanos*	Oficina Tipográfica de la Secretaría de Fomento, 229 pages + vi pages (Appendix)	Daniel and Genaro García	The book is dedicated to Porfirio Díaz.
1898	Spanish	*Descriptive Sociology*, volume II, 1874	*El Antiguo Yucatán*	Oficina Tipográfica de la Secretaría de Fomento, 156 pages	Daniel and Genaro García	The book is dedicated to Porfirio Díaz.

TABLE 7.2 *Some examples of published notes on Mexican newspapers by the end of nineteenth century*

Newspaper	Date	Commentary
El Mundo Científico y Literario (Sunday edition of *La Libertad*)	12/05/1878	Comments on M. Cazelles' translations on Spencer.
La Libertad	13/09/1879	A letter in Spanish from Spencer to Emile Alglave is published.
El Minero Mexicano	9/10/1884	A brief critique of Spencer's ideas of generation and development.
La Patria	10/04/1885	A speech by José María Vigil speech in which he criticizes Spencer.

TABLE 7.2 Cont.

Newspaper	Date	Commentary
El Siglo XIX	19/03/1886	A text on Bismarck's socialism criticized by Spencer. This note appeared again in *La Convención Radical Obrera* (*The Radical Labour Convention*, December 18, 1887).
El Economista Mexicano	08/04/1886	A comment on social stability and *Social Statics*.
El Tiempo	05/10/1887	A critique of positivism by Ramón Valle.
La Juventud Literaria	25/11/1888	On hypnotism, critiques from different scientific and philosophical positions, including that of Spencer.
México Intelectual	01/01/1889	On physiology and hygiene in the schools.
El Foro	13/07/1889	A speech on the regulation of wills and the role of the State.

CHAPTER 8

The Rise and Fall of Spencer's Evolutionary Ideas in Argentina, 1870–1910

Adriana Novoa

The publication of Charles Darwin's *Origin of Species* in 1859 changed the way nature and man's place in it were regarded.[1] In Argentina, European ideas were generally incorporated in two ways. First, there was the adaptation of systems that had already a corpus that determined with clarity the outcome and scope of its ideas; the uses of Kant's work by the beginning of the 20th century is a good example. Second, there was the adaptation of theories that were under construction, and still in flux. Darwinian evolutionism fell into this second category when it began to be known in the 1870s, which explains the apparently confusing way in which Argentine intellectuals understood Spencer's work. The local reaction to evolutionism was not at all passive and, in fact, demanded the collaboration of local thinkers to develop a scientific environment to adapt foreign ideas to local experience.

We might say that this process was similar to the act of building something new while repairing it simultaneously, and without any previous experience, or even a solid foundation to sustain the new structure. This complicated task is relevant to understanding the reasons that led to pessimism about the ability of science to lead the progress of civilization by the 1890s. This chapter deals with this problem, and attempts to present a detailed account of the arc that goes from the swift adoption of Spencer as a thinker who perpetuated treasured European ideas on progress, morality, education, and the importance of civilization, to the dismissal of his work as a betrayal of true civilized ideas defined as the representation of spirituality, universality and harmony. I will first analyze how the synthetic philosophy of Herbert Spencer was

* I am very grateful to Bernard Lightman and Juan Manuel Rodríguez Caso for their excellent comments and editorial expertise. This essay has been greatly improved by their recommendations.

1 Charles Darwin, *On the Origin of Species by Means of Natural Selection* (London: Murray, 1859). On the popularization of science before and after Darwin, see: Bernard Lightman, *Victorian Popularizers of Science: Designing Nature for New Audiences* (Chicago: University of Chicago Press, 2009).

 | DOI 10.1163/9789004264007_010

initially used to correct what was perceived as the intrinsic materialism, and philosophical poverty, of Darwin's theory in order to sustain the continuity of what was becoming the ideology of modern Argentina. I will concentrate particularly on his moral philosophy and ideas on education to understand the role they played in the acceptance and later rejection of his system. I will focus on the last three decades of the nineteenth century when Spencer's work became broadly known and incorporated into the liberal ideology. Obviously, I will not discuss all the reasons that explain why Spencer's work was so attractive to Argentine intellectuals, contenting myself only with a few clear examples to illustrate why his work made sense to them, and why, after a short period of time, it lost its appeal.

1870s: Education and Government

The majority of the intellectuals born in Argentina in the early nineteenth century trusted that civilization was a moral force that would redeem the primitive post-colonial world from centuries of ignorance. The poet Esteban Echeverría (1805–1851), and the writer and politician Domingo F. Sarmiento (1811–1888), like most of the liberals of their generation, believed that progress and transformation required a moral foundation that should be instilled in children through education. Both wrote instructions on how morality should be taught in school, and both longed for the superior morals that would accompany the advent of civilization and help to govern the new republic. Even Juan B. Alberdi (1810–1884), who defended a more materialistic position than the other two, agreed on the need to take matter and spirit into consideration. According to him, a democratic morality would promote "the unitary and synthetic nature of all humankind," through the simultaneous development of the material, moral, and intellectual interests that were present in all human beings.[2] Such utterances help explain why Spencer's work was received by this generation as a source that confirmed their views on the meaning of civilization and the role of education to civilize societies.

James Secord has demonstrated how modern mass culture was fostered by the expansion of reading in England. According to him, the steam-powered printing machine, "machine-made paper, public libraries, cheap woodcut, stereotyping, religious tracts, secular education, the postal system, telegraphy,

2 Juan B. Alberdi, *Obras Completas de J.B. Alberdi*. vol. 1 (Buenos Aires: Imp. de "La Tribuna nacional," 1886), 180. Original text, "la naturaleza unitaria, y sintética de la humanidad entera, por el triple desarrollo de la faz material, moral, e intelectual de la humanidad."

and railway distribution played key parts in opening the floodgates to an increasing reading public" in the second quarter of the nineteenth century.[3] By the 1870s a similar process was underway in Argentina. During this decade there was an increased number of periodicals published in Spanish, including in scientific fields, which broadened the popularization of evolutionary views beyond specialized works coming from outside the country.

This process had begun in 1852, after the military triumph of the liberals over the conservative governor of Buenos Aires, Juan M. de Rosas. Shortly after this victory printing houses and bookstores were opened to mark the importance that a reading culture would have among those who had seized power. One example is the printing shop founded by typographer Carlos Casavalle in 1853, where many books written by local liberal writers were published immediately after it opened for business.[4] Many foreigners established themselves in Buenos Aires to create the printing shops that would end up becoming the main publishing houses of the country a few decades later. One such was José Alejandro Bernheim. He arrived in Buenos Aires in 1854 and established a print shop specializing in the publication of local newspapers, including those of the French and English communities in Buenos Aires.[5] In 1864 Guillermo Kraft arrived from Germany to create a printing shop that would become one of the most respected establishments in Spanish America, particularly known for its use of new technology. He introduced the steam lithograph machine, known as "La Adela" the year he founded his shop.[6]

William Garrett Acree Jr. has studied this period, concluding that the establishment "of public education systems in the Río de la Plata between 1870 and 1910 was the keystone to converting reading and writing into public concerns."[7] In 1869 Angel Estrada, member of one of the traditional families of Buenos Aires, founded his publishing house and imported new machines brought

---

3 James Secord, *Victorian Sensation: The Extraordinary Publication, Reception, and Secret Authorship of Vestiges of the Natural History of Creation* (Chicago: Chicago University Press, 2000), 30.

4 See: Fabio Eduardo Ares, and Marina Garone Gravier. "Letras Argentinas: una Mirada a la Industria Tipográfica del Siglo XIX a través de la Fundición Nacional de Tipos para Imprenta de la Familia Estrada," *Letras Históricas* 9 (Fall/Spring 2014): 115–146.

5 Ibid., 121.

6 María Eugenia Costa. "Tradición e Innovación en el Programa Gráfico de la Editorial Guillermo Kraft: Colecciones de Libros Ilustrados (1940–59)," *Primer Coloquio Argentino de Estudios sobre el Libro y la Edición* (La Plata: Memoria Académica, 2012) <http://coloquiolibroyedicion.fahce.unlp.edu.ar>.

7 William Garrett Acree Jr., *Everyday Reading. Print Culture and Collective Identity in the Rio de la Plata, 1780–1910* (Nashville: Vanderbilt University Press, 2011), 85.

from England and the United States, such as the Minerva and Liberty. In 1872 he signed an agreement to represent the New York publishing house D. Appleton & Co. through 1895, which allowed him access to Appleton's important scientific catalog, including the works of Spencer.[8] Estrada also acquired the rights of other publishers, quickly becoming the main provider of textbooks for schools in desperate need of teaching materials. The expansion of the printing business ran parallel to the education campaign that increased literacy by the 1870s.[9] This process made publishing more profitable, and the number of small shops grew throughout Argentina in the 1870s.

The educational campaigns and the growth of a printing culture helped to spread the knowledge of Darwin's and Spencer's work. An article published in 1883 by Ernesto Quesada (1858–1934), one of the most prominent followers of the latter, mentions the expansion of scientific publishing in the country. According to Quesada, by 1877 Argentina had 148 periodicals, one for every 15,700 inhabitants, which ranked the country fourth in the world in terms of this type of publication.[10] During this decade six journals devoted to science were published; eighteen devoted to literature, and three devoted to philosophy.[11] The total of titles published in 1879 was 472. The increasing number of publications at all levels illustrates widespread interest in disseminating new ideas, particularly those related to science and education. The reception of Spencer's work was aided by this context. In 1869, during one of a series of talks delivered at a convention of elementary school teachers, there is specific mention of Spencer's work in the context of a discussion of basic principles of education. The main ideas cited are connected to the "vital force" of children, their need for physical activity, and for teachers to pay attention to their organic needs.[12]

---

8 Fabio Eduardo Ares, and Marina Garone Gravier. "Letras Argentinas," 123.

9 Fabio Eduardo Ares. "Barrio de Montserrat: Territorio Tipográfico (1780–1871)," *Cordis* 8 (Jan./June 2012): 7–31.

10 According to Quesada, in England the average was 1 for every 23,000, and in Germany one for every 26,000. See Ernesto Quesada, "El Periodismo Argentino (1877–1883)," *Nueva Revista de Buenos Aires* 3, no. 9 (1883): 72–101.

11 The leading scientific publications were *Revista Médico-Quirúrjica*, that began publication in 1864; *Anales del Círculo Médico Argentino* (1877); *Anales de la Sociedad Científica Argentina* (1876); *Revista Científica Ilustrada*; *Boletín de la Academia Nacional de las Ciencias* (1876); *Acta de la Academia Nacional de Ciencias Exactas* (1875); *Periódico Zoológico* (1874).

12 "Conferencias de Maestras," in *Anales de la Educación Común en la Republica Argentina*, edited by Juana Manso, 250–268 (Buenos Aires: Imprenta Americana, 1869), 265. This seems to come from Spencer's *Education*, particularly the first chapter, "What knowledge

This same year, the editor of the *Anales de la Educación Común en la República Argentina*, Juana Manso (1819–1875), translated a chapter of a book written by Norman Allison Calkins (1822–1895), a well-known North American expert in pedagogy. Since Manso was working under the protection of Sarmiento, who was President from 1868 to 1874, she was also very influenced by educational models created in the United States that were viewed as helpful in developing a republican culture. It was through this process that indirect access to Spencer's educational principles became important. Manso's translation of Calkins's work, particularly the chapter on "Object Lessons, Their Nature and Design," made Spencer's name known through its extensive quotations from *Education*, also published in 1861.[13]

In a key passage of his *Education* Spencer wrote that during human progress, "every science is evolved out of its corresponding art. It results from the necessity we are under, both individually and as a race, of reaching the abstract by way of the concrete, that there must be practice and an accruing experience with its empirical generalizations, before there can be science. Science is organized knowledge; and before knowledge can be organized, some of it must be possessed."[14] Spencer's words were not so far removed from what Sarmiento had written in 1855, affirming that the study of science was one of the great challenges confronting new American nations. His definition of "modern science" centered on the importance of "knowing, seeing," and constituting a science that was "practical and useful."[15] By the end of the decade, Spencer's pedagogical ideas were incorporated into the national system of education because they fit the developing liberal ideology, mainly through the elite's plans to create a mandatory system of education that secured their political supremacy.

Since for liberals education was bound up with the teaching and learning of science, and science was the foundation of civilization, in the 1870s there was also a growing interest in the work of Spencer as an evolutionary thinker. The

---

is of most worth?," which circulated broadly in Spanish America. See, for example, "De la Educación Intelectual, Física y Moral," *Revista de Cuba* v, no. January (1879): 52–62.

13 Norman Allison Calkins,. "Lecciones sobre Objetos para los Maestros y los Padres por N.A. Calkins," in *Anales de la Educación Común en la República Argentina*, edited by Juana Manso (Buenos Aires: Imprenta Americana, 1870–1871). Taken from Norman Allison Calkins, *Primary Object Lessons* (New York: Harper & Brothers, 1861).

14 Herbert Spencer, *Education: Intellectual, Moral, and Physical* (London: Williams and Norgate, 1861), 77.

15 This was part of a letter written in New York in 1866. "Irradiación de Civilización," in Domingo Faustino Sarmiento, Luis Montt, and Augusto Belin Sarmiento, *Obras de D.F. Sarmiento: Ambas Américas. 1899*. vol. 29 (Buenos Aires: Belin Hermanos, 1899), 134.

publication of Darwin's *Descent of Man* in 1871 had triggered a debate, and the true meaning of this brand of evolutionism was in dispute.[16] Spencer's understanding of biology was very closely related to the liberal ideology in Argentina, because it had become associated with the work of French thinkers instrumental in developing the idea of the modern republic. Spencer's ideas were blended into an eclectic mix of European sources that transformed his system into a source of philosophical and scientific knowledge that demonstrated the supremacy of liberal ideology. His work was used to further develop the synthetic imperative that had characterized evolutionary ideas in Argentina. As a result of this adaptation, the more materialistic aspects of his system were de-emphasized to promote morality, a civilized sensibility, and the existence of universal ideas of progress.

In the process of adapting Spencer's work to local reality, his system was considered more synthetic than Darwin's in its bringing together of science, politics, culture, and economics, and its possible connections with other European philosophies. His adoption of Lamarckian ideas on inheritance was also very important since it thereby sustained the supremacy of education and moral values. For most liberals one of Spencer's most appealing ideas was the defense of the concept that men's desires were chiefly inherited; while "their beliefs" were "chiefly acquired," and depended on "surrounding conditions;" which, simultaneously, depended "on the social state which the prevalent desires have produced."[17] Obviously, at this time there was not such a clear sense of how Lamarckism was separated from Darwinism, but the perception was that natural selection was harsher and somewhat contradictory to the inheritance of acquired traits, particularly of those traits that were acquired by humans through education.

Spencer's emphasis on a world "governed or overthrown by feelings, to which ideas serve only as guides," provided a sense of continuity that was very important for the liberal intellectuals who had followed Romanticism. In 1880 Domingo F. Sarmiento, a fervent evolutionist, and one of the biggest promoters of education and the growth of print culture in the country, published a report in which he made clear the national relevance and applicability of Spencerian sociology, which by now was more about biology than education. According to him, at the time, "the science of the Englishman Spencer, and recent French

16 Adriana Novoa, "The Dilemmas of Male Consumption in Nineteenth-Century Argentina: Fashion, Consumerism, and Darwinism in Domingo Sarmiento and Juan B. Alberdi," *Journal of Latin American Studies* 39, no. 04 (2007): 771–795.

17 Herbert Spencer, *The Classification of the Sciences: To Which Are Added Reasons for Dissenting from the Philosophy of M. Comte* (New York: D. Appleton, 1864), 37.

writers," called sociology, was scarcely known. This discipline dealt with the "propensities, elements, and human needs that led to the organization of societies as tribal, national, and as such to the form of government that satisfies, for good or ill, these propensities or needs."[18] Sociology would become an extremely relevant discipline in the next decade because it linked the political needs of nation building with scientific principles that lent legitimacy to plans already in place.

1880s: The Apparent Triumph of Continuity

The 1880s was the decade in which Spencer's work was widely read in several fields, and his ideas were quoted as if they were a godsend solution to all sorts of problems. The appropriation of his work continued to be connected to the growing number of publications in several fields.[19] Up to this year there were twelve journals devoted to science; ten devoted to literature, and six devoted to philosophy. The number of works published in 1879 was nearly doubled by 1881.[20] This increase in printing during the 1880s aided the popularization of Spencer's ideas. This follows the same pattern analyzed by Bernard Lightman in his study of England and the United States (see chapter 6). He noticed that among those who practiced science Darwin was the dominant figure, but "the ranks of the Spencerians and neo-Lamarckians contained numerous journalists and scientific writers. It was one of the largest groups of popularizers of evolution in terms of size, and it was influential because of its numbers."[21]

During this decade, the interpretation of Spencer's work was also mediated by the French literature that connected him with French Positivism, which had

18 Consejo General de Educación, *Informe III del Estado de la Educación Común durante el año de 1879 en la Provincia de Buenos Aires. Presentado al Consejo General de Educación por D.F. Sarmiento, Director General de Escuelas* (Buenos Aires: M. Biedma, 1880), 34.

19 In 1882 there were published 224 periodicals, one for each 13,509 inhabitants, which placed the country in third place in the world. In England the average was 1 for each 23,000, and in Germany one for each 26,000. See: Ernesto Quesada, "El Periodismo Argentino (1877–1883)," 72–101.

20 Since Argentina's economy was connected to the cattle and agricultural industries, these publications were also important in the discussion of evolutionary ideas. They began publication in the 1880s: *Revista de Ganadería*; *El Periódico del Estanciero*; *The Argentine Live-Stock and Agricultural Review*; *Noticiero Agrícola*; *Boletín del Departmento Nacional de Agricultura*.

21 Bernard Lightman, "Darwin and the Popularization of Evolution," *Notes and Records of the Royal Society* 64, no. 1 (2010), 12.

been very important in Argentina in the first part of the nineteenth century. Spencer himself rejected the notion that his ideas followed the writings of August Comte (1798–1857), and worked to dispel this misunderstanding. He explained the confusion by reference to the fact that those opposed to science branded as 'positivists' all those scholars interested in scientific ideas. At the same time, according to Spencer, Comte's followers tended to exaggerate the influence of their teacher, and saw continuity where in fact there was none. There was no doubt that Comte had "given a general exposition of the doctrine and method elaborated by Science," but it was not true that all those interested in these problems, were "disciples of M. Comte."[22] Moreover, if people like himself could be called positivists it was only a consequence of the fact that "all men of science have been more or less consistently 'positivists,' but not because they shared the same understanding of science."[23]

Spencer's complaints about the use of positivism as a term connected with the study of science, but not descriptive of a precise methodology or philosophical system, can be applied to Argentina at this time. It was common among the country's intellectuals to provide a synthetic view of science that glossed over the important divisions that might have existed among different thinkers. Science was in fact understood as a synthesis of the diverse philosophical traditions that were important in Europe and, as such, Spencer's system became reduced to provide a link between pre- and post-Darwinian science. Argentine intellectuals, and in general those in Latin America, regarded Darwin as a descriptive scientist, but not a synthetic one. An article published by the Geographical Institute illustrates this perception. In it science's progress is associated with the work of Emile Littré (1801–1881), John

22 Spencer, Herbert. *The Classification of the Sciences*, 28. A translation of this book was published in 1889, but its ideas had been already been circulated by other sources in the Spanish speaking world. For example, Emile Littré, who was read by almost all positivists, had attacked Spencer as, at the same time, one of the English followers of Comte and a traitor to his philosophy. See: Herbert Spencer, *Clasificación de las Ciencias*. Translated by Eduardo Zamacois y Quintana. Biblioteca Económica Filosófica, vol. 45 (Madrid: n/a, 1889); Emile Littré, *Auguste Comte et la Philosophie Positive* (Paris: Librairie de L. Hachette, 1863); Enrique José Varona, "Littre," in *Revista De Cuba*, edited by Jose Antonio Cortina, 71–74 (Habana: Imprenta Viuda de Soler, 1881); Ramón Manterola, *Estudios Científicos y Filosóficos: Ensayo sobre una Clasificación de las Ciencias* (Mexico: Imprenta del Gobierno, 1884). Ernesto Quesada wrote the most detailed analysis of the differences between Spencer and Comte among the intellectuals of Argentina. See: Ernesto Quesada, *Herbert Spencer y sus Doctrinas Sociológicas* (Buenos Aires: Talleres de la Penintenciería Nacional, 1907).

23 Ibid.

Stuart Mill (1806–1873), Auguste Comte, and, by the 1880s, Herbert Spencer, who was "making today the new synthesis according to the current state of the study of science," which meant at the time the incorporation of post-Darwinian evolutionary ideas.[24]

In an article about national high schools published in 1884, Spencer is cited to reflect the link between education and morality: "Science is the best instrument of intellectual and moral discipline."[25] But by the 1880s there was a much better understanding of evolutionary science, in part because *Descent of Man* discussed in detail the evolution of humans, thus having an impact on ideas related to morality, heredity, and education. The main problem was that it was not clear how Darwin addressed the need for a moral system that would structure both society and evolution. First, the main engine of Darwinian evolution was not morality, but survival. Second, the cruel description of an evolution based on the disappearance of less fit groups, and on competition and struggle, seemed quite a departure for Argentine intellectuals. This in part explains the importance that Spencer had in this decade; his system reaffirmed the importance of a synthetic system to address scientific, philosophical, and social problems in concert. His idea of progress, and evolution, was related with moral forces that had been very important in Argentina, a reason that explains why his idea of progress was perceived as a direct continuation of earlier principles. Argentina's conception of progress was discussed widely in the 1880s. A citation from a journal of education gives a good example of the prevalent view at this time:

> THE IDEA OF PROGRESS- (Its main characteristics) There are many ways in which we can establish that the conditions for progress in all creatures, including humankind, are three: unity, variety, and harmony, and that in man, more than in any other being ... there is more intelligence (harmony) ... for he unites all three conditions.[26]

24 Carlos de Mello, "El Siglo XIX y la Geografía," in *Boletin Del Instituto Geográfico Argentino* (Buenos Aires: Jacobo Peuser, 1882), 121–22.

25 "La Enseñanza en los Colegios Nacionales," *Nueva Revista de Buenos Aires* 4, no. 10 (1884): 268–84. 281. This quotes seems to come from Spencer's *Education*, but probably read in translation. The original passage is the following: "Not only, however, for intellectual discipline is science the best; but also for moral discipline. The learning of languages tends, if anything, further to increase the already undue respect for authority." In: Herbert Spencer, *Education: Intellectual, Moral, and Physical* (New York: D. Appleton, 1865, 88).

26 "La Idea de Progreso," *El Monitor de la Educación Común* 3, no. 44 (1883): 100–01.

The understanding of Spencer by the mid 1880s was that, unlike Darwin, he based his evolutionary ideas on moral and ethical concerns that seemed less contradictory with ideas of the past, in part because he insisted on a connection between ethics and education. Darwin's lack of any synthetic philosophical system was viewed with great distrust. Argentine intellectuals were not alone in this perception, since this was a widespread concern. Robert Richards has explained that for those suspicious of excessive materialism, Spencer "proposed that the moral goals of happiness (with its instrument of justice) were achieved through evolutionary means. Darwin's ethical theory, by contrast, derived more directly from biological considerations." According to Richards, this meant that Spencer's "procedure does not vitiate his moral theory, but it might suggest doubts about his biological theory."[27]

In the 1880s, Spencer's sociology became the perfect tool for civilizing a new nation. Spencer, in this way, facilitated the assimilation of Darwinian evolution in Argentina, and in general in Latin America. He was used to set a limit to the changes produced by natural selection, and made the inheritance of acquired traits fundamental to the understanding of how cultural and social traits, central to the creation of a civilized nation, were passed to future generations. In other words, in contrast to Darwin, Spencer's evolutionary ideas made it possible to believe that humans could shape their evolutionary future through social reform, which was an essential political tool for those liberals who saw in a "defective" population the biggest obstacle for the introduction of civilized culture. The implication was that Darwin's work was less important in itself, but truly revolutionary in the ideas it had inspired. This made Darwin more Spencerian than he really was. One example is the essay "The Modern Heterodoxy" written by the Spanish historian Marcelino Menéndez y Pelayo (1856–1912). This work was cited in several Spanish American publications and in it Darwin was defined as "a simple naturalist," but one whose ideas were the origin "of a system of natural philosophy in the *Anthropogenia* of Haeckel, and of a biology and sociology in the work of Herbert Spencer."[28]

This decade was also the one in which Spencer's work begun to be more closely tied to biology, aligning itself with new scientific research. This was in turn related to the role that soft inheritance played in the work of both Spencer

27 Robert J. Richards, *Darwin and the Emergence of Evolutionary Theories of Mind and Behavior* (Chicago: University of Chicago Press, 1987), 309. I am aware of the objections that Richard's work has provoked in some scholars, but his analysis fits very well the development of evolutionary ideas in Argentina.

28 "La Moderna Heterodoxia," in *El Repertorio Colombiano. Volume Xi*, 468–74 (Bogotá: Librería Americana y Española, 1882), 471.

and Haeckel. Even when Darwin accepted the inheritance of acquired traits, his emphasis on natural selection, and the importance that this mechanism of evolutionary change had acquired by the 1880s, had left his view of heredity muddled. As Richard De Witt has explained, blending inheritance and Lamarckianism "were not the only views of inheritance at the time, but they were two of the most common, and serve to illustrate the ways in which the existing views of inheritance did not, in general, point toward natural selection as the main mechanism behind evolution."[29]

Spencer had assimilated Darwin's theory to his earlier ideas, in turn linking them to the new evolutionism, sustaining the importance of blending inheritance, which tied evolution to designed social change implemented through education. But a continuous stream of new scientific evidence needed to be accommodated, and since scientists were conducting their research with ever more independence from philosophical constraints, the synthesis of all scientific knowledge was becoming more difficult and confusing. The tension about the significance of evolutionism for morality and education appeared clearly by 1889. An example of this situation is a congressional debate about the railroads, in which a senator declared that "Spencer's doctrine can be perfectly applied to the nation, and what's more, I believe that it is perfectly applicable," affirming the supremacy of Spencer's analysis.[30] But another member of the chamber who also supported the proposal being discussed asserted that, contrary to his colleague's view, Spencer belonged to a group of men who "do not participate in government, pessimist philosophers," "theorizers" who were concerned only with thought and not action, and supporters of a defeatist morality that paralyzed the nation.

This exchange shows the increasing distrust of the materialism and determinism implied by evolution according to some. This increasing hostility to some aspects of evolutionary theory coexisted with Spencer's influence in the fields of education, psychology, and criminology, which was overwhelming. The study of the criminal mind and the enactment of policies to facilitate the evolutionary process in Argentina were connected with the hygienist movement that characterized the culture of this decade. A growing sense of Spencerian morality was also fundamental to the judiciary, where Spencer's ideas circulated widely.

---

29 Richard DeWitt, *Worldviews: An Introduction to the History and Philosophy of Science* (New York: John Wiley, 2011) 303.

30 *Diario de Sesiones. Senado de la Provincia de Buenos Aires* (La Plata: La Discusión, 1889), 137.

1890s: The Crisis of Materialism and the End of Spencer's Supremacy

This decade was characterized by growing discomfort with Spencer's work on the part of many intellectuals and scientists. Clearly, many of the charges against Spencer that circulated during this decade were unfounded, but in Argentina the contradictions in his writings became more obvious when his work became less useful to create a synthetic system that liberals needed to support their rule. Mark Francis has analyzed Spencer's contradictions carefully, and has concluded that the philosophical ideas that he had advanced at mid-century "had vanished shortly after the publication of *First Principles*" in 1862.[31] This meant that one part of his system was constructed at the same time as new scientific evidence was being published, and continuous evolutionary debates brought new issues to light. By 1864 there were already contradictions in his writings. According to Francis, this "was a continuous problem: successive parts of Spencer's 'A system of Synthetic Philosophy' always interpolated extra and variant arguments into his systems. The results troubled his metaphysical foundations." This might not seem a problem early on, but later "his efforts to avoid unpleasant disputes would confuse later philosophers, whose increasing professionalism made them hypercritical of inconsistency."[32] This section will deal with how intellectuals from Argentina confronted the problems described by Francis, and how these problems made him a less interesting and useful resource for resolving local problems.

In Argentina, August Weismann's work became known in the 1880s, and in the 1890s it was considered in the context of possible theories of heredity. In 1890 the *Revista del Museo* published, in translation, a talk given the year before by the famous English naturalist William H. Flower (1831–1899), which featured an updated analysis of what had been happening in evolutionary science.[33] The article acknowledged many issues that remained unclear and unresolved in the understanding of Darwin's theory. They might become clearer with time, but "until that time comes," a "frank confession of ignorance" was the only "honest position we can assume when questioned on these

31 Mark Francis, *Herbert Spencer and the Invention of Modern Life* (Cornell: Cornell University Press, 2007), 9.

32 Ibid., 11.

33 William H. Fowler, "Los Museos de Historia Natural," in *Revista del Museo de la Plata* 1 (1890–1891): 2–27. The quoted text comes from the original English version. "Museum Organisation. Presidential Address to the British Association for the Advancement of Science. Newcastle-on-Tyne Meeting, 11th September 1889," in *Essays on museums and other subjects connected with natural history* (London: Macmillan, 1898), 27.

subjects."[34] Science was no longer a precise source of certain information, but the source of doubt, which Flower relates to the imperfect state of scientific knowledge, and the difficulty of making accurate scientific observations. But regardless of these claims, at the end of his essay, he affirms the evidence that natural selection "or survival of the fittest, has, among other agencies, played a most important part in the production of the present condition of the organic world, and that it is a universally acting and beneficent force continually tending towards the perfection of the individual, of the race, and of the whole living world."[35]

The crisis described by Flower was also evident in the work of Argentine intellectuals. The naturalist Pedro Scalabrini (1848–1916) offers us a good example of how Spencer's work was viewed by those critical of what was being called materialism. Scalabrini was a leading positivist. He had born in Como, Italy, in 1848, but he moved with his family to Argentina when he was young. He graduated from the University of Buenos Aires in 1869, and from 1872 to 1894 was the director of the Normal School of Paraná, where teachers were trained. In a departure from the practice of just a few years prior, Scalabrini distinguished between Positivism and the work of Herbert Spencer. The latter was an expression of the materialist philosophy that dominated the College of Medicine, and the departments of mathematics, astronomy, physics, and chemistry at the University of Buenos Aires.[36] But according to Scalabrini, materialist philosophy was narrow, lacked "breadth, was almost exclusively analytical, descriptive, dispersed," because it was only concerned with facts, leaving aside 'the spirit of science,' for which the laws established by the facts were more important than the facts themselves."[37]

Scalabrini emphasized that Positivism (Comte) was opposed to materialism (Spencer) "in its defense of a science that is more vast, more humane;" in its definition of a more complete method, and its insistence on a philosophy that was "more real, more systematic." It would be desirable for materialism to abandon "its pretensions of universal dominion through doctrinal objectivism." According to Scalabrini, Positivism should be understood as a precursor to and collaborator with Darwinism.[38] Progress was, following Comte, "the

34 Ibid., 28.

35 Ibid., 29.

36 Pedro Scalabrini, *Materialismo, Darwinismo, Positivismo: Diferencias y Semejanzas* (Paraná: Tip. y Encuadernación "La Velocidad," 1889), 6. Some associated this "medical materialism" with the work of Ludwig Büchner (1824–1899).

37 Ibid., 12.

38 Ibid., 13–14.

work that perfected national beings, in a material, intellectual, and moral way," and positivist morality was "as pure, as disinterested as Christian morality." Positivists' politics were not "based on Darwinian struggle, but on Comte's altruism."[39]

In 1892 Antonio Dellepiane (1864–1939) published a book in partial response to a similar crisis in penal law related to the uses of the concepts of morality and freedom in the wake of all the evolutionary literature on the subject. "The crisis of penal law has the same exact cause as the crisis of morality itself. Both have identical tendencies, and respond to the contemporary philosophical movement initiated by Comte, spread by Littré, his disciple in France, and by Mill and Spencer in England. It is characterized by the aspiration to extend the positive method, proper to the physical and natural sciences, to the study of the human and social facts." This movement had spread everywhere, "'it was in the air,' that people breathed, and in the intellectual environment of the time we live in."[40] This pervasive influence was also provoking a reaction against what was called "materialism," and its influence on accounts of morality and freedom.

The criticism that Spencer received in places like Argentina resembled what happened to other intellectuals in Europe or the United States. Following Francis's analysis, it is clear that he became an outdated thinker because he favored "data over logic," while philosophically there was renewed interest in "non-empirical lines" and the work of Immanuel Kant (1724–1804) and Georg Wilhem Hegel (1770–1831).[41] So, over the 1890s there was growing displeasure with his work as a representative of materialism and individualism, in stark contrast with Sarmiento's earlier praises of Spencer as more spiritual and concerned with ideals of harmony and morality.[42] The existing links between racial determinism and materialist philosophies were perceived as a threat to the nation's future existence, which made very difficult to sustain the local interpretation of Spencer as a synthetic thinker. By 1890s an increasing number of studies emphasized the emergence of natural selection as the most important mechanism in evolution. While in Europe the dominance of natural selection became more evident in the 1920s, in Argentina this idea was dis-

39 Ibid., 65. For a wider analysis of this split between Darwin, Spencer, and Comte, see: Adriana Novoa and Alex Levine, "Darwinism," in *A Companion to Latin American Philosophy* (New York: Wiley, 2009): 95–109.

40 Antonio Dellepiane, *Las Causas del Delito* (Buenos Aires: P.E. Coni, 1892), 239.

41 Mark Francis, *Herbert Spencer*, 11.

42 See: Domingo F. Sarmiento, *Conflicto y Armonías de las Razas en América* (Buenos Aires: La Cultura Argentina, 1915).

cussed since the 1880s. This was in part because being a country devoted to the production of cattle and agricultural goods, the relevance of artificial selection was very obvious early on. In 1888 a study by Juan Llerena (1825–1900), affirmed that after considering the evidence provided by artificial selection, "which was supported by nature," it was not possible "to deny the existence of natural selection."[43]

Due to their reliance on soft inheritance, and their visibility, Spencer's and Haeckel's systems were logical targets among those who had begun to place more emphasis on natural selection. But as Robert J. Richards has analyzed for the European context, the survival of the fittest "was too crude a mechanism to yield up delicate mental structures, refined social adaptations, and a keen sense of justice—-especially since these highly evolved traits had no survival value."[44] In Argentina all these developments were recognized by the 1890s. Carlos O. Bunge (1875–1918), a leading scientist, wrote in 1899 that the "English science of Darwin and Spencer tended to lower men to the level of the beasts;" while the German science "lifted men to the level of God."[45] The two alternatives emphasized the problem that intellectuals had in dealing with multiple understandings of evolution.

In 1898, law professor Baldomero Llerena (1852–1914) was the commencement speaker at the University of Buenos Aires School of Law. In his speech he denounced "the corrupted materialism," that dominated certain areas of education. He called for an enlightened patriotism that would further the cause of the "fatherland" without slowing its development. Political institutions were "the exterior manifestation" of a "public spirit," and without it there would be no "stable wealth, consolidated freedom, and nobody would respect [the nation's] people as educated and civilized;" in times of peace it would be a materialist people, without religion, art, and administration. Its only ideals would be determined by "the bourgeois luxury and the ostentation of bad taste."[46] Spencer's ideas on education and morality were abandoned, and nearly all references to his work were connected to his more recent evolutionary writings that seemed to contradict his early ideas on the subject.

---

43 Llerena, Juan, "Fisiografía y Meteorología de los Mares del Globo," *Anales de la Sociedad Cientifica Argentina* 25, no. 1 (1888), 8.

44 Robert J. Richards, *Darwin and the Emergence of Evolutionary Theories of Mind and Behavior* (Chicago: University of Chicago Press, 1987), 293.

45 Carlos O. Bunge, "El Espiritu de la Educación a través del Tiempo," in *Revista De Derecho, Historia y Letras*, edited by E.S. Zeballos, 264–81 (Buenos Aires: Jacobo Peuser, 1899), 281.

46 Baldomero Llerena, "Colación de Grados," *Anales de la Universidad de Buenos Aires* 13 (1899), 169.

Rodolfo Rivarola (1857–1942), who studied Kant and taught moral philosophy at the University of Buenos Aires, wrote a 1904 essay that questioned whether Spencer had really read Kant when he criticized him. Complaining of Spencer's contradictions, Rivarola decries the sloppiness of his work, and the superficiality of his treatment of the Prussian philosopher.[47] Quoting from the translated original article, he points out that the sense of universality present in Kant's work was part of Spencer's criticism.

> The knowledge of Man in that large sense which Anthropology expresses had made in Kant's day but small advances. The books of travel were relatively few, and the facts which they contained concerning the human mind as existing in different races, had not been gathered together and generalized. In our days the conscience of Man as inductively known has none of that universality of presence and unity of nature, which Kant's saying tacitly assumes.[48]

Intellectuals who believed in evolution, such as Rivarola himself, moved away from Spencer in search of thinkers who provided a renewed sense of the importance of moral philosophy. This explains why Kantian ethics had gained much more visibility. A 1907 report by Rivarola is a good example of this development. In it he argued for changes in the teaching of moral philosophy. According to Rivarola, the direction of scientific thought had changed and was at present organized around "the criticism of knowledge after the successful period in which positivism" had attempted to lay a foundation for all science in the objective reality of everything revealed through perception.[49] Rivarola justified his proposal by arguing for a pedagogy based on the "critical philosophy" that would help in researching the problems of the day, particularly those related to moral philosophy, since "the problem of the moral is connected to

47 Rodolfo Rivarola, "¿Había Spencer Leído a Kant?" *Revista de la Universidad de Buenos Aires* 1, no. 2 (1904): 153–61.

48 Qtd. by Rivarola, Ibid. The text included here is the original in English, and not the translation made by Rivarola, which is very close to the original. This article had originally appeared in the *Fortnightly Review* of July 1888. About this article Spencer said: "This essay was called forth by attacks on me made in essays published in preceding numbers of the *Fortnightly Review* – essays in which the Kantian system of ethics was lauded as immensely superior to the system of ethics defended by me." Herbert Spencer, "The Ethics of Kant," in *Essays, Scientific, Political, and Speculative* (New York: D. Appleton and company, 1899), 192.

49 Rodolfo Rivarola, "Nota de la Presentación y Programa de Ética y Metafísica," *Revista de la Universidad de Buenos Aires* 4, no. 8 (1907): xxix–xxxi.

the metaphysical problem, even in a relativist morality."[50] This philosophical platform explains the substance of his curricular proposal, which included six areas of study for moral philosophy, including the work of Kant, Spencer, Schopenhauer and Nietzsche (called amoralism), with one last section completely devoted to the *Critique of Pure Reason* by Kant.[51]

Those who admired and followed Spencer also recognized that his influence was a thing of the past. The most respected among his admirers was Ernesto Quesada, whose analysis of Argentina's printing culture was discussed above. He belonged to a traditional *porteño* family, and his father had been an important figure of Sarmiento's generation, an active participant in the cultural and political circles of his time. Quesada was probably one of the best-educated men of his generation, professor at the university and editor of an important journal. He recognized that Spencer's reputation had suffered a tremendous blow by the end of the previous century, but he also insisted that this crisis was related not so much with Spencer, but with the confusion that Darwin had created with his theory. The desire to extend everything that Darwin had written to sociological conclusions was responsible for many mistakes. The exaggerations that populated the writings of some Darwinians inspired the skeptics to criticize their understanding of morality, and the place of morals in society.

By the time of this writing, Quesada observed, the storm had calmed with the help of Idealism, and everybody was subjecting Darwin's theory to a "patient examination," that was slowly demonstrating its ability to explain "an immense series of phenomena of the most diverse orders."[52] But the reality, "even when it is sad to say so," was that Spencer's work was "mortally wounded in its doctrine and in its use."[53] It was outdated, and at the present it was "almost a scientific and philosophical anachronism."[54] In Quesada's analysis, the merit of Spencer's work consisted in having provided a philosophical ground for Darwin's theory, making it general enough to apply in all fields, not only the natural sciences. Spencerian sociology was responsible for the success that Darwinian evolution had enjoyed among philosophers, for example. In the process, "the history of civilization became the history of the social struggle

50 Ibid., 29.

51 In 1883 José del Perojo, a philosopher born in Cuba, but trained in Spain and Germany, had published a translation of this work. See: Immanuel Kant, *Crítica de la Razón Pura: Texto de las dos Ediciones. Precedida de la Vida de Kant y de la Historia de los Orígenes de la Flosofía Crítica de Kuno Fischer* (Madrid: Gaspar, 1883).

52 Ernesto Quesada, *Herbert Spencer y sus Doctrinas Sociológicas*, 29.

53 Ibid., 29.

54 Ibid., 69.

and of selection," though not exclusively "the brutal selection of the animal world, but also the selection of the social in the civilized world, whose progress is precisely the effect of such selection."[55]

Quesada believed that this historical interpretation explained how the ethical could become simple rational calculation transformed into organic habit – a frequent target of Spencer's critics. Spencer's rejection of metaphysics and emphasis on the individual explained the problems that his system eventually encountered, which, for Quesada, made evident what had made sociology "such an easy target for critics, who had demolished without any pity the majority of its doctrines." But people like him continued to admire the "brilliant synthetic attempt," though day by day it was becoming a mere "shadow of what it was…"[56]

Conclusion

This essay has analyzed how a system of thought that was originally appropriated as an optimistic doctrine of progress, ended up being considered exactly the opposite. The adoption of Spencer's ideas in Argentina was in its origins linked to the interest that local thinkers had in moral philosophy as applied to the creation of a civilized nation, a process related to the philosophical and scientific reasoning of the first part of the nineteenth century. This emphasis on the links between morality and education coincided with the creation of a mandatory elementary school system, which explains why, as Acree argues, "enrollments and literacy rates rapidly rose," and also how this wider reading culture "solidify the relationship between print, the public, and politics; it was when print media became 'popular' with state backing and officially part of daily life."[57]

This popularization of print media explains why slowly, through the 1870s, Spencer became the philosopher of evolutionism, and the bridge between the ideas that emerged before and after Darwin. But, as he himself had to navigate this transition, the impossibility of his synthetic philosophy became obvious, and by the end of the century, with the apparent triumph of those who attacked him for his defense of the inheritance of acquired traits, it was clear that his work was too contradictory to be considered truly synthetic. So, by the beginning of the twentieth century there were two currents that pushed for his

55 Ibid., 28.

56 Ibid., 87.

57 William Garrett Acree Jr, *Everyday Reading*, 120.

replacement. First, there were the literary and artistic movements initiated by the French symbolist poets; second, the philosophical renewal that tried to create a new synthesis out of neo-idealism and neo-Kantian ideas. The ideology of progress, then, allowed for the reconciliation of material and spiritual concerns, and for a new organization of the relationship between philosophy and science.

The fact that Spencer's work experienced such success and decline in a short period of time only demonstrates the degree of anxiety that the transformation of science, and the gradual separation of scientific inquiry from philosophy, were creating. In Argentina, Spencer was the perfect intellectual because of the connection he forged between the scientific/philosophical and the social, and his rationalization of the biological evolution of society. For this reason, as Quesada noted, what happened to Spencer was quite tragic: his work ended up being misunderstood, and he was unfairly accused of everything that had been wrong with the application of evolutionism to the social sciences and government. Quoting from a section of *Facts and Comments*, entitled "Some Regrets," Quesada attempted to show that Spencer recognized the moral difficulties of some of his concepts, and even rejected some of the popular ideas based on them, suggesting the possibility that given more time he would have been willing to make the needed changes with the help of metaphysics and a new conception of society.[58] Tragically for Spencer, and those who followed him like Quesada, that had never happened.[59]

---

58 Herbert Spencer, "Some Regrets," in *Facts and Comments* (New York: D. Appleton and Company, 1902), 6–11.

59 Ernesto Quesada, *Herbert Spencer*, 87.

CHAPTER 9

Spencerism in Brazil: An Introduction

Heloisa Maria Bertol Domingues

> The history of a people should provide a complete explanation of its evolutionary march (...) Of all the theories proposed, that of Spencer is what comes closest to the target.
>
> SILVIO ROMERO, *1888*

∴

With the above phrase, Silvio Romero,[1] historian of Brazilian literature and teacher of philosophy in Colégio Pedro II, an important school in Rio de Janeiro, introduced his book *História da Literatura Brasileira*, published in five volumes in 1888 – the year slavery was declared illegal in Brazil. Stating that the best theory for the history of Brazil was that of Spencer, Romero explained that "a theory of the historical evolution of Brazil has to clarify the action of the physical environment in all its aspects; study the ethnological qualities of the races which constituted Brazilian people at the beginning of the conquest; take into consideration the biological and economic conditions in which migrants found themselves; determine which old habits had disappeared, like atrophied organs due to a lack of function; follow the appearance of mixed populations; describe the new incentives of national psychology which began in the social organism and determined its future."[2]

* I would like to thank Bernard Lightman for the invitation to participate in the Global Spencerism Symposium, in the 24th International Congress of History of Science, Technology and Medicine, in Manchester, July 2013, and Gowan Dawson for the pertinent comments. I also would like to thank Thomas Glick for the trust in my work.

1 Silvio Romero (1851–1914) was born in a small town called Lagarto, in the Brazilian state of Sergipe. He studied in the Faculty of Law in Recife where he was a student of Tobias Barreto, the great propagator of the evolutionism of Haeckel and Spencer, as well as the positivism of Auguste Comte. He returned to Sergipe where he held political positions. In 1879 he moved to Rio de Janeiro, where he worked as a journalist, initially collaborating in *Revista Brasileira*. He specialized in literary criticism. He was also a teacher of philosophy in Colégio Pedro II <http://basilio.fundaj.gov.br, consult. 22/06/2014>.

2 Silvio Romero, *História da Literatura Brasileira*, 5 vols (Rio de Janeiro: Livaria José Olympio Editoria, 1943), vol. 1, 43, 55.

 | DOI 10.1163/9789004264007_011

He wrote this at an important political and social moment for the representation of Brazilian nationality, because of the transition from the imperial regime to a republican one, the problem of citizenship that had emerged following the end of slavery, and a crisis between Church and State. A few years before controversial positivist thought had been introduced that gave ample space to science and new theories. Darwin's and Haeckel's theories had been widely discussed in scientific circles since the 1870s, as had Auguste Comte's ideas. Spencer's work started to circulate at the same time, but would only really have an impact in the following decades. These ideas appeared in books, newspaper articles, public courses,[3] medical and engineering schools (in the natural sciences), and, most especially, law courses where intellectuals were educated in what were later called the social sciences. The new theories also had an impact on scientific institutions, such as the National Museum (1818), the National Observatory (1827), and the Advanced Schools of Engineering and Medicine (1808), which were undergoing a series of reforms at that time. Law schools introduced new theories for the social environment, similar to what medical schools had done for biology and engineering schools for the natural sciences. This period was called the Scientific Era in Brazil.[4]

According to the Brazilian philosopher, Cruz Costa, Spencerism in Brazil represented the broadest theory of progress and proved the "law of progressive differentiation," which satisfied what the "new nobles" wanted.[5] The historiography of the social sciences in Brazil also shows that sociology was, at the beginning, based mainly on Spencer and Comte.[6] Thomas Glick observed that in Southern Brazil there also emerged a literature based on positivist ideas and on evolutionism in which theories of Darwin, as well as Haeckel, Buckle and Spencer, were the object of debates, partisanship and publications.[7] History,

3 In 1876 the *Cursos Públicos [do Bairro] da Glória* (public courses of [the neighbor] of Glória) commenced in Rio de Janeiro. See Terezinha A. Collichio, *Miranda de Azevedo e o Darwinismo no Brasil* (Belo Horizonte, Itatiaia, São Paulo: EDSP, 1988), 50.

4 João Cruz Costa, *Contribuição à História das Ideias no Brasil* (José Olympio Editora, 1956), 298-300; Maria Amélia M. Dantes, Institutos de Pesquisa Científica no Brasil in *Ciências no Brasil*, ed. Mário G. Ferri and Shozo Motoyama, 3 vols (São Paulo: EDUSP, 1979/81), vol. 1, 341; Angela Alonso, *Ideias em Movimento* (São Paulo: Paz e Terra, 2002), 37.

5 Cruz Costa, *Contribuição*, 301. The 'new-nobles' to whom Cruz Costa referred were the republicans who took power after the end of the imperial regime and the overthrow of Pedro II in 1889. João Cruz Costa (1904-1978), philosopher and Professor of São Paulo University, studied the History of Philosophy in Brazil.

6 Sergio Miceli et al., *História das Ciências Sociais no Brasil* (São Paulo: Vértice, 1989).

7 Thomas Glick also shows that in the south of Brazil Darwin's theory gained scientific recognition as shown by Darwin himself, drawing on the works of Fritz Muller ("O positivismo brasileiro na sombra do darwinismo," in *A recepção do Darwinismo no Brasil*, ed. by Heloisa M. Bertol Domingues, Magali Romero Sá and Thomas Glick (Rio de Janeiro: Editora Fiocruz, 2003), 181–189.

as a sort of genre of literature, was, according to the sociologist Antonio Cândido, one of the great propagators of Spencer's ideas in Brazil.[8]

This paper will consider the transversality of Spencerism in the historiographic production of two Brazilian historians, the aforementioned Silvio Romero and Capistrano de Abreu.[9] Spencer was also important in the formulation of the Political Science program from the São Paulo School of Law by Alberto Salles.[10] Like the first Brazilian sociologists, historians also graduated from law schools. The books written by these intellectuals were published between the end of the nineteenth century and the beginning of the twentieth and even today are representative of their respective specialties.

The three authors gained recognition due to their participation in the movement of ideas at the time. Silvio Romero is considered one of the icons of the so-called 1870s Generation. Alberto Salles is also recognized as part of this 'generation.'[11] Capistrano de Abreu, who arrived in Rio de Janeiro in 1878, is usually not included in it. He was a journalist, soon recognized as a leading intellectual who wrote literary criticism and history under the theoretical inspiration of positivism and Spencerian evolutionism.[12] In their repeatedly republished books Capistrano de Abreu and Silvio Romero, who had parallel trajectories, presented different interpretations of the same facts in the history

8 Antonio Cândido, *Iniciação à Literatura Brasileira* (Rio de Janeiro: Ed. Ouro Sobre Azul, 2004), 63, 64. Antonio Cândido (1918), important sociologist, professor of São Paulo University, and expert in literature history in Brazil.

9 Capistrano de Abreu (1853–1927), was born in Maranguape, in the state of Ceará. He studied law for two years in Recife, then returned to Ceará, where he remained for some years working as a journalist, afterwards moving to Rio de Janeiro where he was a teacher in Colégio Pedro II. In 1879 he entered the National Library, making it his laboratory. His works contributed greatly to the history of Brazil and to a radical change in the historiography, bringing the hinterland, the *sertão*, and the 'paths of settlement' constructed by simple people, into the stage of history (José Honório Rodrigues, "Capistrano de Abreu e a Historiografia Brasileira," in *Correspondência de Capistrano de Abreu*, vol. 1, 2ª edição (Rio de Janeiro, Civilização Brasileira, Brasília, INL, 1977), Introduction XXVII–LVI.

10 Alberto Salles (1857–1904), born in Campinas, São Paulo, son of coffee planters, graduated from São Paulo School of Law, where he worked as a professor. He was important in journalism and a great publicist of politics. He was considered an ideologue of the federalist republic, a regime he ardently defended, having written in 1885, the Republican Catechism, which had enormous repercussions, with up to 10.000 copies being printed. His writings still have an impact today, as shown by the facsimile edition of his book one hundred years later. He was the brother of the Brazilian president Manuel Ferraz de Campos Salles, elected in 1898, whose mandate lasted until 1902 (Fernando Cardim, "Preface," *Alberto Salles, Sciencia Política* (Brasília: Senado Federal, 1997), III–XXIV).

11 Alonso, *Ideias em Movimento*, 23.

12 Rodrigues, *Capistrano de Abreu*, XXXIX.

of Brazil based on theoretical principles drawn from Spencer's concept of social evolution. For Romero, as for Spencer, Brazil was part of the march of civilization. According to Romero, the Brazilian people originated from three races, but the white conquerors imposed their habits and culture, as could be seen in the history of Brazilian literature. For Abreu, Brazilian historical evolution was expressed by the social habits and culture of the simple men that lived in the countryside. He reinterpreted Spencer's evolutionism making an analogy with the process of the evolution of intelligence, as a part of the nervous system, saying that the intelligence of the Brazilian people was formed in the struggle and adaptation to colonial life. This shows that historical representation is not exempt from the injunctions of the present as to how a theory, in this case Spencer's evolutionary theory, served as scientific argument. Romero represented the colonization ideology, while Abreu, an anti -colonialist, draws a type of cultural nationalism. Political positions were translated that were subject, above all, to the publishing industry, creating representations for the future. More than Romero and Abreu, Salles publicized politics, his journalistic specialty. Can they be considered as the 'popularizers of sciences' like the British figures identified by Bernard Lightman?[13]

It can be noted that although Spencer's thought had provided a foundation on which intellectuals could build their ideas, Spencerism was not a term used at that time, unlike Darwinism '*tout court*,' 'Social Darwinism' and on a lesser scale, Hackaelianism. Positivism was more than an idea. The positivist church had been active in Brazil since the end of the nineteenth century. The adepts of positivism were divided into orthodox, liberal, and republican.[14] Romero, who was well educated, called himself a Darwinist, though he looked to Spencer for theoretical guidance. Salles included himself among the non-orthodox Comtean positivists in relation to the non-autonomy of the field of cultural production.[15] Abreu abstained from self-classification, and although he emphasized the ideas of Spencer, he also acknowledged the impact of the positivism of Buckle and Comte on his ideology. Spencer's ideas served to corroborate different representations of the same historic facts, as in the cases of political organization and the ethnic formation of Brazilian society. Appropriating Spencer's ideas gave these authors scientific credibility. As a

13 Bernard Lightman, *Victorian Popularizers Science. Designing Nature for New Audience* (Chicago: University of Chicago Press, 2007). See also James A. Seccord, "Knowledge in Transit," Isis (2004): 654–672.

14 Ivan Lins, *História do Positivismo no Brasil. Coleção Brasiliana* (São Paulo: Companhia Editora Nacional, 1967).

15 About non-autonomy of a field of cultural production see Pierre Bourdieu, *Campo Intelectual e Projeto Criador. In Problemas do Estruturalismo* (Rio de Janeiro: Zahar Ed., 1966).

result, their appropriations oriented Brazilian social thought and became influential as historical accounts of the development of the nation.

Spencer's Ideas and the Context of Circulation

Acknowledging the importance of what James Secord said, quoting Robert Westman, that "books and letters," and not *isms*, "passed hands," and considering that libraries are sites where books circulated, a survey was carried out of the records of Spencer's books in the Brazilian National Library and in the Minerva Database of the libraries of the Federal University of Rio de Janeiro. This includes all the libraries of that time from the medical and engineering school, as well as that of the National Museum [of natural history]. It also includes libraries of social science courses, which were established during the twentieth century. With these records it was possible to observe the dates of book editions, if they were original, and to discover who were the publishers.

It was found that Spencer's books are scattered among various university libraries, as well as the National Library. Some can be found in original editions, in English, though the majority of first editions are in French.[16] It can also be seen from the publishing dates that the circulation of Spencer's works intensified between 1880–1890, when the first Portuguese translations of his books appeared in Brazil (Table 9.1). This means that Spencer's ideas gained greater importance with the republican government and the emergence of the question of citizenship following the abolition of slavery.[17] The importance for the history of Brazil by Silvio Romero and Capistrano de Abreu, on the one hand, and the theoretical approach of Alberto Salles on the other, will also be looked at here. They are representatives of that political moment. Spencer continued to shape social thought in Brazil throughout the twentieth century. In the libraries consulted, his publications ranged from the 1920s to the 1940s, especially those about chemistry and biology.

A dynamic book trade flourished in Brazil at the end of the nineteenth and beginning of the twentieth century. Many newspapers circulated, and

16 The importance and popularity of French in Brazil can be seen in Romero's note about the *Classification of the Sciences*, stating that he had translated the book because there was no previous French translation, leaving it understood that Brazilian readers dominated this language. See attached table 9.1.

17 Herbert Spencer, *Educação intellectual, moral e physica* (Education: Intellectual Moral and Physical) (Rio de Janeiro: Ed. Laemert, 1901); *A Justiça* (Rio de Janeiro: Ed. Francisco Alves, 1891); *O que é a moral?* (Lisboa: J. Bastos, 1879); *O progresso, sua lei e sua causa* (Lisboa: Inquérito Ed., s/d).

TABLE 9.1 *Translations of Herbert Spencer's works into Brazil (1879-19??)*

Years	Language	Original	Title	Publisher	Translator	Comments
1878	*French*	*Education: Intellectual, Moral, and Physical* (1861)	*Educação intellectual, moral e física (1879)*	Paris		BN
1901	*Portuguese*			Rio de Janeiro, Livraria Laemmert	From English	Library MN
1879	*Portuguese*	*What is moral?*	*O que é a moral?*	J. Bastos, Lisboa		
1939	Portuguese	"Progress: Its Law and Cause" (1857)	*Lei e causa do progresso*	Ed. Inquérito		PGG/IFCS/UFRJ Library
1891-98 1899	French	*The Study of Sociology* (1873)	Principes de Sociologie	F. Alcan, Paris	M. E. Cazelles	Obras Raras da Escola de Belas Artes/UFRJ Library
1871	English	The Classification of the Sciences	*Classificação das Ciências*	Rio de Janeiro, Jornal do Commércio		BN Published in the newspaper shared in parts
1885	Potuguese			Paris, Alcan	Silvio Romero	
1897	French		*Classification des Sciences*		F. Rethore	Library MN , Input record 1965
1886	English	Principles of Biology		New York, D. Appleton,		CCS, Health Sciences Center/ UFRJ Library
						Library MN
1893	French		*Principes de Biologie*			
1880	French		*Les Bases de la Morale Evolucioniste*	Paris, G. Ballière		Library MN Also Ed. 1901 and 1905, ed. Paris, Alcan; Input record 1910
1896	French		*La morale des differents peoples et la morale personelle*	Paris, Guillaumin	E. Castelot, Etiènne Martin Saint-Léon	Library MN Input record 1958

TABLE 9.1 Cont.

Years	Language	Original	Title	Publisher	Translator	Comments
1907		*First Principles* (1862)	*Les Premiers Principes*	Paris, F. Alcan	M. E. Cazelles	Library MN
1879	French		*Essaies Scientifiques, suivue de réponses aux objectives sur les Premiers Principes*	Paris, G. Ballière	M. A. Burdeau	Library MN
1905	French		Principes de Psychologie			Obras Raras/UFRJ Library
1909-1912				Paris, Alcan	Th. Riboi end A. Epinas	
1888	French	*The Man Versus the State*	*L'Individu contre l'État*	Paris, Alcan	Gerschel	Library MN
1904	English	*An Autobiography*				Faculdade de Letras/UFRJ Library
1907	French		*Une Autobiographie*	Paris, Alcan		Library MN
1875	French		*Introduction à la Science Sociale*	Paris, G. Ballière		Instituto de Filosofia e Ciências Humanas/UFRJ and MN Library, 1908 edition
1909	Spanish		Las ceremonias de la Vida			Faculdade de Ciências Sociais/UFRJ

Note. BN – Biblioteca Nacional / National Library; MN – Museu Nacional/National Museum (Museum of Natural History); UFRJ – Universidade Federal do Rio de Janeiro (Federal University, Rio de Janeiro). The National Library has many books by Spencer about biology published in the twentieth century.

publishers printed works by Brazilian authors.[18] Leuzinger, a Swiss company in Rio de Janeiro since 1832, was one of the great book producers during the Empire. Leuzinger published Capistrano de Abreu's book, even after his death, for *Sociedade Capistrano de Abreu*.[19] Livraria Laemmert was also in Rio de Janeiro since 1827. The German owner of this publishing house, Eduard Laemmert, had studied in Karlsrhue, where there were bookselling schools. Laemmert was the publisher of numerous Brazilian intellectuals. In 1901 he published the first translation of a book by Spencer in Brazil, *Education: Intellectual, Moral, and Physical*. At the end of the century he published various journals, including *Revista Brasileira*, which included many articles by Silvio Romero. In 1909 he closed down his business, selling the copyrights he owned to Francisco Alves, who would be one of the most important Brazilian publishers in the twentieth century, alongside José Olympio.[20]

Although the original books of Spencer, Comte, Darwin, and so many other theorists circulated through the country, the new ideas gained popularity principally through newspapers and cultural magazines. Silvio Romero published many articles at the end of the century in magazines, including *Revista Brasileira*, and in newspapers, such as *Jornal do Commércio* from Rio de Janeiro, where he published the first Brazilian translation of *The Classification of the Sciences* in 1885.[21] Capistrano de Abreu published numerous journalistic articles, which after his death were collected in five volumes, entitled "*Ensaios e Estudos*."[22]

The great impact of Spencerian evolutionism could be felt when it was considered that Brazil was one of the important sites for Charles Darwin's own

---

18 Laurence Hallewell, *O livro no Brasil, São Paulo* (São Paulo: EDUSP, 1985), 158. Specialized scientific journals and those from scientific societies, such as *Instituto Histórico e Geográfico Brasileiro* (IHGB–1838), *Sociedade Auxiliadora da Indústria Nacional* (SAIN–1825), and *Revista do Imperial Instituto de Agricultura* (1860), or scientific institutions, such as *Archives do Museu Nacional* (1874) or *Anais do Observatório Nacional* (1876), journals which were exchanged with numerous similar international institutions, stuck to the scientific work and specialties of each institution; no references to Spencer call attention.

19 Hallewell, *O livro no Brasil*, 160.

20 Ibid.

21 Silvio Romero, "Herbert Spencer," *Jornal do Commércio* (October 27, 1892): 7 and 15. *Jornal do Commércio* published many articles about the sciences, naturalist journeys in the hinterland of Brazil, the theory of evolution, and even the descent of man from apes.

22 Abreu initially published in Ceará, where he was born, in Maranguapense newspaper and in *Escola Popular do Ceará*. In Rio de Janeiro he published in various magazines, notably *Gazetinha, Revista da Academia Brasileira de Letras*, and *Revista Brasileira*. He published in newspapers such as *Gazeta de Notícias, O Jornal, O Globo* and especially *Jornal do Commércio*, whose articles form the second volume of *Ensaios e Estudos*.

voyage, for Alfred Russel Wallace's and Henry Bates' expeditions to the Amazon, as well as for the research trips of the anti-Darwinist Louis Agassiz, who wanted to prove *in loco* that Darwin' theories were scientifically flawed. However, Brazil was not the stage for the application of Darwin's theory. One of the first works that proved the theory of natural selection was 'born' in Brazil. It was written by a German immigrant and naturalist, Fritz Müller, whose book, entitled *Für Darwin*, was published in Europe in 1863, and translated into English in 1869, by recommendation of Darwin himself.[23] The same book was translated in Brazil only in 1907, in a cultural magazine, *Kosmos*, by a zoologist from the National Museum, Alipio Miranda Ribeiro, under a pseudonym.[24] Spencer's books were already circulating in Brazil, including the Portuguese translations published on 1878 and the first Brazilian translation from 1901. Considering these facts it is possible to say that, although Spencerism was not talked about, it was actually behind the so-called Darwinism that oriented social thought.

Spencer and Brazilian Historiography: Silvio Romero and Capistrano de Abreu

By the end of the nineteenth century, when scientific ideas had become an integral part of the intellectual milieu, the Brazilian Historical and Geographical Institute (*Instituto Histórico e Geográfico Brasileiro* – IHGB) had already built up a tradition in Brazilian historiography. IHGB was created in 1838 as a private scientific society, whose most important 'patron' was Emperor Pedro II. It was created to write the history of Brazil and in this way give shape to the image of the nation and 'found' the Brazilian nationality.[25] The first question raised was to define who the first inhabitants of the country had been. In other words, what was the founding historical landmark of Brazil? The indigenous past or the arrival of Portuguese colonizers in the sixteenth century? The first Brazilian history written from the perspective of IHGB was *Historia Geral do Brasil* by Adolpho Varnhagen (1854). It was a history of 'civilization' in Brazil, where the Portuguese were the principal actors, with blacks playing a supporting role.

---

23 Domingues, Sá and Glick, *A recepção*, 98; About Fritz Muller: Patrick Tort, *Pour Darwin* (Paris: PUF, 2001), Introduction; David A. West, *A Naturalist in Brazil* (Blacksburg, Virginia: Pocahontas Press, 2003).

24 Domingues, Sá and Glick, *A recepção*, Introduction.

25 Domingues, H. M. Bertol, *A noção de civilização na visão dos construtores do Império* (UFF, Niterói, 1991, Masters Thesis). Chapter 1.

Indians were ignored.[26] This vision emphasized colonization and provoked great debate within the IHGB. Even the Emperor spoke against it and the author made some changes in later editions. In Brazilian literature at this time the *Indianism* movement emerged. At the end of the century, when the equilibrium of political forces had changed, the new scientific ideas served to reinterpret society and consequently history. Silvio Romero and Capistrano de Abreu were representatives of this 'new' vision of Brazilian History. In the 'evolutionism wave,' indebted to Spencer's ideas of different social classes, they considered the ethnic diversity of Brazilian society in the historical evolution of the political and social organization of the country.

Although identified by the same theoretical approach, Romero and Abreu appropriated Spencer's theories for their own distinct political and social positions. In discussing the origin of Brazilian society they adopted a political position towards the republican state, imperialism (in relation to colonialism), and to their peers regarding the reality of the 'three races' (Indians, blacks, and whites). Although each had different positions, they shared the theme of the origin and formation of Brazilian society. Furthermore, both published articles entitled "Herbert Spencer" in the mainstream press.

In his unsigned article,[27] Abreu acknowledged the importance of Spencer although he was somewhat critical of him.[28] He commented that Spencer gained recognition only after the appearance of Darwin's theory.[29] Abreu ended this article comparing Spencer and Comte: "[Spencer] was one of the greatest spirits of the century, one of those whose ideas most agitated and who exerted most influence. His place is alongside the great lights of the human spirit, together with Comte, who formulated the subjective synthesis, while Spencer provided the architecture for the objective synthesis; the springs from which there originate all the philosophical currents of our times." Nevertheless, he leaves it understood that he did not agree with all of Spencer's ideas (Fig. 9.1).

---

26 F. Adolpho de Varnhagen, *História Geral do Brasil*, 1st edn (Rio de Janeiro: Livraria Laemmert, 1854). The first edition of his book was more radical in relation to his view of Indians and was modified in later editions, although his personal position did not change.

27 Capistrano de Abreu, "Herbert Spencer," *Gazeta de Notícias*, 4 June 1887, in Abreu, *Ensaios e Estudos*, 1976, 4th vol., 173–176. This volume of *Ensaios e Estudos* includes a compilation of various articles by Abreu, some of which were unsigned, and whose authorship was attributed by scholars of his work and by his son.

28 According to Abreu, Spencer began to develop his philosophy before Darwin. In 1852 he published *Social Statics*, which did not have a great repercussion and Spencer spent a long time without income.

29 Abreu, *Ensaios e Estudos*, 4th vol., 174.

FIGURE 9.1
Capistrano de Abreu. Obras de Capistrano de Abreu Correspondência, *Volume 1, 2ª edição (Rio de Jeneiro: Editora Civilização Brasileira, 1977), book cover.*

Silvio Romero, on the other hand, wrote a series of articles in *Jornal do Commércio* in 1893, entitled "Herbert Spencer e a Classificação Positivista das Ciências." The first article was the translation of Spencer's book *The Classification of the Sciences*. However, the final articles in this series were commentaries with harsh criticism of the orthodox vision of Brazilian positivists and of Littré, which he called *systematic idiotification.* In one of these articles he wrote, "Anyone who has read Spencer on the classification of the sciences translated by me and published here, must have noted the perfect lucidity of the argumentation of the great British philosopher (...)."[30] (Fig. 9.2.)

Silvio Romero and Capistrano de Abreu studied in the same university, Olinda School of Law (in Pernambuco). Both were journalists and professors.[31] Romero presented his ideas about Brazilian history in his book *História da Literatura Brasileira*, published in five volumes (1888).[32] In this work he

30 Silvio Romero, *Jornal do Commércio*, Rio de Janeiro, Aug. 7, 1893.

31 Among the Law Schools in Brazil the most traditional in the nineteenth century were the school in Olinda, Pernambuco, followed by São Paulo, both created in 1827. In the middle of the nineteenth century Law Schools were created in Porto Alegre (Rio Grande do Sul), Rio de Janeiro, and Salvador (Bahia).

32 By tracing a history of romanticism and its end, Romero explained that "only the agnostic concept of the universe, which is the great achievement of modern science, a concept which is based on the triple supports of the positivism of Comte, the evolutionary ideas of Spencer, and German religious criticism, can, in my opinion, be the inspiration for the actual art" (Romero, *História da Literatura Brasileira*, vol. 4, 139).

FIGURE 9.2
Silvio Romero. Silvio Romero, História da Literaturta Brasileira, *5 vols., 3ª edição aumentada (Rio de Janeiro: Livraria José Olympio Editora, 1943), Tomo I, p. 1.*

developed the idea of a mixed nation of whites and blacks, excluding Indians. Abreu published various books, notably *Chapters of Colonial History* (First Edition 1906),[33] *First Peoples in Brazil,*[34] and *Caminhos Antigos e Povoamento do Brasil* (First Edition 1899),[35] in which he described the lives of 'simple' men in the Brazilian countryside and their 'struggle for survival' in the inhospitable hinterlands, making them a part of a history which previously had only been seen from the perspective of the Portuguese conquest.

Romero's and Capistrano's books were republished continuously during the first half of the twentieth century. An association was formed to remember Capistrano de Abreu (*Sociedade Capistrano de Abreu*). It organized, edited, and republished all his work, including his correspondence and articles published in journals, in which his Spencerism emerged.[36] Romero, however, was more eloquent in showing his Spencerism. At the beginning of *A História da Literatura Brasileira* he said that social studies should begin with Herbert Spencer:

33 Capistrano da Abreu, *Capítulos de História Colonial, 3rd edn* (Rio de Janeiro: F. Briguiet 1934 – Published by Sociedade Capistrano de Abreu).

34 Capistrano de Abreu, *Primeiros Povoadores do Brasil* (São Paulo: Companhia Editora Nacional, 1954).

35 Capistrano de Abreu, *Caminhos Antigos do Povoamento* (São Paulo: Companhia Editora Nacional, 1975).

36 Capistrano de Abreu, *Correspondência, vol. 1* (Brasília: INL, 1977);---- *Ensaios e Estudos: crítica e história* (Brasília: INL, 1975); 2ª Série, 1976; 4ª Série, 1976.

> We have to start with Herbert Spencer, because he is excellent for showing the general foundations of society, its internal and external factors, and how the fundamental laws of evolution were applied. After reading *Principles of Sociology* we understand how society was formed and how it had created its various manifestations, customs, ceremonies, religion, family, politics, law, professions, economic institutions, etc."[37]

Paraphrasing Spencer, Romero saw history as the supreme receptacle of the march of all ideas – the evolution of thought – which was also modified by the moral sciences, which come from the physical and natural sciences.[38] In his *Education: Intellectual, Moral and Physical*, Spencer stressed that history should be concerned with the phenomenon of social progress. It is necessary to know how society was created and organized itself with its own forces. For Spencer, historians needed to make wide-ranging descriptions of the manifestations of life in society, politics, economy, beliefs, aesthetics, music, and science. The facts had to be observed as mutually interdependent parts of the whole and be presented in such a way that the existing consensus between them could be rapidly traced, in order to discover which social phenomena coexisted. And lastly, to connect the whole, should exhibit the morals, theoretical and practical, of all classes: as indicated in their laws, habits, proverbs, and deeds. All these facts, given with as much brevity as possible in combination with clearness and accuracy, should be so grouped and arranged that they may be comprehended in their *ensemble* and thus may be contemplated as mutually dependent parts of one great whole.[39]

Spencer believed that the "highest office which the historian can discharge, is that of so narrating the lives of nations, as to furnish materials for a Comparative Sociology; and for the subsequent determination of the ultimate laws to which social phenomena conform."[40] However, without the generalization of biology and psychology, the rational interpretation of social phenomena would be impossible. Romero followed these ideas when he analyzed the history of Brazil based on three aspects of Spencer's ideas. First, he considered the history of literature intertwined with the history of Brazil as a concatenated march of humanity, subject to the law of development of civilization – that was Spencer's concept of social progress toward the complexity of social relations. Romero wrote, "All and any historical and literary problem has to have two principal faces in Brazil: one general and the other particular, one influenced by the European moment and the other by the national environment... Literature throughout the Americas has been a process of the adaption of

37 Romero, *História da*, vol. 5, 406.

38 Romero, *História da*, vol. 2, 307.

39 Spencer, *Education: Intellectual, Social, and Physical*, 69.

40 Ibid., 70.

European ideas to the societies of the continent (...) From tumultuous imitation and old mental servilism, we want to move on to choice, to literary and scientific selection." [41]

According to Romero, and expressed in Spencerian language, "literary selection, in analogy with biological, was the result of the Law of Vital Competition divided into adaptation and heredity. The latter represented stable static elements, energies of races, fundamental predicates of people; it was the national side in literatures. Adaptation expresses the mobile, dynamic, genetic elements, transmissible from people to people; it was the general, the universal face of literatures. Two forces blend themselves both indispensable, natural products of the physical and social environment."[42] Understanding hereditariness as a static element, as did Spencer, Romero annulled one of the principles dear to Darwinian evolution. At the same time, by accepting adaptation [to the environment] as a mobile and dynamic transmission element, he followed Spencer, as well as Lamarck. Romero believed there was a continuity between nature and society taking into account the survival of the fittest. According to Patrick Tort this idea of continuity between nature and culture was anti-Darwin.[43]

Secondly, another important idea of Spencer – the 'struggle for existence' – appeared in the work of Romero. Taking as a backdrop the ideas of civilization and progress, Silvio Romero considered the 'struggle for existence' in the history of Brazil as a struggle in which "the Portuguese superseded the *caboclo* (culturally and biologically mixed workers of the land). The black served as weapon and support – this was his great historical title in the New World." In turn, the Portuguese represented the connection with European civilization ("though his Iberian-Latin origin caused him prejudice" because, according to him, Portuguese was considered a backward country in Europe).[44] Following Spencer, Romero explained that the superiority of the white race was inevitable, since it had expanded at the cost of the inferior races.[45]

In the second chapter of the first volume of *History of Brazilian Literature*, entitled "Theories of Brazilian History," Romero stated that literature in Brazil, and in all of the Americas, had been a process of adaptation of European ideas to the societies of the continent. However, this 'adaptation' of ideas had been more or less unconscious in colonial times, resulting in a tumultuous imitation of the old mental servility, which, following independence in 1822, came to be

41 Romero, *História da Literatura Brasileira*, vol. 1, 45.

42 Ibid.

43 Tort, *Pour Darwin*, 32.

44 Romero, *História da Literatura*, vol. 1, 91.

45 Citing Spencer, *The Principles of Sociology*, 1876–1896.

an adaptation of choice, representative of the 'natural literary and scientific selection' of the newly created political regimes. Therefore, according to Spencer, the thought of the Old World had to be known, in addition to a precise idea of the past and current state of the nation. In other words, it was necessary to know the whole to understand the process of adaptation of doctrines and European thought to the Brazilian social and literary environment.[46]

Third, in relation to the 'biological' origin of the people, Romero recognizes the fusion of races as a characteristic of the evolutionary formation of Brazil, but does not admit monogenism, which was fully accepted by Darwin.[47] Although Spencer did not talk about this, Romero, followed his approach in explaining the whole mechanism of society. The theory of the history of Brazil had to compute the whole and "elucidate the action of the physical environment, in all its aspects." History, according to Silvio Romero, should show the reasons for the origins of a people, their particularities, and not be restricted to saying what a certain people has in common with others.[48]

During the 1870s, Brazilian anthropology, as a natural science, also interpreted the formation of Brazilian society – whether indigenous or white European – and the inclusion of populations scientifically classified as inferior.[49] In the National Museum [of Natural History] anthropometric anthropologists discussed the origin of Brazilian and/or American man. Naturalists specialized in anthropology, João Batista de Lacerda and Rodrigues Peixoto, drew on the study of craniums discovered by Lund in Lagoa Santa, in the state of Minas Gerais (cited by Darwin in *Descent of Man*), and contrasted the polygenism of Agassiz to monogenism.[50] This debate echoed the discussion in the Academy of Science of Paris where it was discussed by Quatregafes de Bréau, archenemy of the theory of natural selection by Darwin.[51] Romero

46 Romero, *História da*, vol. 1, 45, 46.

47 Although Spencer was not concerned with this question, this idea was a watershed.

48 Romero, *História da*, vol. 1, 54.

49 Lilia K. M. Shwarcz, "Usos e abusos da mestiçagem e da raça no Brasil: uma história das teorias raciais em finais do século XIX." *Afro-Ásia* 18 (1996): 77–101; Heloisa M. Bertol Domingues and Magali Romero Sá, "Controvérsias evolucionistas no Brasil," in *A recepção*, 103.

50 Agassiz had visited Brazil in the 1860s to prove that Darwin was wrong. See Domingues and Sá, "Controvérsias evolucionistas no Brasil," in Domingues, Sá and Glick, *A recepção*, 97-123; Gastão Galvão de C. e Sousa, "Conferências de Agassiz após seu retorno da Amazônia," in *Darwinismo, meio ambiente, sociedade*, edited by Heloisa M. Bertol Domingues, Magali Romero Sá, Miguel Angel Puig-Samper, Rosaura Ruiz Gutierrez (Rio de Janeiro: MAST, São Paulo, Via Lettera, 2009), 101–112.

51 Domingues and Sá, "Controvérsias evolucionistas," in Domingues, Sá, Glick, *A recepção*: The scientists from the National Museum, João Batista de Lacerda and Rodrigues Peixoto

said that according to them it was necessary to end the habit of reducing the American races to a single type. – "The American races are a product of the American environment."

Based on the arguments of Brazilian anthropologists, Romero concluded that the origin of the Brazilian came from three different races, which in its evolution became mixed and was heading towards whitening, due to growing European immigration – essentially Germanic, which had begun in the 1820s. He concluded that Brazil was becoming mixed, but emphasized that the socio-historical domination of the Portuguese was undeniable. He thereby adapted Spencer's ideas. For Spencer, those born in a particular social class never moved into a different social class. Romero agreed when he spoke about Portuguese dominant inheritance in literature, language and power. Cross-breeding, a contingency of the environment, resulted in white domination: biologically Germanic and socially Portuguese. But conceiving of the social heritage as a static element, as claimed by Spencer, was rejected by Romero, because he didn't ignore the different colors of Brazilian people. The "mestiçagem" was a reality: that wasn't in any of Spencer's writings.

Silvio Romero and Capistrano de Abreu differed about the historical interpretation of the three races, which formed the Brazilian 'people.' Applying Spencerism, Abreu produced a political interpretation of the historical relationship between the three races. Unlike Romero, Abreu criticized the idea that the Portuguese were responsible for the development of the history of Brazil. To explain the question of Brazilian social formation and (Portuguese) colonial domination the idea of race had to be used and for this Abreu drew on Spencer. An article dated 30 November 1875, in which he discusses Brazilian nationality, starts with an epigraph from Spencer, taken from a French translation of *Introduction to Social Science*. Here Spencer explained that social progress is the action of 'great men' – for Abreu, the conqueror. Spencer wrote, "Si vous supposez qu'un Newton puisse naître d'une famille Hottentote, qu'un Milton surgit au milieu des Andamans, qu'un Howard ou un Clarkson puisse avoir des Fidjenspour parentes, alors vous réussirez facilement à expliquer le progress social comme amené para les actions du 'grand homme.'"[52] He makes

were polygenists. They concluded that Indians that they studied had "such intellectual inferiority that they were close to anthropomorphic monkeys." Their work was presented in the Academy of Sciences in Paris, and was commented on by Quatrefages.

52 H. Spencer, *Introduction à la Science Social*, 35–36. "If you assume that Newton could be born a Hottentot family, a Milton arises among Andamans, a Howard or Clarkson could have Fidjenspour parent, then you easily manage to explain the social progress brought as the actions of the great man."

an analogy between ethnicity and the evolution of the nervous system stating, as Spencer did, that intelligence was manifested through the nervous system, implying predispositions, and was thus an intellectual, super-organic factor.[53] According to Spencer, "Those however, who recognize the truth that the structural changes are the slowly accumulated results of the functional changes, will readily draw the corollary, that a part cause of the evolution of the nervous system, as of other evolution, is this multiplication of effects which becomes ever greater as the development becomes higher."[54]

To study the different people that formed Brazilian society Abreu began with the Portuguese. They had conquered kingdoms, founded empires, discovered worlds, which had led to the "heaven of History." However, they had declined and the fall was terrible, creating an "emotion of inferiority" and it was this emotion which characterized the primitive times of Brazil, and which had established "our trajectories." There were fights and the interests of the kingdoms and colonies divided, becoming antagonistic. "Little by little the emotion of superiority erupted, grew, and gave us 7 September" (Brazilian independence).[55] The colony created a feeling of superiority due to their struggles, independence movements, and cultural manifestations such as popular songs, inspired by contempt of the oppressor. Independence modified that feeling and had to be considered as the translation of awareness of superiority over Portugal. He explained that this established the unity of the history of the country, showing the role of the people in its evolution. He then stated that: "When the movement is so general and persistent, it is not improvised, it is not ordered: it emerges slowly, like a coral reef in the secular architecture of the people. The influence of great men is forcefully limited and directed by the influence of the social environment." He stated this citing of Herbert Spencer at length in a phrase which ended: … "an interpretation of social phenomena, while insisting on the importance of the changes consummated by great men, forgets the vast accumulation of latent force which serves as a valve and the immense number of prior facts from which comes the strength and the great men."[56]

Abreu also said that the independence movement was political, hence its strength and its insufficiency, since a revolution could have modified the relations between the states and the conditions of the governed would not have modified the social structure. Although 7 September 1822 transformed the

53 Abreu, *Ensaios e Estudos*, 1ª Série, 48.

54 Spencer, *First Principles*, vol. 1, 43–44.

55 Abreu, *Ensaios e Estudos*, 1ª Serie, 47–48.

56 Citing Spencer, *Introduction to Social Science, in French translation*, 37.

colony into a sovereign nation it did not abolish the other more profound, industrial, mental, and social dependency on Europe. "Our current status reproduces a feeling that flourished in primitive times. Then Brazil deemed itself inferior to Portugal. Now it deems itself inferior to the other European countries." He insinuated that this was the feeling of a colonized people and that the same had been observed in the United States. Abreu was an anti-colonialist.

In Abreu's vision, the originality of Brazil was due to the climate and the Indians, not the Africans, as Romero argued. The influence of the environment was helped by another factor: the great mass of the Tupi population who had been incorporated by the Portuguese colonists and their descendants.[57] He saw the Indian as being segregated, despite the cultural legacy in language and customs. He also observed that the Indians were numerous when the Portuguese arrived in Brazil and that they were unstructured, which confirmed Spencer's idea that only an organized society could survive.[58] However, in relation to this point it was Romero who disagreed, claiming that indigenous society was not complex: "There being no complex society, the community could not be big." To which Abreu responded: "Replace the words *great people* with the words *small but numerous tribes*, which were mutually hostile, irreconcilable, without a core of common resistance, and all difficulty disappears; Spencer's law is saved, the law of anthropology is respected and historical truth is obeyed."[59] Spencer was the theoretical instrument for Abreu to define independent Brazil, free of colonization, as an autonomous country. However, such interpretation of facts was absolutely contrary to the key idea of Spencer, in the sense that people never change their social status. An aristocrat will always be in the upper class, and people of the inferior class never will be in a superior class.

One of the political objectives of Abreu's historiography, based on Spencerian evolutionism, but also on Buckle and Comte, was to combat the colonialism built into the vision of the history of Brazil. Spencer's ideas also served to corroborate the history of the political dominion of the Portuguese. Similar to Romero's view, Brazil was mixed and evolving toward whitening, but for Abreu

57 Abreu, *Ensaios e Estudos*, 3ª Série, 104–108.

58 Ibid., 109.

59 Anthropology was also immersed in the debate about Brazilian social formation, discussing evolutionism and the ideas of Indians and blacks, racial inferiority, and superiority See: Lilia K. M. Schwarcz, "Usos e abusos da mestiçagem e da raça no Brasil: uma história das teorias raciais em finais do século XIX," *Afro-Ásia*, 18 (1996): 77–101; Domingues and Sá, "Controvérsias evolucionistas," in Domingues, Sá, Glick, *A recepção*, 97–123.

this served to show that diversification was the mark of an independence which could not be reputed to the colonizer, but to the people who had evolved and won this. Spencer actually had the merit of facilitating debate about the ethnic diversity of Brazilian society and although these historians had not absorbed the term 'social class,' as Spencer did (in his *Principles of Sociology*), the idea permeated the concept of 'race' they used.

Spencer and a Proposition of Political Science by Alberto Salles

From a different perspective than Romero and Abreu, Alberto Salles also relied heavily, and uncritically, on Spencer. In 1891, he published a book entitled *Political Science*, one of the first studies of political science in Brazil (Fig. 9.3). Salles sought inspiration in Spencer to compose his own political theory, inscribed in liberal capitalism.[60] *Political Science* appeared at the beginning of the republican regime in Brazil, but it has a very intriguing feature in that a commemorative edition was published on its centenary in 1997, organized by a Professor of Political Science from the National University of Brasília (UNB). The publisher was the Federal Senate Press. According to the organizer of this facsimile edition, Salles' book was pioneering, as it served as a basis for political science in the São Paulo law school, when this science was also being organized in other countries.[61] Law courses were the gateway for the entrance of new scientific ideas in Brazil, which allowed the social to be thought about and structured the first social science courses in the country.[62] Disciplines like Sociology, Political Economy,[63] and Political Science[64] were introduced in these courses as scientific interpretations of society, thereby spreading the so-called "social Darwinism"[65] as a theoretical orientation of the Brazilian social sciences. Although this book cannot be considered a historiographical book, it was a textbook in a school which produced many who became historians, writers or journalists, as was the case of Romero, Abreu and even Salles.[66]

60 Fernando Cardim, "Presentation of Political Sciences," in Alberto Salles Salles, *Sciencia Política*, XIII.

61 Ibid., III–XXIV.

62 Micelli, *História das Ciências Sociais no Brasil*, Introduction.

63 Aprígio Guimarães, *Estudo de Economia Política* (Recife, 1902).

64 Salles, Alberto, *Sciencia Política*, 1ª Edition, São Paulo, Teixeira & Irmão, 1891 (Edição Fac Similar: Brasília, Gráfica do Senado, 1997), 297.

65 Richard Hofstadter, *Social Darwinism in American Thought* (Boston: Beacon Press, 1955).

66 Law courses existed since 1827, in Olinda, Pernambuco, and in São Paulo and since the middle of the century in Rio de Janeiro, Porto Alegre and Salvador. Many of those who

FIGURE 9.3 *Alberto Salles'* Sciencia Politica. *Alberto Salles, Sciencia Politica (Brasília: Editora do Senado Federal, 1996), book cover.*

In the prologue of his book Salles presented his vision of the state, saying that democracy should be its foundation. "Democracy organized itself in the influxes of the *natural law of equivalence*. It is equality in reciprocity. The whole reacting harmoniously to the parts and these to the whole, in an admirable set of functions which are equilibrated, without being excluded...."[67] The organization of government, according to Salles, is formed from the will of the people ('desires'), reflection on it ('opinion') and implementation of this, which should be distinct and independent: "There is a scale of desires, as there is a scale in social culture and in the needs, which is an elementary phenomenon. Opinion, however, is by its nature complex and special, because it refers to thought (synthetic concept)."[68] Therefore, "it was absurd that the masses wanted to reason. Opinion has to want, publicists have to propose the means of implementation and the government has to implement (...) In summary, when politics had become a positive science, the public had to necessarily give to publicists the same amount of trust that they currently give to astronomers in relation to astronomy, to doctors in relation to medicine, etc."[69] He cites the positivism of Pierre Laffite to reaffirm the need for a [political] 'doctrine' as an opinion forming element.[70]

In relation to the most adequate means for the preparation of this policy, Salles has no doubt that it was the organization of public education, as also explained by Spencer in his book on education. Showing himself to be a liberal, he stated that economic prosperity and education are intertwined, since 'social capital' increases to the extent that it operates the economic evolution (of industry and agriculture). In other words, "intellectual and moral aptitudes are perfected to the extent that civilization progresses, in such a way as to develop education and to compete directly for the elimination of prejudices and to prepare ever more the spirits for the establishment of a doctrine."[71] He appropriated Spencer's idea of fixed heredity and applied it to men, seeing them as an economic instrument: -"Man is a fixed capital capable of producing utilities, exactly like a field, a machine... His superiority consists of being a utensil susceptible to being improved by himself. The more improved he is the greater is his capacity for utilities; as a consequence the greater is his value."[72]

---

graduated from them worked in what are now called the human and social sciences.

67 Salles, *Sciencia*, 7.

68 Ibid., 13.

69 Ibid., 17. It is interesting to observe that the term 'publicist' refers to those who translate the will of the public, the masses, in other words it defines political science intellectuals.

70 Ibid., 19. Emphasis added.

71 Ibid., 23.

72 Ibid., 24.

Perfection, which was an economic problem of labor, came from the valorization of human aptitudes and efforts; it depends on the resolution of the problem of education, concretely professional teaching. Economic prosperity was linked to the level of intelligence and the use of labour, "It is a civilizing mission."[73]

Salles criticized beliefs of all sorts, religious, superstitions, etc., calling them the inhibitors of the emancipation of the human spirit. In this point he managed to share somatometric anthropology and the social thought of Spencer. According to Salles, these prejudicial manifestations of the human spirit were nothing other than atavisms of primitive anthropomorphism, "sustained by naïve incredulity and exploited with pertinence by clerical imposture."[74] In this field it was necessary to carry out the fight, or to involve education. Like Spencer, he did not enter the discussion which divided monogenists and polygenists. He accepted the conclusions of the polygenistic anthropology of Quatrefages, who stated, like some Brazilian anthropologists, that there were differences between inferior and superior races, measured by cranial capacity, however, he disagreed when he said that the latter had no capacity for improvement by external stimulus, such as by example of education. Salles then accepted another line of anthropological interpretation, citing Paul Broca whose research led him to state that the "evolution of the social environment is parallel to the anatomical evolution of the brain, which becomes heavier and rougher as civilization progresses."[75] Salles also stated that teaching could not dispense with the knowledge of the moral and intellectual capital of society. He wrote, "What general anatomy does for animal organisms, politics does for national organisms (...) There is political and social evolution, as well as anatomical and biological."[76]

With these premises Salles then formulated his theory of social organization based on Spencer, saying that political organization is born out of the "struggle of one society against another." Still inspired by Spencer he concluded that contemporary nationalities emerge from social evolution, which, even with the refinements of civilization, were born in simple, savage and primitive hordes, capable of founding a nationality – constructing a factor of social cohesion which varied from one society to another – the first stage of political evolution.[77] "An empire, which is usually painted with such strange colors,

73 Ibid., 25.
74 Ibid., 30.
75 Ibid., 33.
76 Ibid., 88.
77 Ibid., 116–126.

enlightened by the scientific criteria of political evolution, becomes a natural fact, perfectly explicable by the laws of evolution."[78] Repeating Spencer he stated: "a nationality only becomes possible through the feeling that the units nurture towards the whole. It is this sentiment which constitutes the principal strength of a nation…."[79] With this statement he shows that Spencer's discussion of the idea of the nation was included in the thought of the theorists of the nation-state.[80]

Nationality therefore had to be studied through social and political evolution, looking at its historical past, and focusing on the factors of the law of evolution: environment, race, mental constitution, and the struggle for existence. First, the environment, which included the physical and biological conditions that affected the general process of political integration and differentiation, i.e., political evolution. In this case climate, alimentation, physical aspects, health, orographic soil relief and the special condition of the region.[81] Second, race "has an uncontestable relationship with the formation of nationality; it is a preponderant social and political factor of evolution."[82] Races, according to Salles, are produced by the environment, by crossbreeding and by heredity. He quoted Spencer to say that the mixing of two different races produces a worthless mental type. To the contrary, people of the same origin produce a superior mental kind – an English 'sub-race' was an admirable case of perfect homogeneity, because in England crossbreeding happened among similar groups, "fusion by crossbreeding."[83] When crossbreeding happened between very different individuals it did not allow a general fusion of characters, consequently a unique nationality was not formed, but diverse regional types. However, mental constitution allowed the progress of civilization.

In relation to the third political and social factor of evolution, mental constitution, Salles considered linguistic and ethnic differentiation. Salles, agreeing with Littré, stated that "all peoples are more and more susceptible to intellectual and moral culture, however, in mental evolution some people can precede others. A background of ideas and sentiments that forms the national character of each people according to the mental capacity and to the past."

78 Ibid., 132.

79 Ibid., 138.

80 See George Gusdorf, *Les principes de la pensée au Siècle de Lumières* (Paris: Payot, 1971); Jean Plumyene, *Les Nations Romantiques* (Paris: Fayard, 1979).

81 Salles, *Sciencia*, 145.

82 Ibid., 159.

83 Salles, *Sciencia*, 1897, 157.

Salles then repeated Spencer in criticizing the common belief that the governmental forms by themselves were capable of promoting the progress of a nation. For Spencer, Salles said, only those forms that sprouted naturally from the national character were efficient.[84]

In the fourth and final factor of political evolution Salles evoked Spencer's idea of the struggle for existence, seeing this as a 'law' which all organized beings obey. In this struggle, it is always the strongest, the best constituted, those who best adapt to the general conditions of existence, who survive. In short, war is a means of improvement, which has led to civilization. Repeating Spencer, he stresses that it is a truth found at all times and in all places that without war neither the formation of large agglomerations of individuals would not be possible, nor a developed industrial state.[85] The conquering of a semi-civilized or savage race by another signified a step towards civilization. Furthermore, the organization of labor has to be taken into account, so that the performance of any task given to the incapable at the cost of the capable signifies multiplying the former at the expense of the latter, lowering the level of the social and organic capacity of the race.[86] The state itself is subject to the factors of evolution.

These Spencerian interpretations, both of history and of political science were based on the inequality of races. However, the emphasis on war, on the 'struggle for existence,' stressed by Salles, did not appear in the same way either in Romero or in Capistrano de Abreu. Spencer's aristocratic doctrine, predominant in the social sciences since the nineteenth century, was founded on this idea, seeing race as peoples or society.

Conclusion

The writings of Romero, Abreu and Salles cannot be considered as the popularization of science, from the perspective of Lightman. They used scientific ideas to popularize historical, political and sociological ideas that shaped the social imagination on one hand, and to lay the groundwork for a political science.

As a materiality of ideas, books generate cultural representations guided by the authors and also by various interests of agents that leave traces of their interventions in the bibliographical resources, in the texts and beyond the

84 Ibid., 166, 167.

85 In relation to this, see Antonello La Vergata, "Darwinismo, evolução e guerra," in Domingues et al., *Darwinismo, meio ambiente*, 237.

86 Salles, *Sciencia*, 170.

texts.[87] Spencer's ideas and their strong influence remained invisible in the representations of the social and political imagination of the country. They were so invisible that Spencerism never was a coherent term; it survived in the shadow of Darwinism in Brazil.

Evolutionism oriented the natural and social sciences, which were also intertwined in the debate about historical and social formation. Spencerism introduced into the thought of those intellectuals common questions about the evolution of intelligence, resulting from the stimulus of the environment, from education, and principally providing a foundation for the debate about racial (actually social) inequality. However, each author used Spencerism to make different interpretations of the same historical facts, affirming, with this, political positions. In the case of Romero and Abreu the aim was the construction of the history of Brazil, in the case of Salles, it was the rationalization of politics. All of them analyzed the same historical facts or social situations, applying the ideas of Spencer, which they used in part or integrally according to their own point of view of the facts or of their social and political ideologies. This was the case of the interpretation of social evolution of the Brazilian people. Although Salles did not refer to the history of Brazil, it is possible to say that all of them agreed that the history of society resulted from a process of social evolution, as Spencer stated. In the case of Brazilian society, formed by different "people," the political position and ideology determined the way Spencer's ideas were appropriated. The publications in question emerged in the context of important political and social moments of the representation of Brazilian nationality.

Despite having remained invisible in representations which emerged, Spencer's ideas constituted cultural baggage which still intervened in the image which Brazilian society created of itself. These ideas corroborated social differences and, although he made education a way of instituting 'social evolution' this would never erase the differences between one or other ethnic group.

---

87 Roger Chartier, *Autoria e História Cultural da Ciência*, Ed. by Priscila Faulhaber and José Sergio Leite Lopes (Rio de Janeiro: Azougue, 2012; Gustavo Sora, *Brasilianas: José Olympio e a gênese do mercado editorial brasileiro* (São Paulo: EDUSP, 2010); The third and final edition of *História da Literatura Brasileira* by Silvio Romero is from 1943. According to a list Romero himself prepared, between 1880 and 1913, he published 17 books about literary criticism, including *História da Literatura Brasileira*; four about folklore; four on ethnography; seven about political and the social condition [of Brazil]; three about philosophy; two of poetry, and 17 which he called *Pamphlets*. Abreu's books were republished until the 1970s by *Sociedade Capistrano de Abreu*. While he was alive his books on the history of Brazil were republished three or four times. Salles wrote about politics and while he lived he published many pamphlets which sold well, as mentioned. However, his book only seems to have been republished on its centenary in 1997.

The full scope of the impact of Spencer's ideas on Brazil is immeasurable, because when the country in the 2000s instituted *Racial Quotas* to minimize the persistent racial issue as an impediment to higher education for many young people, regardless of economic differences, it was unthinkable that these differences were due only to [wild] Capitalism. There was an ideological reason for recognizing the social inequality as racial inequality. Spencerian thought was widespread and remained intertwined in Brazilian social thought.

CHAPTER 10

The Importance of Being Quantified: Herbert Spencer in Liberal Italy, and Beyond

Paola Govoni

After the great success of the 1880s, a rapid fall in popularity overtook Herbert Spencer's work in Italy, and the image of a Spencer who had disappeared from the Italian public sphere as rapidly as he had arrived begun to circulate. However, with the new century, there was actually a renewal of interest in Spencer, which was to last until the 1920s: a phenomenon which seems to have escaped the notice of the numerous historians of ideas who have worked on Spencer south of the Alps. When Spencer died on December 8, 1903, the British press reported that Italian newspapers were giving great prominence to the news. In the Italian Parliament, the jurist and university professor Agostino Berenini (1858–1939) delivered a commemorative speech on Spencer, warmly appreciated by the whole assembly.[2] On the same occasion, philosopher Cesare Ranzoli (1876–1926) observed that the cables containing condolences reaching England from south of the Alps had been recognized by the British press as more numerous than those arriving from any other country.[3] It was young philosopher Guglielmo Salvadori (1879–1953), believing the *First Principles* to be "the most important work of the 19th. Century,"[4] who was one of the main originators of what I call the second wave of the Spencer effect in Italy (Fig. 10.1).

---

1 Many thanks are due to Bernie Lightman for comments on the initial draft and, for useful discussions, to all the Spencerians attending the 2013 Manchester International Congress of History of Science, Technology and Medicine. The research for this project was supported by the University of Bologna (2011–2012 RFO funds).

2 Information drawn from two albums of cuttings from the international press, MS 791/356, 1 and 2, Spencer Papers, Atheneum and University Library, Senate House, London (hereafter HSP).

3 Cesare Ranzoli, "La fortuna di Erberto Spencer in Italia," *Rivista di filosofia e scienze affini* I/1–2 (1904): 97. See the second and third parties of Ranzoli's essay in ibid. I/3–4 (1904): 219–35, and I/4–5 (1904): 440–67.

4 William Salvadori [Guglielmo Salvadori] to Herbert Spencer, 25 April, 1901, Spencer Papers, MS 791/281 (HSP).

 | DOI 10.1163/9789004264007_012

FIGURE 10.1 *Cover of* L'evoluzione morale (*Moral Evolution, Turin, Bocca, 1909*) *by Herbert Spencer, Italian translation by Guglielmo Salvadori, from the first volume of* Principles of Morality (*private collection*).

Spencer had reached Italy late, the first translation – from *Education* – being published in 1876. However, publishers and scholars promoting Spencer knew how to make up for lost time. In the 1880s, scientists were decisive for the first wave of Spencer's success, but endorsing Spencer were also those economists and politicians who saw themselves as part of a so-called positivist culture. The national context after Italy's unification (1861) was characterized by considerable tension between the liberal elites and the Catholic Church. Despite being in the presence of sometimes very different positions,[5] to call oneself a 'positivist' was the equivalent not only of declaring science one's cultural guide, but also one's political orientation, typically anti-Catholic and republican. It was in that milieu, often called the Liberal Age,[6] that Spencer's thought had a great success, was appropriated, and adapted to political positions which, as we shall see, in some cases horrified Spencer himself. Heterogeneity of political positions, indeed, characterized also his most hostile critics, like the Marxist philosopher Antonio Labriola (1843–1904), the liberal neo-Idealist and anti-Fascist philosopher Benedetto Croce (1860–1952), or the other champion of neo-Idealism, the 'philosopher of Fascism' Giovanni Gentile (1875–1944). Despite the authoritative status of his detractors, there can be no doubt that Spencer's admirers were the more numerous.

Economist Tullio Martello (1841–1918), a supporter of free trade and a follower of Adam Smith, bestowed important official honors on Spencer. Already mentioned MP Berenini, a member of the Socialist Party, on the other hand, and other well-known Marxist writers such as Enrico Ferri (1856–1929), acclaimed Spencer. Spencer's most famous writings were used – if often not quoted – by the anthropologist and popularizer of international renown, Paolo Mantegazza (1831–1910), a conservative, and likewise by Cesare Lombroso (1835–1909), another 'high priest' of the cult of positivism, but with socialist sympathies. If Spencer's admirers were typically anticlerical, he was taken seriously even by novelist Antonio Fogazzaro (1842–1911), a convinced believer. What the various scholars usually shared in their interest in Spencer (though not in the case of Martello) was in the first place his philosophical evolutionism, which seemed to offer theoretical support for undertakings of every kind.

5 On Italian positivism, see at least: *Scienza e filosofia nella cultura positivistica*, ed. Antonio Santucci (Milan: Feltrinelli, 1982); *Il positivismo e la cultura italiana*, ed. Emilio R. Papa (Milan: Angeli, 1985); *L'età del positivismo*, ed. Paolo Rossi (Bologna: il Mulino, 1986). For further bibliography, see: *Il positivismo italiano: Una questione chiusa?*, ed. Giuseppe Bentivegna et al. (Rome: Bonanno, 2008). On the impact of Spencer's works in Italy: Mario A. Toscano, *Malgrado la storia: Per una lettura critica di Herbert Spencer* (Milan: Feltrinelli, 1980), 16–20.

6 By the Liberal Age is meant the period from the political unification of the country (1861), to the eve of fascism (1922).

But the political aspect was just as important. A system of thought based on the cult of science, and politically reassuring, seemed to offer answers to many at a crucial time in the construction of national unity.[7] The utilitarian nature of Spencer's thinking also held so many attractions, plus the centrality of 'liberty' in his intellectual universe. The Anglophile tendency in important sectors of the Italian elites also played an important role in Spencer's appropriation. A cultural tradition consolidated in the eighteenth century, in the nineteenth century Anglophilia became fashionable once more thanks to John Stuart Mill (1806–1873), and of course also thanks to Darwin and Spencer himself. Mill was the friend and correspondent of Pasquale Villari (1827–1917), one of the most interesting figure among Italian positivists.[8]

The data emerging from the publishing market seems to contradict what philosophers and historians of ideas have been saying about him. Well before Talcott Parsons in *The Structure of Social Action* in 1937 announced to the world that interest in Spencer had ceased, in the 1920s in Italy some philosophers had already begun to declare that Spencer had "outlived his work."[9] Quantitative data contradicts this affirmation, which suggests that the statement "Spencer outlived his work" is a myth. As far as the specialized readers go, it may be seen that the number of times Spencer's name was cited in the Italian literature collapsed towards the end of the nineteenth century, but picked up again with the new century. Besides, publishers' data indicate that except for the fascist parenthesis, when the regime censored English language culture in all its forms (books, music, and cinema), Spencer's work continued to appear in new translations and reprints in the twentieth century (Tables 10.1 and 10.2 at the end of this chapter). In the absence of specific information on the sales of the Italian translations of Spencer's books, the data on the number of translations and reprints allows us to state that, if the experts stopped taking Spencer seriously, the Italian well educated general public (a restricted one, as we shall see) continued to read him. This is a further confirmation that, when analyzing the impact of an author on a specific national context, it is important to examine

7 Eugenio Garin, "Il positivismo italiano alla fine del secolo XIX," *Giornale Critico della Filosofia* 1 (1980): 1–27.

8 On Anglophilia in relation to Spencer, see Angelo Crespi, "Herbert Spencer e il Socialismo," *Critica Sociale* 13 (1903): 376; Paolo Orano, "Spencer e Darwin in Italia," *La Vita Britannica* 3 (1918): 228. More in general, see Maurizio Isabella, *Risorgimento in Exile: Italian Émigrés and the Liberal International in the Post-Napoleonic Era* (Oxford: Oxford University Press, 2009); Paola Govoni, "The Power of Weak Competitors: Women Scholars, 'Popular Science' and the Building of a Scientific Community in Italy, 1860s-1930s," *Science in Context* 3 (2013), 409–10.

9 Zino Zini, *Spencer* (Milan: Athena, 1926), 5.

also the networks of publishing production, circulation, and consumption, and the general public.

In the pages that follow, I shall offer some general information on the Italian context in the Liberal Age, and data that will help to 'quantify' the presence of Spencer in Italy over a long period: from the first wave of its success (1870s-1890s), through the second one (1900–1920s), to the present. I shall then offer examples of the different expert readers who took Spencer seriously, and of others who appropriated his thought during the first wave of its success south of the Alps. The extent of that phenomenon sustained a second wave of the Spencer effect in Italy, from the very end of the 1890s to the beginning of the 1920s. During this period fashionable neo-idealist philosophers strongly attacked Spencer, while fascism rose to power (1922). Under the dictatorship, Spencer disappeared from publishers' catalogues, incompatible as he was with the regime's obsession with the role of the state and the denial of the individual liberties. As I shall recall in the conclusion, it has been the importance of individual liberty in Herbert Spencer's thought that periodically attracted the attention of Italian readers in the second half of the twenty century, and all too recently.

1 Context and Data

Data from surveys in the first decades after Italian political unification gives us an image of Italy as one of the most backward countries in Europe; a context characterized by a huge gap between the elites and the majority of Italians. The publishing industry, mostly concentrated in the north, began very slowly to extend to the rest of the country in the 1860s; slow development in the postal service, in the network of roads and railways, in the diffusion of telegraphy, all provide evidence of what was a dramatic economic situation. In 1861 the rate of illiteracy was 75%, and by 1911 it was still at 40%.[10] Only one newspaper, *Il Secolo*, was successful around 1872, with 20,000 copies sold. In 1881, when the Spencer effect emerged, what have been recently described as the middle classes were about 11% of the population:[11] non-occasional readers were likely to belong to that restricted élite. Because of the social situation rapidly sketched

10 On illiteracy, see Carlo M. Cipolla, *Literacy and Development in the West* (Harmondsworth: Penguin, 1969); for the economic conditions of the country, V. Zamagni, *The Economic History of Italy, 1860–1990: Recovery after Decline* (Oxford: Clarendon Press, 1997).

11 Marzio Barbagli and M. Pisati, *Dentro e fuori le mura: Città e gruppi sociali dal 1400 a oggi* (Bologna: il Mulino, 2012), 295.

out, cheap science books in the period 1870–1890 were unable to achieve a truly extensive circulation. The only case uncovered so far of a cheap popular science book capable of reaching 10,000 plus copies remains the almanac on hygiene by Mantegazza, published annually from 1865 to 1905. Within the mentioned literate public, the success of popular science in post-unification Italy cannot be doubted.[12] In the name of a 'progress' written in nature, and given orientation in society by 'positive science,' publishers and scientists, often involved in politics, from the 1870s supported the success of 'vulgarized science,' popular or 'for all.' Scientist popularizers thought that the 'truths' unveiled by science would lead to new ways of improving education and social development, healing the scars of the past.[13] In that context, Spencer's work offered a means to extend an evolutionary perspective to any social or cultural dimension. In 1876 the first two volumes of Spencer to be translated were taken from *Education*: Spencer's most translated and reprinted book in Italy far into the twentieth century (table 10.2). In 1881 two other translations followed: one from the *Principles of Sociology*, the other from *First Principles*.[14] The two volumes appeared in the Italian branch of the International Scientific Series, and the introductions were by the evolutionist anthropologist Giuseppe Sergi (1841–1936), soon to become one of the most academically powerful scientists of the age.[15]

Spencer's success extended beyond what we now call the age of progress, as both qualitative and quantitative data show. Reporting direct testimony from readers, a survey of "The most widely read books by Italian people" shows that in 1906 Spencer was still appreciated by "workers and shop keepers," by "white collars and professionals," as well as Catholic priests, who read Spencer's books

---

12 Govoni, "The Power of Weak Competitors."

13 For a general history of the period, see *Italy in the Nineteenth Century*, ed. John A. Davis (London: Oxford University Press, 2000). On the popularization of science in XIX-century Italy, see Paola Govoni, "The Historiography of Science Popularization: Reflections Inspired by the Italian Case," in *Popularizing Science and Technology in the European Periphery, 1800–2000*, ed. Faidra Papanelopoulou et al. (Aldershot: Ashgate, 2009), 21–42.

14 Herbert Spencer, *Introduzione allo studio della sociologia*, pref. Giuseppe Sergi (Milan: Dumolard, 1881); Id., *La base della morale*, pref. Giuseppe Sergi (Milan: Dumolard, 1881).

15 The Italian Series was published by Dumolard between 1875 and 1891. See Govoni, *Un pubblico per la scienza: La divulgazione scientifica nell'Italia in formazione* (Rome: Carocci [2002], 2011), ch. 3. For further bibliography on the Series, see Bernard V. Lightman, *Victorian Popularizers of Science: Designing Nature for New Audiences* (Chicago: Chicago University Press, 2007), 378–88.

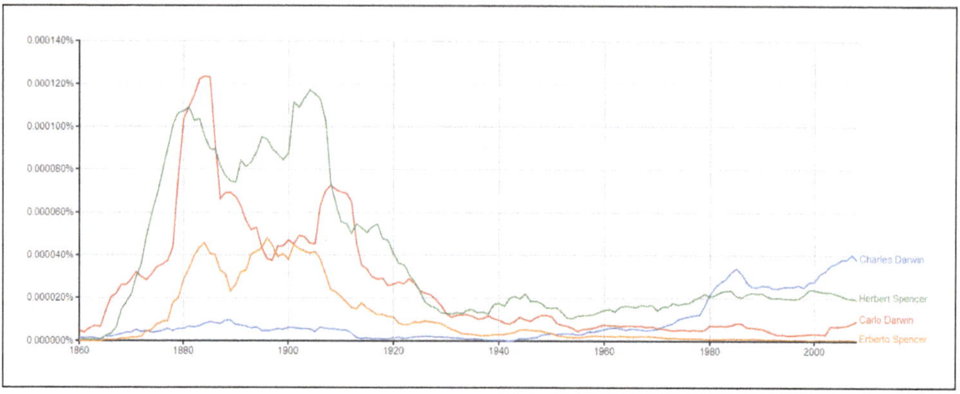

GRAPH 10.1 *Research carried out with Google Ngram Viewer of the phrases 'Charles Darwin,' 'Carlo Darwin,' 'Herbert Spencer' and 'Erberto Spencer,' from the corpus 'Italian books,' period 1860–2008 (January 30, 2014).*

"in order to refute his theories from the pulpit."[16] On the quantitative side, interesting figures emerge – through Google books via Ngram Viewer – from a comparison between the presence of Spencer in the Italian literature, and the presence of Darwin, an author often raised to mythical status by many Italian scientists and reformers of the time.[17] A dual survey carried out by making use of the Italianized names of Darwin and Spencer (Erberto and Carlo) shows that Spencer (Herbert) outdid Darwin (both Charles and Carlo) in the period 1890–1910, while only in 1980 did 'Charles Darwin' outdo 'Herbert Spencer' (Graph 10.1). Those data also confirm the two waves of success of Spencer in Italy: a first one, between the 1870s and the 1880s, and a second wave of success, from the 1900 to the First World War. We don't have data on the number of copies of Spencer's works sold in Italy, unfortunately, but the number of translations and reprints testify to readers and publishers' interest in Spencer up until the 1920s (Table 10.1).

After this rapid but needed contextualization, it is time to see who, and how, among Italian specialists – in some cases through a process of appropriation, or even misunderstanding – contributed most to the Spencer effect in Italy.

---

16 Società Bibliografica Italiana, *I libri più letti del popolo italiano* (Milan: Società Bibliografica Italiana, 1906), 23, 33, 13.

17 Giovanni Landucci, *Darwinismo a Firenze: Tra scienza e ideologia (1860–1900)* (Florence: Olschki 1977); Pietro Corsi and Paul J. Weindling, "Darwinism in Germany, France and Italy," in *The Darwinian Heritage*, ed. David Kohn (Princeton: Princeton University Press, 1985), 683–730; Giuliano Pancaldi, *Darwin in Italy: Science Across Cultural Frontiers* (Bloomington: Indiana University Press, 1991).

2 The Spencer Effect: The First Wave (1870s-1890s)

To understand the extent of the Spencer effect south of the Alps an overview of his Italian expert readers, and an example of the debate emerging after the reading of his books, may be useful. It is true that scientists contributed the most to Spencer's success in Italy. Anthropologist and psychiatrist Enrico Morselli (1852–1929), one of the most interesting Italian scholars of the period, was an enthusiastic admirer of Darwin, but on commemorating him after his death he placed Spencer beside him, calling them together the "founders of modern evolutionism [who] complement[ed] each other."[18] Spencer's work not only fascinated scientists, but in some cases led them into dialogue with philosophers over new research projects. The debate on the relationship between mind, brain and consciousness, for example, has been seen by experts as an interesting example of how, inspired by Spencer's work, various Italian scholars devoted themselves to new directions in their studies.[19] At the end of the 1870s the debate involved anthropologist Giuseppe Sergi, physiologist Alexander Herzen (1812–1870), and other scientists, but also philosophers such as Roberto Ardigò (1828–1920), and Felice Tocco (1845–1911). Along with Carlo Cattaneo (1801–1869), and Villari, Ardigò was one of the most interesting positivist theorists. He was never convinced by Spencer as a philosopher in general,[20] and in particular by his pedagogical thinking; to Ardigò, Spencer's interpretation of learning processes were too determined by hereditary factors, compared to the social and cultural conditioning which from birth, Ardigò believed, played a crucial role.[21] But his reading of Spencer played a role in his exploring deeply questions of education in relation to mind, as it did for the anti-metaphysical philosopher Tocco who, contrary to Ardigò, continued to find inspiration in Spencer, on whom he was still giving university courses in 1904.[22] Anthropologist Sergi too, in 1904, still admired in Spencer his "utilitarian ethic [...] superior to that [of] Bentham and Mill," as well as his merit in having helped scientists to understand a good many phenomena, including

18 Enrico Morselli, "Carlo Darwin," in *Carlo Darwin e il Darwinismo nelle scienze biologiche e sociali: scritti vari*, ed. Enrico Morselli (Turin: Dumolard, 1892) 47–8 (originally published in 1882 in the *Rivista di Filosofia Scientifica*).

19 Landucci, *Darwinismo a Firenze*, 191 and ff .

20 For example, Spencer's 'theory of the unknowable' he called 'mistaken,' without any beating about the bush. Roberto Ardigò, *Opere Filosofiche* (Padua: Draghi, 1898), 352.

21 Tiziana Pironi, *Roberto Ardigò, il positivismo e l'identità pedagogica del nuovo stato unitario* (Bologna: Clueb, 2000).

22 See *Felice Tocco: Le carte e i manoscritti della Biblioteca della Facoltà di Lettere e Filosofia dell'Università di Firenze*, ed. Anna Olivieri (Florence: Olschki, 1991), 242.

those of the psyche.[23] Spencerian theories played an important role for scholars who didn't want to reduce psychology to physiology, nor give room to the "autonomy of the spirit," theorized in the German philosophical tradition.[24] These are just a few of the most influential scholars of that first generation who took Spencer seriously, presenting him in their university courses and thus transmitting an interest in Spencer's work to several generations of students, well into the twentieth century.

It was in this way, for example, that Spencer had a hand in the formation of internationally well-known physician and educator Maria Montessori (1870–1952). Montessori's university education occurred at a time when, according to many, Spencer's star had already faded completely. A pupil of Sergi, Montessori's theories in the field of education were rooted in psychiatry and anthropology. In the 1917 American edition of her famous *Advanced Method*, Montessori recognized Spencer as an important influence: "Experiences are, according to [Spencer] and the English empiricists, the constructive factors of the mind [...] Man is what experiences has made him; hence, in education by preparing a suitable structure of experiences, it is possible *to build up the man*."[25] Similarly significant was how popular science literature used Spencer. Physiologist and anthropologist Mantegazza was one of the most famous Italian popularizers with a considerable international reputation, and he was the one who made the most use of Spencerian information and expressions in articles, or in the popular science books mentioned before.[26] However, the uses Spencer was put to with political aims in mind were also pervasive.

Among the economists who read Spencer with care, an interesting case to explore would seem to be that of the free trader and individualist Tullio Martello.[27] Martello had been a follower of Garibaldi and a republican. A graduate in mathematics, he turned to economics, and in 1867 took part in the

---

23 Giuseppe Sergi, "Herbert Spencer," *Nuova Antologia* 109 (1904): 138.

24 Landucci, *Darwinismo a Firenze*, 191 and ff.; Stefano Poggi, *Introduzione al Positivismo* (Bari-Rome: Laterza 1987), 6.

25 Maria Montessori, *The Advanced Montessori Method: Spontaneous Activity in Education*, trans. Florence Simmonds (New York: Frederick A. Stokes Company, 1917), 159. Italics of the author.

26 Govoni, *Un pubblico per la scienza*, ch. 5.

27 The Tullio Martello Papers are undergoing re-ordering (Archivio Storico, University of Bologna, hereafter ASUB). The papers include interesting material, which will offer a better understanding of this figure and the role played by Spencer for Italian economists of the period.

first Congress for Peace and Freedom in Geneva.[28] A member of the Cobden Club, and in fierce polemics with the so-called Socialisti della Cattedra (Socialists of the Chair),[29] from 1884 he taught economics at the University of Bologna. In a volume on *L'economia politica e la odierna crisi del Darwinismo* (Political Economy and the Current Crisis of Darwinism, 1912; from a collection of articles published in the 1890s), he took on an entire generation of those scientists who had used evolutionism to explain the workings of society, including the economy. He entered into the merits of Spencerian theses, reiterating that between "man and beast" there is a "line of demarcation […] insuperable, an abyss. Spencer […] wanted to prove too much."[30] To Martello, between mankind and the animals there is a *hiatus*, which is the raison d'être of the science of economics. It was not a concession to a religious vision: "Positive science has a thousand reasons for rejecting final causes, because they have a *Providential* character."[31] The point for Martello was freedom of choice, the crucial difference between mankind and animals. Admired by Vilfredo Pareto (1848–1923), for whom "Spencer's positivism is simply a metaphysics,"[32] in 1888 Martello promoted a degree *honoris causa* to Spencer in the context of the celebrations for the eighth centenary of the University of Bologna.[33] This was one of the few occasions when Spencer rejected official recognition with more scruples than usual.[34] If Martello's position was anti-evolutionist, his important public recognition of Spencer in 1888 has to be read from the perspective of politics. Scholars from all over the world converged on

28 Wilhelmus H. van der Linden, *The International Peace Movement, 1815–1874* (Amsterdam: Tilleul Publications, 1987), 685. On Martello see Angelo Bertolini, *Vita aneddotica e opera scientifica di Tullio Martello* (Bari: Laterza, 1917); Franco Della Peruta, "Politica ed economia nelle lettere di Tullio Martello a Enrico Cernuschi (1873–1888)," *Tra Lombardia e Ticino: Studi in memoria di Bruno Caizzi*, ed. Raffaello Ceschi et al. (Bellinzona: Casagrande, 1995), 279–326.

29 The Socialisti della Cattedra were inspired by the German school (*Kathedersozialisten*) and were in favor of state subsidies to the most vulnerable sectors of the national economy.

30 Tullio Martello, *L'economia politica e la odierna crisi del Darwinismo* (Bari: Laterza, 1912), 271–2.

31 Ibid., 285, author's italics.

32 Vilfredo Pareto, *Trattato di sociologia generale* (Milan: Edizioni di Comunità, 1964), 112.

33 Martello, a teacher in the Faculty of Law, obtain the honorary degree for Spencer from the Faculty of Letters and Philosophy. Folder VI Lauree d'onore, Folder Rubrica IV Ringraziamenti dei dottori (ASUB).

34 Herbert Spencer to unnamed addressee (almost certainly the Vice-Chancellor), 19 July 1988, Titolo VI Lauree d'Onore, Rubrica IV, Ringraziamento dei Dottori (ASUB). See also David Duncan, *The Life and Letters of Herbert Spencer* (London: Methuen & Co., 1908), 291.

Bologna to celebrate the university claiming to be the oldest of the western world. For the free trader Martello it was an ideal opportunity to show the Italian academic world "on whose side" an authority like Spencer stood. You could look the other way on evolutionism if you had the chance to celebrate Spencer as the hero of free trade and *laissez faire*.

Confidence in this kind of undertaking was not, apparently, misplaced: not very long afterwards a quite sensational event occurred, seeing Spencer become a Socialist hero. In 1894, Enrico Ferri, a criminologist and lawyer, formerly a pupil of Ardigò and later of Lombroso, published *Socialism and Positive Science* (*Darwin-Spencer-Marx*)[35]: this is very likely one the most striking cases of appropriation of Spencer's thought internationally. A "syncretistic muddle,"[36] but written in a brilliant style, this little volume on the three biggest names of the age earned Ferri worldwide fame: it was translated into French, English, German, Spanish and Lithuanian. Ferri argued that Darwinism was the "scientific foundation of Marxist socialism," which in its turn was the "complement in social life of that modern scientific revolution, inaugurated several centuries back by the revival of the experimental method [which] has triumphed in our days, thanks to the labors of Charles Darwin and Herbert Spencer."[37] Stoutly anti-socialist, Spencer reacted badly. A tough letter dissociating himself from Ferri made the rounds of Italian conservative newspapers, who could hardly believe their luck to be able to expose the Socialists to ridicule in this way.[38] But the Ferri episode was not an isolated one. Spencer enjoyed the great admiration of various writers appearing in *Critica Sociale*, the journal of the Socialists who, when Spencer died, saluted him as "an intel-

35 Enrico Ferri, *Socialismo e scienza positiva: Darwin, Spencer, Marx* (Rome: Casa Editrice Italiana, 1894).

36 Mario Sbriccoli, *Storia del diritto penale e della giustizia: Scritti editi e inediti (1972–2007)*, vol. 1 (Milan: Giuffré, 2009), 869. On the Ferri's case, see Lanaro, *L'evoluzione*, 23 and ff., Noemi Beck, "The Diffusion of Spencerism and its Political Interpretations in France and Italy," in *Herbert Spencer: The Intellectual Legacy*, ed. Greta Jones and Robert A. Peel (London: The Galton Institute, 2004), 37–60; Ead., "Enrico Ferri's Socialism: A Marxist Interpretation of Herbert Spencer's Organic Analogy," *Journal of the History of Biology* 38 (2005): 301–25.

37 I quote from p. 23 of the English edition of 1909, which in 2013 was reproduced in the Forgotten Books Series: accessed January 30, 2014, https://www.google.com/url?sa=t&rct=j&q=&esrc=s&source=web&cd=5&ved=0CEgQFjAE&url=http% 3A%2F%2Fwww.forgottenbooks.org%2Fdownload_pdf%2FSocialism_and_Positive_Science_Darwin-Spencer-Marx_1000017837.pdf&ei=1su1UarJDY2shQfU_YDwDw&usg=AFQjCNFmV3EnFWxyddsgUDHCt_tzW31QPg&bvm=bv.47534661,d.ZG4.

38 Giuseppe Rensi, "Un sillogisma socialista (Spencer – Romagnosi – Marx),' *Critica Sociale* 22 (1894): 344–7.

lectual giant."[39] That same year, successful Socialist writer Paolo Orano (1875–1945) explained that Darwin's work had found "in the masculine doctrine [...] of Marx [...] a further integration," and celebrated with emphasis the "great and vigorous" work of Spencer.[40] For writers like these, Spencer in his *Social Statics* had provided influential international publicity for the subject of equal rights for all in the use of land, and the denunciation of workers' exploitation. The prospect of a society in which individuals would be fully realized in agreement with others was entrusted by Spencer to an evolutionary progress received enthusiastically by 'dilettante' or superficial socialists like Ferri, Orano and others. In a somewhat buccaneering manner they combined Spencerian evolutionary progress with Darwinian evolutionism, which they likewise appropriated, simplifying it for their own purposes. Thus Spencer, with Darwin and Marx, became one more useful instrument in anti-Catholic and anti-conservative political campaigning. In evident conflict with the theoretical principles of socialism, which entrusted the liberation of the working class to the action of history and the class struggle, there were those who tried to provide clear arguments for improbable links between Spencer and Marx.

Antonio Labriola, a philosopher and one of the most interesting of Italian Marxists, was upset by the appropriation of Spencer by Ferri. Labriola was an admirer of Darwin, and supporter of the need to give more room to the sciences in educational curricula, including those of the humanities in universities.[41] However, to him Spencer was a "weak, vacuous, prolix and tedious reasoned,"[42] who seemed "sometimes a Kantian but unaware of this, and sometimes a caricature of Hegel."[43] Labriola's judgment of Spencer had been made more critical by the interest in him in the Socialist press, as he wrote to Friedrich Engels in 1894, of the "generalizers of Darwinism and the admirers of the great Eunuch Spencer."[44] According to Labriola there shouldn't be confusion between Spencer's 'altruism,' and the socialist praxis of the struggle for

---

39 Crespi, *Herbert Spencer*, 376. However, in the journal Spencer's anti-socialist positions, and those on the inferiority of women, were criticized. See, Giuseppe De Franceschi, "Il pregiudizio sessuale nello Spencer: Note sulla pretesa inferiorità della donna," *Critica Sociale* (1892), 173–5; A. Lissoni, 'Dalla schiavitù alla libertà. Risposta al libro "Dalla libertà alla schiavitù" di Erberto Spencer,' ibid (1893): 282–4.

40 Paolo Orano, "H. Spencer e l'avvenire," *Rivista di Filosofia e Scienze Affini* 1–2 (1904): 95.

41 Landucci, *Darwinismo a Firenze*, 15.

42 Antonio Labriola, *Saggi sul materialismo storico*, ed. Valentino Gerratana and Augusto Guerra (Rome: Editori Riuniti, 1964), 51 (1st ed. 1895).

43 Ibid., 232.

44 Antonio Labriola to Frederick Engels, June 13, 1894 in Antonio Labriola, *Scritti filosofici e politici*, ed. Franco Sbarberi, vol. 1 (Turin: Einaudi, 1973), 393.

justice.[45] But the *triunità*[46] (Marx, Darwin, and Spencer) that Labriola was laughing at had by now become a dogma for many Italian socialists. In 1909, the ambitious young editor in chief of *Il Popolo*, a Socialist paper, Benito Mussolini (1883–1945), wrote an article in commemoration of Darwin in which he referred to Spencer as the intelligent interpreter of that "period of transition from the old society based on individual and collective oppression, and the new society ordered according to criteria of justice."[47]

Clear traces of Spencer also ended up in the writings of the most famous Italian novelists. Socialist Edmondo De Amicis (1846–1908), journalist and educationalist, was the author of a best-seller also famous abroad, *Cuore* (Heart, 1886). On May 1st, 1896, in a speech to workers, De Amicis could take for granted that his audience knew Spencer, "our most formidable adversary."[48] Conservative Italo Svevo (1861–1928), a writer who became famous internationally between the wars, wrote a satirical short story on Spencer in 1897,[49] referring to him derisively again in his last short novel of 1928.[50] Antonio Fogazzaro, a successful mystical writer who adhered to the movement of so-called modernism and was excommunicated, used both Darwin and Spencer to support his synthesis "between the evolutionist hypothesis and the religious idea."[51] Thus in 1904 the (then) Marxist Orano read Spencer as the author who had reduced to silence "every kind of religiousness,"[52] whereas in 1899 the Catholic Fogazzaro admired "the sense of mystery" which for Spencer surrounded the origins of life, as an "act of Evolution."[53]

Chosen as samples from among the many available for the variety of their professions and political positions, the mentioned authors testify to the extent of the Spencer effect in the period 1870s-1890s. A generation of Italian scholars and politicians were involved, who were the protagonists of the political unifi-

45 Labriola, *Saggi sul materialismo*, 19, 53.

46 Ibid., 232. Author's italics.

47 Benito Mussolini, "Centenario Darwiniano," *Il Popolo*, 11 febbraio 1909, in *Opera Omnia*, ed. Eduardo e Duilio Susmel, vol. 2 (Florence: La Fenice, 1951), 10.

48 Edmondo De Amicis, *Il 1º maggio. Discorso tenuto all'Associazione Generale degli Operai la sera del 1º Maggio 1896* (Turin: Libreria Editrice Socialista del Grido del Popolo, 1896), 21.

49 Italo Svevo, *Tutte le opere: Racconti e scritti autobiografici* (Milan: Mondadori, 2004), 654.

50 Italo Svevo, *Confessioni di un vegliardo* (Pordenone: Studio Tesi, 1987), 57.

51 Antonio Fogazzaro, *Ascensioni umane: Teoria dell'evoluzione e filosofia cristiana*, ed. Paolo Rossi (Milan: Longanesi, 1977), 49. On this interesting case, see Paolo Rossi, "Introduzione," *ibid.*, 7–46, and Stefano Bertani, *L'ascensione della modernità: Antonio Fogazzaro tra santità ed evoluzionismo* (Soveria Mannelli: Rubbettino, 2006).

52 Orano, "H. Spencer," 95.

53 Fogazzaro, *Ascensioni umane*, 50.

cation of the country, and of the so-called age of positivism south of the Alps. Those authors carried with them a well educated public – restricted but curious about science – in their interest for the British polymath. It was that widespread interest in Spencer among different readers that paved the way to a second wave of his success in Italy in the following two decades.

3 The Spencer Effect: The Second Wave (1899–1920s)

New translations and reprints are evidence of an interest in Spencer on the part of the general public, which continues until after the First World War (Tables 10.1 and 10.2). Young philosopher Guglielmo Salvadori was one of the main originators of the new wave of the Spencer effect in Italy. In 1899 the first of six volumes of Spencer translated by Salvadori was published,[54] including, in 1904, a new translation of *Le basi della morale* (from *Principles of Morality*), reprinted several times until 1920, and then again in 1946 and 1951. Salvadori finished his translation of the first volume of *Principles of Morality* in 1909, with *L'evoluzione morale* (Moral Evolution). Having graduated with a thesis on Spencer, in 1900 he published his first book on him, and in 1901 he translated the *First Principles*, which by 1906 had already been reprinted six times. In 1906 he also published a second volume, in which he reported a fragment of a letter from Spencer: "You have grasped certain aspects of my work considerably more clearly than many of my countrymen. […] Great generalizations escape the English intellect."[55] From the little correspondence that has survived, it can be deduced that Salvadori beavered away busily at the publishers to promote the work of his *Maestro*, as he called Spencer, who showed generosity to the much younger man.[56] Salvadori met Spencer at least once, and when in 1903 Salvadori decided to marry the Anglo-Italian writer and traveller, Giacinta Galletti di Cadilhac (1875–1960), he asked the opinion of the *Maestro*. Spencer, of course, was not in favor, but Salvadori and Galletti married anyway. Ironically, this was not the first time he had been approached by Italians on questions

54 Herbert Spencer, *Istituzioni professionali; Istituzioni industriali*, trans. and ed. Guglielmo Salvadori (Turin: Unione Tipografico-Editrice Torinese, 1899). From parts VII and VIII of *The Principles of Sociology*.

55 Guglielmo Salvadori, *La scienza economica e la teoria dell'evoluzione: Saggio sulle teorie economico-sociali di Herbert Spencer* (Florence: Lumachi, 1901), 167. See also, Id., *Herbert Spencer e l'opera sua* (Florence: Lumachi, 1900).

56 On at least one occasion, Spencer sent back to Salvadori the check for the payment of his Italian royalties. Guglielmo Salvadori to Herbert Spencer, January 2, 1903, MS 791/299 (HSP).

concerning male-female relations.[57] When in 1902 a debate on divorce was reopened, the senators Carlo Porro (1854–1939) and Melchiade Gabba (1874–1952), on behalf of the Committee against Divorce,[58] asked Spencer what he thought: Spencer replied that he was "decidedly in favour of allowing divorce, when adequate reasons exist."[59]

When he died, Spencer was celebrated by the Italian popular press and by the Accademia dei Lincei. The latter was one of the few institutions whose membership he accepted.[60] Cesare Ranzoli, author of what has remained one of the most thorough studies of Spencer's influence on Italian philosophy, wrote that the English author "exercised an influence on our philosophical speculation and scientific thinking certainly no less than that of [...] Kant and [...] Hegel."[61] *Nuova Antologia*, the most widely read periodical of general culture, published some articles on Spencer, including one by Angelo Mosso (1846–1910). An evolutionist whose researches into the physiology of respiration and pain had impacted internationally, for Mosso "Spencer did not discover anything new and does not leave behind him [in science] any memorable trace," but nevertheless his "merit was to have introduced the data of the experimental sciences into the field of moral and social studies."[62] This was what Spencer had represented for the generation of Italian scientists born between the 1830s and 1850s.

In the twentieth century *Education* continued to be the book of Spencer's that most interested the public. In 1884 a new translation had been published, edited by the anthropologist Paolo Riccardi (1854–1924). The book was planned for educators and parents, to whom Riccardi pointed out that Spencer's educational writings had been those most appreciated by the English, people "who educate [...] independent beings, lovers of individual freedom and jealous of it."[63] Riccardi's edition was reprinted at least five times, and up until 1915. In

---

57 On Spencer's complex personality and relationships with women, see Mark Francis, *Herbert Spencer and the Invention of Modern Life* (Ithaca: Cornell University Press, 2007).

58 *Civiltà Cattolica*, also referring to Gabba and Porro, noted that also secular people, and even anti-Catholics, had taken the field in opposition to divorce. See "La follia del divorzio," *La Civiltà Cattolica*, VI: 1248 (1902): 168.

59 Herbert Spencer to Carlo Gabba and Melchiade Porro (written in a third hand; not Spencer's), February 24, 1902, MS 791/21 (HSP).

60 Giacomo Barzellotti, "Commemorazione del Socio straniero Herbert Spencer," *Rendiconti della reale Accademia dei Lincei*, XVIII (1904): 3–21.

61 Ranzoli, "La fortuna di Erberto Spencer in Italia," 97.

62 Angelo Mosso, "Herbert Spencer nella sua Autobiografia,' *Nuova Antologia* CXI (1904): 577.

63 Paolo Riccardi, "Introduzione," in Herbert Spencer, *L'educazione intellettuale, morale e fisica*, trans. Luigi Cussini, intr. Paolo Riccardi (Milan: Trevisini, 1884), v.

1894 Paravia had published another translation of *Education*, which also went through innumerable reprints until the 1920s. In the 1920s Spencer's works continued to attract the interest of non-specialist readers, especially school teachers,[64] or the odd politician.[65] Some chapters of the *Principles of Sociology* were published in 1922, together with the umpteenth reprint of *Education*. But then fascism arrived, and the combination of the fascist campaign against 'perfidious Albion'[66] with the success of neo-idealism meant that, apart from a reprint in 1935,[67] no more works of Spencer would be published.

4 Between Neo-Idealism and Fascism

In 1920, in an often reprinted volume, the famous neo-idealist philosopher Croce provided a two dimensional image of the Liberal Age, in the years when "the reputation of Herbert Spencer towered above us, with his train of French and English positivists."[68] A Spencer who, between the 1870s and 1890s, was derided by Croce as an "idol of the time, read quite considerably and revered and followed in Italy by historians and philosophers;"[69] an "Erberto Spencer, who everyone read and referred to as a great authority, as well known in Italy as Marx, was no longer either read or referred to, and had disappeared into complete oblivion."[70] The embarrassing case of Ferri gave the lie to this. Fascist Giovanni Gentile, no less than liberal Croce, strengthened that stereotyped image. Long before becoming the fascist regime's undisputed authority in educational fields, his opinion of Spencer was well known. In 1903, reviewing a

---

64 Maria G. Benso, *L'Educazione secondo E. Spencer, A. Bain, R. Ardigò* (Biella: Coop. Biellese, 1919); Leonardo Catalano, *Rileggiamo la scienza dell'educazione di Erberto Spencer* (Avellino: Labruna, 1923).

65 Manuel Devaldes, "Herbert Spencer," in Herbert Spencer, *Il diritto d'ignorare lo stato* (Rome: Edizioni di "Fede!," 1920), 3–17.

66 Fascism supported a policy of autarchy in economics but also in the world of culture; among its various initiatives was the censorship of the use of individual English words, as well as the translation of books written in English.

67 Erberto Spencer, *Dell'educazione intellettuale, morale e fisica*, trans. Angelo Valdarnini (Turin: Paravia, 1935).

68 Benedetto Croce, *Storia d'Italia dal 1871 al 1915* (Rome-Bari: Laterza 1973), 130; digital edition accessed January 30, 2014 <http://www.scribd.com/doc/93789202/Benedetto-Croce-Storia-d-Italia-dal-1871-al-1915>.

69 Benedetto Croce, "La storiografia in Italia: Dai cominciamenti del secolo decimonono ai giorni nostri," *La Critica*, 18 (1920): 262.

70 Croce, *Storia d'Italia*, 154.

collection of "many scant writings" of Spencer, he had liquidated them with "there is not one of any philosophical value."[71] It was in that neo-idealism context that the myth of a forgotten Spencer took root in Italy: "a chill shadow wrapped itself around that name, and enveloped that philosophy like a shroud."[72] This is from a 1926 popular book, entirely devoted to Spencer, by philosopher Zino Zini (1868–1937). Quite ironically, Zini's book appeared in the series *Pensatori d'Oggi* (Thinkers of Today), meeting with some success, and was reprinted in 1932. The little volume did Spencer justice, although Zini was a neo-Idealist who did not like "the excess of mechanical thinking [and] a contrived simplification and a monotonous schematism" in Spencer. A collaborator of *Ordine Nuovo*, the paper edited by Antonio Gramsci (1891–1937), Zini went out of his way to stress that "economic factor of historical facts [...] deeply separates Spencer from Marx."[73]

In the meantime, Ferri now adhered to Fascism. In 1929 he published a book in which he demonstrated his loyalty to the ambiguous anthropology of the Lombrosos that he applied now in the new autarchic context: "Mussolini [...] is a true Latin type. Such exceptional personalities can be met with less frequently among the Germanic and Anglo-Saxon peoples."[74] The Marx of German origins, and the Anglo-Saxons Darwin and Spencer, were water under the bridge. Indeed, this was the evolution of other Socialists, such as Orano, and, of course, Mussolini himself.[75]

Salvadori, having in the meantime become a university teacher, remained faithful to Spencer. If Spencer's conception of history was certainly partly at fault – as he wrote to Croce in 1931 – Croce's negative judgment seemed "excessive and not fair." Salvadori commented: "his great love of liberty, constantly reiterated by Spencer, should at least render the noble English thinker less contemptible in your eyes, in your judgment."[76] Salvadori was writing to Croce from Switzerland, where he had taken refuge, because as an anti-Fascist activist his life had been in danger in Italy.

Croce's anti-positivist obsession, and that of Gentile made them allies in the campaign to promulgate a neo-Idealism which created very serious problems

71 Giovanni Gentile, "Recensione a H. Spencer, *Fatti e Commenti*, versione dall'inglese del dott. Gugliemo Salvadori, Torino, Bocca, 1093," *La Critica*, 1 (1903): 296.

72 Zini, *Spencer*, 5.

73 Ibid., 72, 106.

74 Enrico Ferri, *Il fascismo in Italia e l'opera di Mussolini* (Mantua: Mussolinia Edizioni Paladino, 1929), 6.

75 Renzo De Felice, *Mussolini il rivoluzionario: 1883–1920* (Turin: Einaudi, [1965] 2005).

76 Guglielmo Salvadori Paleotti to Benedetto Croce, April 25, 1931, Carteggio, Archivio Privato di Benedetto Croce, Fondazione Biblioteca Benedetto Croce, Naples.

for the liberal and utilitarian approach of Anglo-Saxon derivation; but this was perhaps the reason why many people had remained drawn to Spencer and, like Salvadori, continued to defend him in the 1930s. After the "generalizers of Darwinism," those of neo-Idealism imposed a two dimensional image of nineteenth century scientific culture, and of an author like Spencer, which Italian historiography began to question again only in the 1980s.[77]

5 Conclusion

James G. Kennedy began his book on Spencer with a well known witticism: "When a man has written 400,000 words about himself, there is need for interpretation."[78] A salutary invitation to never take oneself too seriously, that witticism invites us to take seriously the concept of 'quantity,' a category which seems central to Spencer. Working on scientific best-sellers, there is one question I ask myself on starting a new project: beyond the intelligence and literary ability of a writer, what other features are required for a work to enjoy popular success, and overcome language barriers? This is destined to remain an open question, of course, because of the variables it involves. In the case of Herbert Spencer (but perhaps the same thing goes for Agatha Christie?), seriality and quantity play a role. The first thing is to find a subject (or a character) which readers can come to depend on: 'evolution' from the 1840s with *Vestiges* was 'the' subject of the century,[79] capable of attracting the attention of an enormous public on an international level. Spencer made of 'evolution' and 'progress' the main characters of his later work; he regularly produced writings on the widest variety of subjects in an un-academic style, yet philosophical, using examples which, if they scandalized many scholars, captured others, and above all impressed non specialist readers. In post-unification Italy, Spencer was the incarnation of those myths (evolutionism and progress) and fears (chaos in society and popular democracy) of the middle classes, crucial issues in the process of nation building. These were subjects that as we have seen, selected and adapted according to requirements, turned out to be useful to readers of a variety of political orientations. To illustrate those processes of

77 *Scienza e filosofia nella cultura positivistica*, ed. Santucci; *Il positivismo e la cultura italiana*, ed. Papa; *L'età del positivismo*, ed. Rossi.

78 Kennedy, *Herbert Spencer*, 11.

79 James A. Secord, *Victorian Sensation. The Extraordinary Publication, Reception, and Secret Authorship of "Vestiges of the Natural History of Creation"* (Chicago: University of Chicago Press, 2000).

appropriation, here I have explored samples of the vast literature of the age on Spencer; whereas it was working on quantities that I was able to discover the existence of a general public interested in Spencer.

Studies that in some cases have served as basic reference points in these pages have often called Spencer "an author willingly neglected;"[80] a Spencer who in the twentieth century saw his fame "practically disappear [with the] collapse of his reputation;"[81] or else "put to one side by the theorists, and not without some basic reasons, [but] wrongly neglected by historians;"[82] a Spencer who was "a classic of the human sciences often misunderstood and today by now forgotten;"[83] a Spencer (in Italy and in France) "immensely popular during the second half of the nineteenth century, yet he quickly fell out of favour."[84] This is the image originally provided by Zini, Croce, Gentile and others, an 'affirmation' that during the twentieth century evolved into a 'fact.' Quantifying the number of translations and reprints of Spencer's books allows us to deconstruct the myth of a forgotten Spencer in the twentieth century: if the publishers encouraged new translations and reprints, this means that, even within limits of the Italian well educated public, Spencer continued to arouse interest.

We can highlight different phases in the circulation of Spencer's works in Italy. A first, successful wave started in the 1870s, and was supported mainly by scientists. That same wave continued until the 1890s with the support of a number of politicians, economists, writers, journalists and the well educated public. Spencer's politically reassuring system of thought, the utilitarian nature of his thinking, and in particular the centrality of personal liberty held so many attractions. The Italians for centuries had had little familiarity with the practice, if not the concept of liberty; a liberty wished for in a country conditioned by foreign domination and by the censorship of the Catholic Church. The evolution over the years of Spencer's political thinking fostered its appropriation both by writers of a liberal or conservative orientation (as with the evolutionist Mantegazza, or the anti-evolutionist Martello), and by writers at that time

80 Giorgio Porcelli, *Identità in frammenti: Prospettive globali di sociologia della conoscenza* (Milan: Angeli, 2005), 31.

81 Toscano, *Malgrado la storia*, 13 and 31.

82 Giorgio Lanaro, *L'evoluzione, il progresso e la società industriale: Un profilo di Herbert Spencer* (Florence: Nuova Italia, 1997), 4.

83 Lino Rossi, *Sviluppo umano e progresso dell'umanità: Herbert Spencer e le scienze sociali* (Naples: Scriptaweb, 2010), from the blurb.

84 From the presentation of Naomi Beck, *La gauche évolutionniste: Spencer et ses lecteurs en France et en Italie* (Presses universitaires de Franche Comté, forthcoming). Accessed January 30, 2014 <http://www.mwpweb.eu/NaomiBeck/publication_745.html>.

socialist (like Ferri and others). A second phase of the Spencer effect, from around 1900 to the 1920s, saw a drop in scholars' interest in Spencer, while lay readers continued to read his books. Then fascism arrived, and Spencer disappeared from the publishers' catalogues. There were several reasons for his disappearance, rooted in the censorship as well as the success of neo-idealism. In addition, people like Ferri, Orano, and Mussolini himself may well have played a role. Ferri had become prominent in internationally famous trials;[85] Orano, now a university professor, had been one of the founders of the Scuola Fascista di Giornalismo (Fascist School of Journalism); to say nothing of Mussolini. For them and their readers, Spencer remained the 'socialist' of whom that press of the 1890s had spoken. This was a Spencer who was now especially incompatible with fascist ideology for reasons ranging from the regime's obsession with the role of the state to its denial of individual liberties. It is no coincidence that in the same period Spencer remained of great interest to Gugliemo Salvadori, who continued to defend his *Maestro* in the name of that idea of personal liberty, which was central to his thought. Unlike Ferri and company, Salvadori could draw on a solid philosophical culture, and the limits of Spencer's thought which had so fascinated him in his youth were now clear to him; what was still precious in Spencer in his eyes, even more during the dictatorship than in previous decades, was that concept of liberty about which he had written so sorrowfully from his Swiss exile in the letter to Croce.

In the second half of the twentieth century, too, an interest in Spencer emerged whenever fears of a widespread, misconceived cult of the national state, or the need to reassert individual liberties, attracted the attention of educated elites. Over the last twenty years, Spencer has been presented again as one of the classics of liberal thought.[86] It will come as no surprise that in an Italian, 2013 reprint of *Social Statics* he was still presented as "the British philosopher, in his own lifetime certainly one of the best known intellectuals of the age, but largely forgotten after his death."[87]

85 Among the many sensational cases Ferri dealt with was Violet Gibson's attempt on Mussolini's life in 1926.

86 Herbert Spencer, *L' individuo contro lo stato*, intr. Antonio Martino, trans. Stefano Mirabelli (Rome: Bariletti, 1989); *La società senza stato: I fondatori del pensiero liberale*, ed. Nicola Iannello (Soveria Mannelli: Rubbettino, 2004). The most recent important reprint is Herbert Spencer, *Principi di sociologia*, ed. Franco Ferrarotti, 2 vols (Turin: UTET, [1967] 1988).

87 *Il diritto di ignorare lo Stato* [The Right to Ignore the State], 2013, ebook version, in the series I Classici del Pensiero Liberale, accessed January 30, 2014 <http://www.ilgiornale.it/news/cultura/spencer-massima-libert-ignorare-stato-928538.html>.

TABLE 10.1 *Number of new translations and reprints traced of Herbert Spencer's works in Italy. Research carried out by cross-checking the bibliographical data in* OPAC SBN *(Catalogo del Servizio Bibliotecario Nazionale) and in* MAI *Azalai (Meta*OPAC *Azalai Italiano).*

Years	New translations	Reprints	Anthologies
1876-1886	7	5	
1887-1897	12	11	
1898-1908	15	12	
1909-1922	8	16	
1935		1	
1937	1		
1945		1	
1946-1963		8	
1967-1989	3	6	5

TABLE 10.2 *Italian translations of some of Spencer's books (including several not mentioned elsewhere in this chapter), chosen among the more than one hundred titles of translations and/or reprints that appeared in Italian in the period 1876-1922.*

Years	Original	Title	Publisher	Translator	Comments
1876	*Education: intellectual, moral, and physical* (1861)	*Educazione intellettuale, morale e fisica*	Tip. Gazzetta d'Italia, Florence, 239 pages	Sofia Fortini Santarelli (professional transl.)	This transl. was later (1883) published by Barbéra, Florence; 13th repr. in 1921.
1876	*Education: intellectual, moral, and physical* (1861)	*La teoria dell'educazione*	Nobile, Caserta, 352 pages	New transl. by Salvatori Pizzi (politician and social reformer)	
1884	*Education: intellectual, moral, and physical* (1861)	*L'educazione intellettuale, morale e fisica*	Trevisini, Milan	New transl. by Luigi Cussini (professional transl.)	Intr. by Paolo Riccardi (anthropologist and pedagogist). A new rev. ed. in 1897; last repr. found in 1915 (not numbered).
1894	*Education: intellectual, moral, and physical* (1861)	*Dell'educazione intellettuale, morale e fisica*	Paravia, Turin, 18 pages	New transl. by Angelo Valdarnini (philosopher and pedagogist)	Intr. by A. Valdarnini. Last (not numberd) repr. found, 1935; new rev. ed. in 1901; 16th repr. of the 2nd ed. in 1916; last (not numbered) repr. in 1922. Other new tranls. followed, with repr.s until the 1970s.
1885	*The Man versus the State* (1884)	*L'individuo e lo stato*	Lapi editore, Città di Castello, 163 pages	Sofia Fortini Santarelli	Intr. by philosopher Giacomo Barzellotti. At least 4 repr.s, last found in 1901.

TABLE 10.2 Cont.

Years	Original	Title	Publisher	Translator	Comments
1888	*First Principles* (1861)	*Primi principi*	Dumolard, Milan, 417 pages	Transl. and ed. by Maria Sacchi (naturalist) and Giacomo Cattaneo (zoologist)	Several repr.s, the last, pref. by G. Egidi, in 1920. New transl. by Salvadori in 1901; several repr.s; last found (not numbered) in 1921.
1881-1889	*Principles of Sociology* (*vol. 1*)	*Principi di sociologia. La sociologia nella storia della scienza, nella religione e nel cosmo*	UTET, Turin, 3 vols	Antonio Salandra (professor of administrative low and politician, conser-vative) and G. Salvadori (philosopher)	Series, Biblioteca dell'Economista, ed. by economists and politician G. Boccardo (liberal, follower of J.S.Mill)
1897	*Domestic Institutions* (1874-75)	*Istituzioni domestiche*	Sandron, Palermo, 303 pages	Ferida Federici	Transl. rev. by philosopher F. Tocco
1905	*The principles of biology* (1864-7; rev. ed. 1898-99)	*Le basi della vita*	Bocca, Turin, 380 pages	Guglielmo Salvadori	1st Italian transl; last (not numbered) repr. in 1922
1922	*Political Institutions* (1882); *Ecclesiastical Institutions* (1885)	*Principi di sociologia: dati e induzioni della sociologia, relazioni domestiche* […]	Casa Editrice Lorenzo Rinfreschi, Padua, 205 pages	Gino Egidi	

CHAPTER 11

Education and Evolution: Appropriations of Herbert Spencer in Scandinavia, 1870–1920

Hans Henrik Hjermitslev

On the occasion of the centenary of the British naturalist Charles Darwin's birth in 1909, the Danish popularizer of science Jens Orten Bøving-Petersen (1864–1937) informed the readers of the radical-liberal newspaper *Politiken* that, "With Darwin and Spencer human thought finally reached the mountain peaks – the open spaces where the air is clean and clear, free from the mist of obscurantism down in the valley." According to Bøving-Petersen and many other Scandinavian freethinkers in the decades around 1900, Darwin and Herbert Spencer were closely linked and seen as scientific heroes and liberators of humankind. From the 1870s, Spencer had been embraced by a new generation of Scandinavian radical-liberal university students who met under the banner of the so-called 'Modern Breakthrough' introduced by the Danish radical-liberal literary critic Georg Brandes in 1871. The radical students challenged idealistic philosophy, conservative values and established faith, and even though Darwin was no doubt more famous than Spencer in 1909, Spencer was still counted among the household gods in radical-liberal circles. In this chapter, I will discuss how Spencer's ideas were appropriated in Denmark, Norway and Sweden. I will demonstrate that in spite of similarities, such as Spencer's appeal to radical students and intellectuals in all three countries, the appropriation of Spencer's ideas in Scandinavia was characterized by a remarkable heterogeneity due to the different national contexts.[1]

The Scandinavian countries are closely linked historically, linguistically and culturally. Norway was under Danish control until 1814 when the country became a relatively independent part of the Swedish kingdom. Only in 1905 did Norway gain full autonomy. During the nineteenth century the Scandinavian countries witnessed a gradual modernization. Democracy and freedom of speech were introduced and school reforms and adult education initiatives resulted in relatively high literary rates. Moreover, economic growth combined

1 Jens Orten Bøving-Petersen, "Charles Darwin," *Politiken*, February 12, 1909, 7–8; Hans H. Hjermitslev, "Danes commemorating Darwin: Apes and evolution at the 1909 anniversary," *Annals of Science* 67:4 (2010): 485–525.

 | DOI 10.1163/9789004264007_013

with technological developments within communication and publishing paved the way for a revolution in print culture. In the last quarter of the century newspapers, periodicals and popular book series were widely circulated in all segments of the Scandinavian societies. Among the most popular publications were inexpensive science book series such as *Frem* (Forward), which was issued in all the Scandinavian countries and reached the astonishing number of 156,000 subscribers in 1898, and the Norwegian *Bibliothek for de tusin hjem* (Library for a Thousand Homes), which published a translation of Spencer's *Education* in 1887.[2]

The Scandinavian languages are quite similar, and in the nineteenth century publications and ideas frequently crossed the Scandinavian borders. This was certainly the case in the period from 1870 to 1920, when Scandinavian intellectuals shared their thoughts on science, society and literature in letters, lectures and periodicals. In spite of these Scandinavian networks of communication, however, the appropriation of foreign ideas such as Charles Darwin's theory of evolution and Herbert Spencer's evolutionary philosophy were to a large extent determined by national and even more local contexts. Thus, different aspects of their work were emphasized and a variety of interpretations were constructed in order to serve specific purposes.

In the case of Spencer, the decisive contexts were indeed very local. The most important arenas were the national universities in Sweden, Denmark and Norway since the vast majority of interpreters of Spencer's ideas were university students, alumni and teachers. In the period under study there existed only one university in Denmark, University of Copenhagen from 1479, one in Norway, the Royal Frederik's University in Kristania (now Oslo) from 1811 and four in Sweden: the traditional universities in Uppsala from 1477 and in Lund from 1666 and the new university colleges in Stockholm and Gothenburg established in 1878 and 1891 respectively. Most of the commentators on Spencer's ideas were connected to philosophy departments. These departments were important, since all university students in Scandinavia at this time were enrolled at the departments of philosophy for compulsory philosophy classes. Philosophy professors were thus influential in disseminating

2 Liv Bliksrud, Geir Hestmark and Tarald Rasmussen, *Norsk idéhistorie, bind IV: Vitenskapens utfordringer* (Oslo: Aschehoug, 2002), 9–21; Tore Frängsmyr, *Svensk idéhistoria: Bildning och vetenskap under tusen år*, vol. II (Stockholm: Natur och Kultur, 2000), 9–209; Casper Andersen and Hans H. Hjermitslev, "Directing Public Interest: Danish Newspaper Science 1900–1903," *Centaurus* 51:2 (2009): 143–67; Hans H. Hjermitslev, Casper Andersen and Peter C. Kjærgaard, "Populærvidenskab og folkeoplysning," in *Dansk Naturvidenskabs historie III: Lys over landet 1850–1920*, ed. Peter C. Kjærgaard (Aarhus: Aarhus University Press, 2006), 345–74.

knowledge to students within a broad range of academic disciplines. Moreover, philosophy professors were often engaged in teaching and writing activities outside the universities, which meant that their views could become familiar to segments outside academia. This included their interpretations and translations of Spencer's work, which were often published by prominent publishing houses and circulated among educators and school teachers. In a few occasions Spencer's ideas also reached a lay audience through periodicals, popular book series, and novels. However, the philosophical traditions at the universities differed and this had a seminal influence on the different appropriations of Spencer's ideas in Denmark, Sweden and Norway. In Denmark, philosophers with positivist sympathies paved the way for Spencer's ideas from the 1870s, while the philosophy departments in Norway and Sweden remained strongholds for idealistic and conservative trains of thought well into the twentieth century.[3]

Introducing Spencer in Denmark

On 2 January 1875 the Danish literary critic Georg Brandes (1842–1927) wrote to the Danish philosopher Harald Høffding (1843–1931) and asked for an article on Herbert Spencer for Brandes' mouthpiece, the highbrow radical-liberal monthly *Det Nittende Aarhundrede* (The Nineteenth Century) produced by the major publishing house Gyldendal. Brandes generously offered Høffding 60 DKK a sheet (16 pages) which equaled the monthly wage of an unskilled laborer. Høffding did not disappoint him. A few months later a 24-page article on Spencer's philosophy, based on a lecture in the Association of Students at the University of Copenhagen, was published in the journal.[4]

3 Hans H. Hjermitslev, "Protestant Responses to Darwinism in Denmark, 1859–1914," *Journal of the History of Ideas* 72:2 (2011): 279–303; Heine Andersen, "Modtagelsen af Spencer i Danmark – i tiden før 1920," *Dansk Sociologi* 19:1 (2008): 103–31; Ulf Danielsson, "Darwinismens inträngande i Sverige," *Lychnos* 1963–64: 157–210, 1965–66: 261–334; Thore Lie, "The Introduction, Interpretation and Dissemination of Darwinism in Norway during the period 1860–1890," in *The Reception of Charles Darwin in Europe*, ed. Eve-Marie Engels and Thomas F. Glick (London: Continuum, 2008), 156–74.

4 Morten Borup, ed., *Georg og Edv. Brandes Brevveksling med Nordiske Forfattere og Videnskabsmænd*, vol. I (Copenhagen: Gyldendal, 1939), 334; Mathias Clasen et al., "Translation and Transition: The Danish Literary Response to Darwin," in *The Literary and Cultural Reception of Charles Darwin in Europe*, vol. III, ed. Thomas F. Glick and Elinor Shaffer (London: Bloomsbury Publishing, 2014), 128–59; Harald Høffding, "Herbert Spencer's Filosofi," *Det Nittende Aarhudrede* 1875: 106–30.

By 1875, both Brandes and Høffding were known as interpreters of British philosophy and science in Denmark. Brandes translated John Stuart Mill's *The Subjection of Women* in 1869 and *Utilitarianism* in 1872. At this time Brandes was introducing what he coined the 'Modern Breakthrough' in Danish cultural life advocating naturalism in literature, positivism in science and liberalism in politics. A number of young Scandinavian writers were part of Brandes' movement and produced some of the most famous works in nineteenth-century Scandinavian literature. With an outspoken self-confidence they declared a clear break with tradition and embraced Darwin and Spencer as providing the chief ideological material for their literary revolution. Among these authors was the botanist Jens Peter Jacobsen (1847–1885), who wrote a series of articles on Darwinism for the highbrow radical-liberal monthly *Nyt Dansk Maanedsskrift* (New Danish Monthly) in the early 1870s and translated Darwin's *Origin of Species* and *Descent of Man* into Danish in 1872 and 1875 respectively. Jacobsen translated *Origin of Species* from the fifth edition of 1869 in which Darwin counted Spencer among his sources of inspiration and applied Spencer's concept of 'survival of the fittest' in his explanation of his theory of evolution by natural selection.[5]

Even though Høffding shared Brandes' liberal and positivist outlook, he never became part of the Brandes circle. Høffding, who achieved a MA in Theology in 1865 before turning to philosophy, was offended by Brandes' strong anticlericalism and later by his Nietzschean individualism and aristocratic radicalism, a position Brandes shared with other Scandinavian authors such as August Strindberg in the 1890s. Høffding defended what he termed democratic radicalism based on utilitarian welfare ethics and public enlightenment.[6]

In the 1870s, however, Høffding and Brandes were united in disseminating modern British thinkers to a Danish audience. In 1873 Høffding lectured on contemporary British philosophers including Mill and Spencer at the University of Copenhagen. This resulted in the work *Den engelske Philosophi i vor Tid* (English Philosophy in our Times) published in 1874, when he also wrote an article on the relationship between philosophy and Darwinism published in a popular weekly (Fig. 11.1).[7]

---

5 Charles Darwin, *Om Arternes Oprindelse ved kvalitetsvalg eller ved de heldigst stillede Formers Sejr i Kampen for Tilværelsen*, trans. Jens Peter Jacobsen (Copenhagen: Gyldendal, 1872); Charles Darwin, *Menneskets Afstamning og Parringsvalget*, trans. Jens Peter Jacobsen (Copenhagen: Gyldendal, 1875); Clasen et al., "Translation and Transition," 105–14.

6 Hans H. Hjermitslev, "Det store menneske – Studier i den tidlige danske Nietzsche-reception," in *Skæbne, magt og vilje*, ed. Pia Rose Böwadt and Ole Morsing (Copenhagen: Anis, 2006), 13–43.

7 Harald Høffding, *Den engelske Philosophie i vor Tid* (Copenhagen: P.G. Philipsen, 1874); Harald Høffding, "Filosofien og Darwinismen," *Nær og Fjern* 3:93–94 (1874); 1–4, 4–7.

FIGURE 11.1
Harald Høffding (1843–1931). As a professor of philosophy at the University of Copenhagen in Denmark, Harald Høffding was the most influential popularizer of Herbert Spencer's ideas in Scandinavia. Even though he remained skeptical of aspects of Spencer's thought, Høffding translated, outlined and commented on Spencer's philosophy extensively from the 1870s and included a substantial chapter on Spencer in his internationally acclaimed A History of Modern Philosophy *published by Macmillan in 1908.*

In the late 1870s Høffding devoted much of his time to translating Spencer into Danish. This was a welcome financial and intellectual supplement to his work as reader of philosophy and school teacher by which he made his living until he was appointed professor of philosophy at the University of Copenhagen in 1883. In 1876 Høffding's translation of *Education* was issued by the middle-sized publishing company Høst & Søn. It was priced at 2.50 DKK, which was a modest price compared to Jacobsen's 1872-translation of Darwin's *Origin of Species*, which Gyldendal sent to the market in 1,500 copies at the price of 8.00 DKK. The print run of the translation of *Education* is not available, but it is clear that Høffding's translation was widely reviewed and disseminated in Denmark, and in 1895 a second edition was issued. With the approval of Spencer, Høffding omitted the first, and in other countries very influential, essay of the original edition of 1861 in his translation. According to Høffding's foreword this was requested by the publisher, who wished to keep the costs and the price as low as possible. In 1877 the publisher used the same strategy when issuing a translation of the German evolutionist Ernst Haeckel's *Natural History of Creation*, which was reduced from the original 688 pages to 279 pages and priced at 3.00 DKK. However, excluding the essay "What knowledge is of most worth?" probably suited Høffding well, since its criticism of traditional education and studies of ancient languages and cultures seemed to undermine Høffding's own profession as a teacher of Greek and Latin and a scholar of ancient philosophy. Høffding's translation of *Education* was very influential among Danish educational thinkers in the last decades of the nineteenth century and became a standard reference in the pedagogical textbooks published at this time.[8]

The cooperation between Høffding and Høst & Søn continued shortly after the publication of the translation of *Education* when Høffding commenced to translate a handful of Spencer's essays from the 1850s and 1860s into Danish. The volume was published in 1878 and included the central evolutionary tracts "The Development Hypothesis" (1852) and "Progress: Its Law and Cause" (1857), originally published before Darwin's *Origin of Species* made transmutation of species and evolution the talk of the town. A biographical sketch sent to Høffding by Spencer, and originally published in the American journal *Popular Science Monthly* in March 1876, supplemented the translations.[9]

While translating Spencer, Høffding developed his version of an empirically-based ethical theory. In 1876 Høst & Søn published his work *Den Humane Ethik* (Human Ethics) in which Høffding drew on Spencer's *Social Statics* and on Darwin's newly translated *Descent of Man* when explaining how sympathy developed through human evolution. Høffding based his ethics on purely naturalistic principles and thus distanced himself from the tradition of Christian ethics prevailing at this time. In his ethical synthesis, Høffding combined Spencer's evolutionary perspective with an original interpretation of Mill's utilitarian ideas including an appeal to social justice through political reform. Thus, Høffding claimed that the ethical aim was to create as much public welfare and progress as possible. Small wonder that Høffding was a source of inspiration for the politicians, public officers and academics who developed the principles of the Scandinavian welfare states in the decades around 1900. In spite of their contrasting assessments of the value of the state, Spencer and Høffding shared an optimistic view of the future of mankind and believed that the egoism manifest in the struggle for existence would gradually be replaced by the harmony of human altruism.[10]

In 1882 Høffding's main work on psychology was issued. It was published in eleven editions, widely read in the Scandinavian countries and translated into several foreign languages. Høffding's psychology was based on British empiri-

---

8 Clasen et al., "Translation and Transition;" Andersen, "Modtagelsen af Spencer i Danmark;" Carl Henrik Koch, *Dansk filosofi i positivismens tidsalder 1880–1950* (Copenhagen: Gyldendal, 2004), 15–87; Ernst Haeckel, *Skabelseshistorien fra et naturvidenskabeligt Standpunkt*, trans. anon (Copenhagen: Høst & Søn, 1877). See Table 11.1.

9 For translations, see Table 11.1.

10 Harald Høffding, *Den Humane Ethik* (Copenhagen: Høst & Søn, 1876). See also Harald Høffding, *Etik* (Copenhagen: Det Nordiske Forlag, 1887) which was published in five editions, sold in more than 6,000 copies and translated into French and German. Koch, *Dansk filosofi i positivismens tidsalder*, 41–69.

cism and Immanuel Kant's transcendental philosophy, but in fact the most cited work was *Principles of Psychology* (1855). This book was bought by the Royal Library of Copenhagen in 1869 and discussed among philosophers in Copenhagen from around 1870. While Høffding drew on Spencer's investigations when explaining specific emotions and expressions such as sympathy and laughter, he was critical of Spencer's evolutionary explanation of the development of the fundamental elements of the human mind as a process of adaptation and inheritance through generations, which Høffding compared with Plato's concept of cognition as memory from preexistence. Following Kant, Høffding argued that there existed a priori conditions on which all experience must be based, and that these fundamental conditions could not be explained by Spencer's evolutionary psychology.[11]

In sum, Spencer's thought was introduced in Scandinavia in the 1870s by the Danish philosopher Harald Høffding, who translated several works by Spencer and applied, discussed and criticized aspects of Spencer's ethics and psychology in the process of designing his own independent philosophy. In this way, Høffding exploited the liberal atmosphere among academics and publishers in Copenhagen in order to strengthen naturalist and positivist ideas (see Table 11.1 at the end of this chapter).

Høffding's Interpretation of Spencer Abroad

In his influential chair as professor of philosophy from 1883 to 1915, Høffding continued to discuss and popularize evolutionary theory. In 1889 he wrote a small introduction to Darwin's life and work published in a popular book series issued by the radical-liberal Society of Students, and in 1908 his views on the philosophical consequences of evolution became accessible for an international audience when his two-volume work *A History of Modern Philosophy* was translated into English. In 1909 his interpretation of evolution was summed up in a chapter on "The Influence of the Conception of Evolution on Modern Philosophy" in the prestigious volume *Darwin and Modern Science* published by the University of Cambridge Press in commemoration of the centenary of Darwin's birth.[12]

---

11 Harald Høffding, *Psykologi i Omrids paa Grundlag af Erfaring* (Copenhagen: P.G. Philipsen, 1882); Andersen, "Modtagelsen af Spencer i Danmark," 107; Koch, *Dansk filosofi i positivismens tidsalder*, 41–60; Danielsson, "Darwinismens inträngande i Sverige," 313.

12 Harald Høffding, *Charles Darwin* (Copenhagen: Studentersamfundets Smaaskrifter, 1889); Harald Høffding, *A History of Modern Philosophy*, vols. I-II (London: Macmillan, 1908);

In his history of philosophy Høffding devoted 14 pages to Darwin and no less than 44 pages to Spencer, while in his commemorative chapter Høffding naturally focused on Darwin with Spencer as runner-up among many other philosophers. In Høffding's eyes, Darwin and Spencer were responsible for making a developmental and evolutionary view of life dominant. Their emphasis on struggle in nature broke down any idealist interpretation of nature and society. Thus Darwin's theory of struggle for existence and Spencer's concept of survival of the fittest incited people to discuss the foundations of human life and moral values in naturalistic terms. According to Høffding, natural selection in nature was not an argument for brutality and cruelty among humans. Criticizing Friedrich Nietzsche's philosophy, Høffding argued that good and evil could not be deduced from nature and that Darwinism, therefore, should not be used as an argument against humanity and morality. In Høffding's view, Darwin and Spencer's natural explanation of the evolution of moral sentiments could not determine the actual validity of moral principles. The validity of moral principles, according to the utilitarian Høffding, should instead be judged by their ability to increase welfare in society. Høffding was cautious to distance himself from materialist and atheist interpretations of Darwinism. He explicitly denied that Darwin and Spencer were materialists, and he supported their reluctance to talk about first causes and the origin of life. Høffding thus honored Darwin and Spencer for realizing the limits of scientific knowledge.[13]

In spite of his positive evaluation of Spencer, Høffding emphasized in an eloge in 1904 that "I have never been a Spencerian."[14] Thus, while appropriating Spencer's evolutionary perspective and applying his empirical methodology when developing his own ideas, Høffding remained critical of certain aspects of Spencer's thinking. As noted earlier, in the sphere of politics, Høffding distanced himself from Spencer's strong individualism and laissez-faire liberalism, and sided with the social democrats in the development of the Scandinavian welfare states. Moreover, in his detailed account and analysis of Spencer's philosophy of science, biology, psychology, sociology and ethics in *A History of*

Harald Høffding, "The Influence of the Conception of Evolution on Modern Philosophy," in *Darwin and Modern Science*, ed. Albert C. Seward (Cambridge: University of Cambridge Press, 1909), 446–64; Hjermitslev, "Danes commemorating Darwin," 494–95.

13 Høffding, *A History of Modern Philosophy*, vol. II, 434–85; Høffding, "The Influence of the Conception of Evolution on Modern Philosophy;" Hjermitslev, "Danes commemorating Darwin," 495.

14 Harald Høffding, "Herbert Spencer," *Tilskueren* 1904: 7. All translations are made by the author.

Modern Philosophy, Høffding repeated some of the philosophical reservations that he had put forth back in the 1870s and again in the eloge in 1904.[15]

According to Høffding, Spencer's concept of 'the Unknowable,' which led him to a reconciliation of religion and science, was illusionary and inconsistent. No adherents to a positive religion would accept that their dogmas should be reduced to a mere mystery that transcends all human experience. Høffding, who coined his own philosophy 'critical monism,' criticized the epistemological dualism between the relative and the Absolute inherent in Spencer's philosophy. If 'the Unknowable,' which Spencer regarded as identical with 'the Absolute,' is the first cause of all phenomena then it must "reveal itself in the forms and laws under which phenomena occur; it cannot really be entirely unknowable," Høffding instructed the readers. He added that "Spencer is wrong in pronouncing the leading concept of experience, *i.e.* the concept of evolution, to have no validity when applied to 'the Absolute.' It would be a contradiction if the laws and forms revealed in all experience were entirely meaningless in relation to that which lies behind or on the other side of experience."[16] Thus, Høffding did not embrace the relativistic currents of Spencer's *First Principles* and did not become an uncritical advocate of Spencer's synthetic philosophy. Even though both Høffding and Spencer advocated positivism, agnosticism, and evolutionism they differed in the philosophical details.[17]

Spencerian Philosophy and Sociology in Copenhagen

Høffding was no doubt the most important popularizer of Spencer's ideas among theologians and philosophers in Denmark. He was, however, not the only one. Two of Høffding's students, the non-believing theologians Søren Hansen and Karl Gjellerup (1857–1919), also had their share in the wide circulation of Spencerian ideas in the university environment in Copenhagen. In 1887 Hansen translated into Danish the first five chapters of *First Principles*, which discussed 'the Unknowable,' at the request of Høffding, who wrote a short foreword to the translation. Høffding expected the translation to be of interest to Danish readers since it focused on the relationship between science and

15 Høffding, "Herbert Spencer's Filosofi;" Høffding, "Herbert Spencer;" Høffding, *A History of Modern Philosophy*, vol. II, 434–85.

16 Høffding, *A History of Modern Philosophy*, vol. II, 462.

17 Høffding, *A History of Modern Philosophy*, vol. II, 434–85; Koch, *Dansk filosofi i positivismens tidsalder*, 42–45.

religion, which was much debated in Denmark at this time. In the footnotes of the translation, Hansen was eager to explain the philosophical vocabulary applied by Spencer, and when in 1903 Hansen defended his thesis *Etikens Begrundelse* (The Foundations of Ethics) the influence from Spencer (and Høffding) was evident. Hansen advocated a naturalistic view on ethics by arguing that ethics should be based on results from the sciences of physiology, psychology and sociology.[18]

Two decades before Hansen's thesis, Karl Gjellerup defended the same line of thought in his prize paper *Arvelighed og Moral* (Heredity and Morality) for which he was awarded the University of Copenhagen Gold Medal in 1881. Gjellerup was a devoted disciple of Brandes and wrote the paper at Høffding's request. In the late 1880s, however, Gjellerup broke with Brandes and distanced himself from naturalism. Instead he turned to Arthur Schopenhauer's pessimistic philosophy, Buddhism, Indian spiritualism and Christian mysticism. He left Denmark for Germany in 1892 and was soon reduced to a marginal figure on the Danish literary scene. Consequently, when Gjellerup in 1917 shared the Nobel Prize in Literature with his countryman, the renowned social realist Henrik Pontoppidan, it came as a surprise to many Danes.[19]

In his prize paper, however, Gjellerup drew heavily on Darwin, Spencer, Francis Galton and his mentor Høffding. Gjellerup argued that both physical and psychological traits were hereditary, and in line with Spencer he emphasized the role of inheritance of acquired characteristics. On the basis of his analysis of recent studies of heredity, Gjellerup defended a deterministic, naturalistic and evolutionary view on morals and backed his arguments by references to Spencer's works *Principles of Psychology* (1855) and *Principles of Biology* (1864–67). Like Spencer, Gjellerup had great hopes for the future. He was convinced that through the progressive laws of evolution eventually the struggle for existence among humans would be replaced by a mutual feeling of sympathy and happiness. In the concluding chapter Gjellerup put forth a concept of 'evolutionary religiousness' by praising these fortunate consequences of the evolutionary process, which he viewed as a scientifically sound alternative to outdated positive religion.[20]

In sum, Spencer was among the most debated foreign thinkers among a new generation of philosophers in Denmark in the 1870s and 1880s, when German

---

18 Søren Hansen, *Etikkens Begrundelse* (Copenhagen: Lehmann & Stage, 1903); Andersen, 'Modtagelsen af Spencer i Danmark,' 114; Table 11.1.

19 Karl Gjellerup, *Arvelighed og Moral* (Copenhagen: Andr. Schou, 1881); Clasen et al., "Translation and Transition," 120–22.

20 Gjellerup, *Arvelighed og Moral*; Andersen, "Modtagelsen af Spencer i Danmark," 112–3.

idealist philosophy was challenged by positivist ideas primarily from Britain. Spencer's ideas were not kept inside the walls of the University of Copenhagen. Høffding thus played a seminal role in popularizing evolutionary theory including Spencer's philosophy through lectures, articles, books and translations. In addition, Høffding's colleagues as professors of philosophy at the University of Copenhagen in the decades around 1900, Kristian Kroman (1846–1925), Claudius Wilkens (1844–1929), and Høffding's successor, Carl Nicolai Starcke (1858–1926), all held Spencer's version of positivism and agnosticism in high esteem and drew on his ideas in their work which besides philosophy also included sociology, psychology and pedagogics. Thus, Danish philosophy around 1900 was in many respects Spencerian in its outlook. In the first Danish introduction to sociology *Samfundslegemets Grundlove* (The Basic Laws of the Social Body) from 1881, Wilkens embraced the Spencerian idea of social progress through adaptation and survival of the fittest and he applied Spencer's basic concepts of differentiation and specialization. Starcke, in turn, discussed Spencer's analyses of the development of family types in *Principles of Sociology* in his main anthropological work *Die primitive Familie* from 1888, which was published in German and translated into English and French.[21]

In his study of the reception of Spencer in Denmark, Heine Andersen has demonstrated how academics promoting the new scientific discipline of sociology drew on Spencer. In the journal *Nationaløkonomisk Tidsskrift* (Journal of National Economics) Spencer's sociological works were outlined and praised from the late 1870s. The appropriation of Spencer focused on his view of society as an organism and on his evolutionary perspective. Spencer's liberalism and individualism was downplayed in favor of a collectivist view of society, which was more in line with the initial ideals of a welfare state that was discussed among social scientists, public officers and politicians at this time. Through political compromises between the conservative government and the liberal opposition the state and the municipals, rather than the market and civil society, became increasingly responsible for social security such as old-age pensions. Thus, Spencer's laissez-faire liberalism was marginalized and did not gain much ground in Denmark. Among many liberal thinkers and politicians, such as the moral philosopher Severin Christensen (1867–1933) and the above-mentioned Starcke, the American economist and politician Henry George's social liberalism including land-value taxation was preferred to

21 Andersen, "Modtagelsen af Spencer i Danmark;" Koch, *Dansk filosofi i positivismens tidsalder*, 31–167; Claudius Wilkens, *Samfundslegemets Grundlove* (Copenhagen: Wroblewsky, 1881); Carl Nicolai Starcke, *Die primitive Familie in ihrer Entstehung und Entwickelung* (Leipzig: Brockhaus, 1888).

Spencer's individualistic liberalism. In fact, George's followers in Denmark, who founded the Justice Party in 1919, referred to Spencer's work *Social Statics* (1850), in which Spencer criticized private property of land, when arguing their case, and they taunted him with his retreat from his original view in polemics with George on the Irish land question in the 1880s and 1890s. Moreover, Spencer's anti-militarism and pacifism were embraced by radical-liberal intellectuals in Denmark such as Georg Brandes, who campaigned against nationalism and militarism in the first decades of the twentieth century, and the highbrow liberal journal *Tilskueren* which honored Spencer shortly after his death by publishing an extract of his anti-militarist piece "Re-Barbarization."[22]

Thus, the Danish versions of Spencer were constructed in order to suit a variety of purposes. While nobody seems to have bought the whole package of Spencer's synthetic philosophy, his ideas were exceptionally useful for philosophers interested in promoting certain topics such as agnosticism, naturalistic ethics, sociology, land-value taxation and anti-militarism. Indeed, academics were eager to pay tribute to Spencer. In 1889 Høffding succeeded in getting Spencer elected a fellow of the Royal Danish Academy of Sciences and Letters – 10 years after Darwin had been awarded a fellowship. To the surprise of the academy, however, Spencer declined the offer in a three-page letter in which he argued that these marks of honor "though intended as encouragements to science and literature, they are in their indirect effects impediments to science and literature." Maybe the academy should not have been so surprised, since it was not the first time Spencer declined an offer to become a fellow of a national academy. He was awarded 32 honorary positions throughout his life, but declined practically all of them, including fellowships of the Royal Society and Institut de France.[23]

---

22 Andersen, "Modtagelsen af Spencer i Danmark," 115–21; Jørn Henrik Petersen, Klaus Petersen and Niels Finn Christiansen, *Dansk Velfærdshistorie I: Frem mod socialhjælpsstaten* (Odense: Syddansk Universitetsforlag, 2010); Severin Christensen, *Herbert Spencer contra Henry George: En Verdenshistorisk Tvekamp*, special edition of *Det ny Aarhundrede*, 1909; Henry George, *Om Erstatning: Uddrag af Henry Georges Svar til Herbert Spencer ("en forvirret filosof")*, trans. P. Larsen, 2nd edition (Copenhagen: Henry George Forlaget, 1919); Koch, *Dansk filosofi i positivismens tidsalder*, 379–93; Jørgen Knudsen, ed., *Georg Brandes – den mangfoldige* (Copenhagen: Gyldendal, 2005); Table 11.1.

23 Andersen, "Modtagelsen af Spencer i Danmark," 106, 112 (quote); Hjermitslev, "Danes commemorating Darwin," 493n.

Spencer among the Swedes

At the time when Høffding and his students and colleagues popularized Spencer's philosophy in Denmark, his ideas were also discussed and his works translated across the sound in Sweden. However, the appropriations of Spencer in Denmark and Sweden took place in different national contexts and had little to do with each other. While a 'Modern Breakthrough' in literature and philosophy, introducing positivist and naturalistic ideas and challenging idealistic philosophy and established Lutheran faith, took place in the early 1870s in Denmark, it only happened in the 1880s in Sweden due to the strong conservatism of the intellectual elite.[24]

During the 'Modern Breakthrough' in Sweden in the 1880s, a young generation of radical students at the University of Uppsala near the capital of Stockholm, including the novelist and playwright August Strindberg (1849–1912) and the philosopher Allen Vannérus (1862–1946), confronted conservative professors by advocating free thought and naturalism. In 1882 they established the society *Verdandi*, which soon became a vibrant meeting place for people eager to discuss modern science and literature and challenge the established order. The radical Uppsala-students were inspired by Georg Brandes in Denmark and like him they discussed and popularized ideas from Britain, such as liberalism and agnosticism, and authors such as Mill, Darwin, Spencer and Henry Thomas Buckle. The debate between conservative academics and radical students generated much interest in Sweden in the 1880s and resulted in a large number of translations of British works, including Spencer's *Study of Sociology*, *Education*, *First Principles*, *Data of Ethics* and a collection of his *Essays*.[25]

However, already in 1873 the Swedes could become acquainted with Spencer through an article in the highbrow journal *Samtiden* (Contemporary Times). At this time, Darwin's theory of evolution was embraced by a number of young zoologists and botanists who popularized evolutionary theory in the periodical press. Furthermore, Darwin's *Origin of Species* and *Descent of Man* were available in Swedish translations from 1871 and 1872 respectively. These publications, however, did not generate the same public debate as did the promotion of evolution by Brandes, Jacobsen, and Høffding in Denmark. While evolution was a relatively widely discussed topic among theologians and philosophers in

24 Sven-Eric Liedman, "Sweden and Victorian Britain," *The European Legacy: Toward New Paradigms* 4:6 (1999): 72–83.

25 Liedman, "Sweden and Victorian Britain," 72–6; Danielsson, "Darwinismens inträngande i Sverige," 317; Table 11.1.

Denmark in the 1870s, the evolutionary debate in Sweden was more or less restricted to professional scientists until around 1880.[26]

In 1879 the Swedish physician and public educator Anton Nyström (1842–1931) published his work *Positivismen* (Positivism) in which he outlined the views of Darwin, Spencer and Comte among others. Nyström sided with Auguste Comte and criticized Spencer for his teleological and evolutionary views. He accused him of clinging to metaphysical tendencies and being a de facto defender of theism. As early as 1864 Nyström had defended Darwinism at a meeting at the University of Uppsala, but in the 1870s he became a devoted disciple of Comte, and in 1879 he established a positivist society in Stockholm in order to popularize a religion of humanity to all segments of society. In 1880 Nyström went on to establish *Stockholms Arbetareinstitut* (Working Man's Institute of Stockholm), which disseminated modern science and arts to the working classes in line with Comte's dictate. Nyström's initiatives were quite successful and made him a well-known and controversial figure among both workers and academics in Sweden. His introduction of positivist science and philosophy thus played an important part in the 'Modern Breakthrough' in the 1880s. In a pamphlet published in 1887, Nyström repeated his criticism of Spencer's agnosticism, which Nyström regarded as an unwelcome safe haven for theologians eager to defend their faith against a modern scientific worldview. True, that year bishop Anders Fredrik Beckman (1812–94) indeed used Spencer's concept of the Absolute as an argument for the existence of a Christian God.[27]

In 1879 the Swedish army lieutenant Richert von Koch (1838–1913) wrote an introduction to Spencer's views on education. Progressive educational thinkers such as the feminist Ellen Key (1849–1926), who received Spencer more positively than academic philosophers, read von Koch's introduction. Key's pedagogical ideas were strongly influenced by Spencer and reached a large audience through her 1900-work *Barnets Århundrade* (The Century of the Child), which was also read by Danish and Norwegian educators and later translated into Danish, German, and English. In line with Spencer, Key argued that important biological and intellectual differences between the sexes had evolved through mankind's struggle for existence, and consequently feminists should focus on advocating the inherited female talent of nursery and

---

26 Danielsson, "Darwinismens inträngande i Sverige," 327–34; Lie, "The Introduction, Interpretation and Dissemination of Darwinism in Norway during the period 1860–1890," 171.

27 Anton Nyström, *Positivismen* (Stockholm, 1879); Anton Nyström, *Positivismen och Herbert Spencer* (Stockholm, 1887); Danielsson, "Darwinismens inträngande i Sverige," 312–14; Frängsmyr, *Svensk idéhistoria*, 119, 137–40.

education instead of attempting to copy male behavior. Small wonder, this argument was seen as reactionary by many of her fellow feminists.[28]

In the 1880s Spencer became a household name among radical students in Sweden. Inspired by Georg Brandes in Denmark they advocated naturalism and positivism and challenged the strong conservative intellectual elite. This debate generated considerable financial interest among publishers, who issued many translations of progressive British thinkers, including five works by Spencer. Due to the campaign by the educational thinker and feminist Ellen Key around 1900, it was primarily his educational ideas that reached a large audience.

The Philosophical Debate in Sweden

Until the 1880s when Swedish philosophers were challenged by their radical students, the philosophical environment and the learned and official classes in Uppsala, Lund and Stockholm were totally dominated by thoughts developed by Christoffer Jacob Boström (1797–1866), who served as professor of practical philosophy in Uppsala from 1842 to 1863. Boström, who went by the name 'the Swedish Plato' but remained unknown outside his homeland, was an extreme idealist. Unlike the German philosopher Georg Wilhelm Friedrich Hegel, who was influential among philosophers in Copenhagen and Kristania at this time, Boström denied any connection between the real world of ideas and the illusionary world of phenomena, and therefore he was very critical of the validity of the results brought forth by the empirical sciences. According to Boström the real world was a hierarchy of ideas with God at the top. The role of philosophy was to explain this hierarchy through a process of deduction based on logical reasoning. One of the results of Boström's deductions was to establish society as a conservative order with God, the State, the Monarch and the public officers at the top and the people at the bottom. Consequently, Boström declined any liberal reforms and remained an opponent of democracy and individual freedom until his death. In the 1860s and 1870s at least three out of four university professors of philosophy in Uppsala and Lund were adherents of Boström. Even though Boström's influence decreased after the 'Modern

28 Richert von Koch, *Uppfostran: En Fremställning af Herbert Spencers Åsigter derom* (Stockholm: Norstedt, 1879); Danielsson, "Darwinismens inträngande i Sverige," 312; Frängsmyr, *Svensk idéhistoria*, 204–7; Ellen Key, *Barnets Århundrade* (Stockholm: Albert Bonniers Förlag, 1900); Anne Katrine Gjerløff and Anette Faye Jacobsen, *Dansk Skolehistorie III: Da skolen blev sat i system* (Aarhus: Aarhus University Press, 2014), 273–94.

Breakthrough' in the 1880s his legacy lived on in Sweden well into the first decades of the twentieth century.[29]

Small wonder that Spencer's liberalism and empiricism, disseminated by students connected to Verdandi, was seen as a challenge to Boströmian idealism at the universities in Uppsala and Lund. In 1883 Boström's successor as professor of practical philosophy in Uppsala, Carl Yngve Sahlin (1824–1917), picked up the gauntlet when he wrote the introduction to a translation of Spencer's chief philosophical work *First Principles* published by the newly established academic press Almqvist & Wicksell in Uppsala.[30] Sahlin was vice chancellor of the university at this time and arguably one of the most influential intellectuals in Sweden. Therefore, his introduction was seen as a conservative and idealist response to positivist and naturalistic ideas embraced by some of his rebel students. However, Sahlin's introduction was surprisingly conciliatory. He offered a balanced account of Spencer's main ideas of evolutionism and relativism, but stressed that real philosophy only begins where Spencer's places the limits of positive knowledge. Spencer's empirically-based philosophy was therefore secondary to Broströmian idealism according to Sahlin, who emphasized that the object of real philosophical thinking was absolute knowledge, which Spencer was only capable of reducing to 'the Unknown.' Thus, in the words of intellectual historian Sven-Eric Liedman, "Sahlin's message may have been that contrary to what some critics might believe, idealism does not compete with science but provides a deeper understanding of the facts and laws that science offers."[31] In this way, Sahlin attempted to transform Spencer's controversial positivism and evolutionism into what historian of science Jonathan Topham has coined 'safe science' in order to defend the idealistic tradition of Swedish philosophy.[32]

Indeed, Sahlin seemed to have been quite successful in his attempt to marginalize British positivism and liberalism among philosophers and other intellectuals in Sweden. One of his students, Vitalis Norström (1856–1916) for example, campaigned against materialism and naturalism, including Spencer, from the late 1880s. When in 1893 Norström became professor at the newly established university in Gothenburg, he became a renowned conservative critic of radicalism and especially of the feminist Ellen Key's pedagogical and

29 Liedman, "Sweden and Victorian Britain," 72–3; Frängsmyr, *Svensk idéhistoria*, 31–6, 162–66.

30 Table 11.1.

31 Liedman, "Sweden and Victorian Britain," 73.

32 Jonathan Topham, "Science and Popular Education in the 1830s: The Role of the 'Bridgewater Treatises,'" *The British Journal for the History of Science* 25:4 (1992): 397–430.

political ideas. Key applied Spencer's evolutionism and liberalism in her struggle for political and educational reforms and thereby challenged the established order that Norström defended. According to Norström, Spencer's naturalistic ethics were dangerous for society since they did not include any absolute principle such as a Christian God or a Kantian notion of duty.[33]

In general, Swedish professional philosophers still embedded in Boströmian idealism ignored or remained very critical of Spencer throughout the nineteenth century. However, Johan Jacob Borelius (1823–1909), professor of philosophy from 1866 at the University of Lund, was a noteworthy exception. Borelius was one of the few Hegelians in Sweden which placed him in opposition to Boström and his adherents. Contrary to the static nature of Boströmian Platonic idealism, Hegelian idealism contained a dynamic and progressive view of the world, which was echoed in evolutionary theories. Borelius lectured on Spencer's *First Principles* in the academic year of 1887–88, and even though Borelius did not himself fully embrace Spencer's evolutionism, one of his students, Johan Bager-Sjögren, published an appreciative study of Herbert Spencer in 1893.[34]

Due to the dominance of Christoffer Jacob Boström's conservative idealism, Spencer's philosophy did not gain a foothold at universities in Sweden. Thus, unlike their Danish colleagues most Swedish philosophers remained very critical of Spencer's thought in the period from 1870 to 1920.

Bjørnstjerne Bjørnson on Evolution, Education, Marriage and Morals

In the third Scandinavian country, Norway, the conservative Hegelian Marcus Jacob Monrad (1816–97) served as an influential professor of philosophy. While Borelius's Hegelian sympathies and opposition to Boströmianism in Lund made room for Spencer's evolutionism, Monrad's right Hegelian outlook made Spencer unacceptable in Kristiania. Thus, Spencer did not generate much debate among Norwegian intellectuals until the 1880s, when the renowned

33 Frängsmyr, *Svensk idéhistoria*, 166–8; Danielsson, "Darwinismens inträngande i Sverige," 316.

34 Danielsson, "Darwinismens inträngande i Sverige," 309–11, 316–17; Johan Bager-Sjögren, *Herbert Spencer och Utvecklingsfilosofien* (Lund: Collin & Zickerman, 1893), a copy of the work in the holdings of the State Library in Aarhus, Denmark includes a handwritten dedication from the author to the Danish philosopher and Spencer expert Carl Nicolai Starcke.

Norwegian author Bjørnstjerne Bjørnson (1832–1910) applied Spencer in his campaign for liberal education. In his 1884-novel *Det flager i Byen og på Havnen* (The Flags are Flying in the Town and at the Harbor) Bjørnson's main character explicitly advocated Spencer's educational program and its focus on healthy living and upbringing, physical training and the teaching of natural science. Moreover, in 1887 Spencer's educational ideas received a large audience in Norway, when a full translation of *Education* was published in the popular book series *Bibliothek for de tusin hjem* aimed at working- and middle-class readers. To some Norwegians, however, Spencer's educational program was too radical. In his study of Spencer's educational thought, the Lutheran minister Christian A. Bugge, for example, warned against the dangers of excluding religion and discipline from moral education.[35]

Bjørnson was an outspoken freethinker inspired by Brandes and the 'Modern Breakthrough' in Denmark. In the 1870s he read the Danish translation of Darwin's *Descent of Man*, and in 1879 he officially announced his conversion from Christianity to Darwinism. However, Bjørnson was not interested in Darwin's biological details, but rather in a general theory of evolution as outlined by Spencer. Bjørnson took to heart the idea of progress through survival of the fittest and preached the evolutionary gospel in poems and letters. He wrote to Ellen Key that the theory of evolution must become a new religion, and in 1881 he told Brandes in a letter that he was much enriched by reading Spencer in English, and in his response Brandes suggested that Bjørnson read *Principles of Psychology*. However, the friendship between Brandes and Bjørnson came to an end in the middle of the 1880s, when Bjørnson used arguments drawn from Spencer in his criticism of promiscuity in a heated moral debate that included intellectuals from all of the Scandinavian countries. In a series of public lectures in Scandinavia, Bjørnson criticized Brandes' vision of free love which in turn made Brandes furious. In this way, Spencer's anthropological research on the evolutionary development of the institution of marriage from polygamy to monogamy played an important role in the break between two of the leading freethinkers in Scandinavia.[36]

35 Bjørnstjerne Bjørnson, *Det flager i Byen og på Havnen* (Copenhagen: Gyldendal, 1884); Christian A. Bugge, *Herbert Spencers opdragelseslære* (Kristania: Cammermeyer & Fabritius, 1889); Table 11.1.

36 Tore Rem, "Darwin and Norwegian Literature," in *The Literary and Cultural Reception of Charles Darwin in Europe*, vol. III, ed. Thomas F. Glick and Elinor Shaffer (London: Bloomsbury Publishing, 2014), 160–80; Morten Borup, ed., *Georg og Edv. Brandes Brevveksling med Nordiske Forfattere og Videnskabsmænd*, vol. IV.1 (Copenhagen: Gyldendal, 1939), 140–88; Bjørnstjerne Bjørnson, *Engifte og mangegifte* (Fagerstrand: Bibliothek for de tusin hjem, 1888); Bliksrud, Hestmark and Rasmussen, *Norsk idéhistorie, bind IV*, 251–56.

FIGURE 11.2
Bjørnstjerne Bjørnson (1832–1910). Bjørnson was among the most famous Norwegian novelists and poets in the second half of the nineteenth century. He began his intellectual life as a Christian disciple of the Danish theologian and educational thinker N.F.S. Grundtvig, but in 1879, influenced by the Danish literary critic Georg Brandes, he converted to Darwinism, which he regarded as a new religion. In the 1880s he became strongly influenced by Herbert Spencer's evolutionary and educational thought, which he advocated in novels and poems, and in letters to other Scandinavian intellectuals. © NATIONAL LIBRARY OF NORWAY. PHOTOGRAPHER: WARREN'S PORTRAITS (BOSTON 1880).

In Norway, no one embraced Spencer's evolutionary and educational ideas with more enthusiasm than the acclaimed author Bjørnstjerne Bjørnson. From 1879, he advocated Darwin's and Spencer's evolutionism, which he regarded as a new religion that should replace Christianity. However, in the 1880s it was Spencer's educational ideas, and not his evolutionary philosophy, that reached the largest audience in Norway through Bjørnson's fiction and a widely circulated translation of *Education* (Fig. 11.2).

Popular Appropriations of Spencer

In an eloge published in the liberal highbrow journal *Det ny Aarhundrede* (The New Century) in 1904, Carl Nicolai Starcke reflected upon Spencer's influence in Denmark compared to other countries:

> In England herself, in France and Germany, the Spencerian ideas are barely yet diffused to the common people, and their advance in the leading circles is constantly in danger of meeting opposition through the

> support which old transcendent and supernatural conceptions hold in common people's consciousness and instincts. At home, however, there is hardly any village teacher who has not had his share of the Spencerian thoughts at the college of education, hardly any man who has not, in his newspaper or his society, seen urgent social questions and the basis of everything that gives his life backbone and glory connected to ideas which are due to Spencer. Whether or not Spencer's name is mentioned, whether or not one agrees with him in the details, the entire spirit of the public debate, the leading views from which one departs, are Spencerian. It is welfare ethics and the belief in man's social destiny. From people's political home rule to opposition to corporal punishment one can see the same belief prevail. The belief that it is not compulsion, but evolution, not governing, but organizing, that matters.[37]

According to Starcke, Spencer's liberal political and educational views and his progressive evolutionary sociology thus strongly influenced the public debate in Denmark. His ideas were widely appropriated, interpreted, praised, criticized and applied in a variety of contexts for a variety of purposes. Even though the occasion of Spencer's death might have made Starcke exaggerate Spencer's direct and indirect influence – many other liberal thinkers also played an important part in the progressive atmosphere at this time – it seems to be a fair assessment that Spencer's ideas were influential among at least some segments of society. As we have seen, he was appropriated by philosophers and sociologists and discussed in highbrow journals and among intellectuals in Copenhagen. However, his name was known outside the chattering classes. Although he does not seem to have been as renowned among the working classes as Charles Darwin, whose portrait and theory made it to the front pages of popular newspapers and periodicals from the 1870s, many workers would have been familiar with Spencer's sociology through the philosopher Oscar Hansen who delivered a lecture series on evolutionary theory before an audience of between 1200 and 1500 workers during a lock-out in the summer of 1899. He repeated his lecture series in the winter of 1900–01 before turning the manuscript into a book published in cheap installments for subscription by the major publishing house Gyldendal.[38]

Among rural people in Denmark other aspects of Spencer's work than his sociological thoughts were emphasized. In 1890, the renowned headmaster of

37 Carl Nicolai Starcke, "Herbert Spencer," *Det ny Aarhundrede* 1903–4: 443–44.

38 Hjermitslev, "Danes commemorating Darwin;" Oscar Hansen, *Udviklingslære* (Copenhagen: Gyldendal, 1901).

the Grundtvigian people's high school in the village of Testrup in rural Jutland, the theologian Jens Nørregård, published a comparative study of the Danish theologian Nicolai Frederik Severin Grundtvig, the German philosopher Hermann Lotze and Spencer in which he argued that Spencer's agnosticism and science in general did not challenge established faith. This was a welcome conclusion for many of his fellow Christian believers who felt threatened by scientific theories such as the theory of evolution popularized by freethinkers such as Georg Brandes and Karl Gjellerup in Denmark and Ernst Haeckel in Germany.[39]

Nowhere did Spencer's ideas have such a deep and lasting impact than among educational thinkers and teachers in Scandinavia. His 1861-work *Education* was translated into all the Scandinavian languages in the 1870s and 1880s and the translations were widely circulated and discussed. The work was generally considered a much readable extract of modern, scientifically-based educational thought, and it provided progressive educational thinkers with useful arguments in their campaigns against traditional education, rote learning and corporal punishment.[40]

In 1879 the Danish-Norwegian pedagogical journal *Vor Ungdom* (Our Youth) was launched by the Danish philosopher and liberal politician Herman Trier (1845–1925) in cooperation with Danish and Norwegian educators. Trier was engaged in public enlightenment and adult education and was closely linked to the group of liberal Copenhagen-philosophers inspired by Spencer. The journal soon became a common Scandinavian platform for discussions of educational theories and politics. In the journal, Spencer was applied in the struggle for a reform of the education of teachers from theological and classical learning to scientific pedagogics based on psychology and physiology.[41]

In 1878, Trier had introduced "Pedagogics as a Science" to a Danish audience in a popular work that drew on Spencer in the sections on psychology and didactics and offered a detailed and sympathetic summary of Spencer's educational thought. This line was followed by the professors of philosophy Heegaard, Wilkens and Starcke, who were all engaged as readers of educational theory and enrolled as consultants and examiners at the commission that controlled

39 Jens Nørregård, *Studier over Spencer, Lotze og Grundtvig* (Copenhagen: Schønberg, 1890); Hjermitslev, "Protestant Responses to Darwinism in Denmark."

40 Vagn Skovgaard-Petersen, "De blev lærere – læreruddannelse mellem 1860 og 1945," in *For at blive en god lærer: Seminarier i to århundreder*, ed. Karen B. Braad et al (Odense: Syddansk Universitetsforlag, 2005), 207–11; Table 11.1.

41 Skovgaard-Petersen, "De blev lærere," 207–11; Andersen, "Modtagelsen af Spencer i Danmark," 121–24.

the Danish colleges of education. From 1880, the Ministry of Culture and Education requested philosophy professors to write pedagogical textbooks aimed at student teachers at colleges of education. Also other educational thinkers such as Oscar Hansen and Niels Bang wrote textbooks, which were an attractive commodity at the scientific marketplace. The textbooks generally emphasized Spencer's idea of natural education with a focus on self-development, utility, everyday experience and object lessons. Moreover, they agreed with Spencer that learning should be based on inclination rather than compulsion and that the teacher should be seen as a friend by the pupils. Spencer remained a standard reference in textbooks well into the twentieth century, even though his laissez-faire liberalism and his emphasis on individual self-development were seen as outdated among many educators from the 1910s when social issues made politicians in Scandinavia lean towards state-sanctioned social engineering with a focus on solidarity, citizenship and democracy.[42]

Through the pedagogical textbooks Spencer's educational thought was disseminated to thousands of Danish primary and secondary school teachers in the period from 1880 to 1920 and thus influenced the education of coming generations. However, Spencer's ideas were not received without criticism. Among authors of textbooks, Spencer's focus on utility and natural science was seen as one-sided and his radical anti-authoritarian pedagogics was met with some reservations. Moreover, theologically-trained teachers at colleges of education, many of whom were connected to the Grundtvigian movement, a liberal faction of the national Danish Evangelical-Lutheran Church, were not too impressed by Spencer's criticism of religious education and his secular ethics. Politically, the impact of the new educational ideas put forth by Spencer and other progressive thinkers had its limits. While scientific pedagogics and didactics was included in a reform of teachers' education in 1894 and object lessons and physical education were strengthened by a primary school reform in 1899, natural science, including evolutionary theory, remained a relatively marginal subject at schools and colleges in Denmark.[43]

---

42 Andersen, "Modtagelsen af Spencer i Danmark," 121–24; Herman Trier, *Pædagogikken som Videnskab* (Copenhagen: Høst & Søn, 1878); Sophus Heegard, *Om Opdragelse* (Copenhagen: Gyldendal, 1880); Kristian Kroman, *Om Maal og Midler for den højere Skoleundervisning* (Copenhagen: Høst & Søn, 1886); Claudius Wilkens, *Grundtræk af Pædagogikken* (Copenhagen: Wroblewsky, 1892); Oscar Hansen, *Opdragelseslære* (Copenhagen: Gyldendal, 1898); Niels Bang, *Opdragelsens og Skolens Historie*, 4th edition (Copenhagen: Gyldendal, 1912); Carl Nicolai Starcke, *Lærebog i Pædagogik* (Copenhagen: Gyldendal, 1914).

43 Skovgaard-Petersen, "De blev lærere," 207–23; Andersen, "Modtagelsen af Spencer i Danmark," 121–24; Gjerløff and Jacobsen, *Dansk Skolehistorie III*, 65–84.

Conclusions

The different intellectual contexts at the national universities resulted in a heterogeneous appropriation of Spencer's ideas in Scandinavia. In Denmark aspects of his philosophical, sociological and educational thoughts were welcomed by liberal professors and students at the University of Copenhagen from the 1870s, while in Sweden the majority of philosophers were loyal to Boström's idealism and conservatism and therefore remained critical of Spencer's positivist and naturalistic ideas well into the twentieth century. In Norway, in turn, the Lutheran and Hegelian influence at the university prevented Spencer's philosophy from gaining ground. Thus, while Spencer's philosophy had a major impact on academic learning in Denmark, it was left to radical students and modernist authors to popularize his ideas in Norway and Sweden. Spencer's educational thought received a particularly large audience through translations and pedagogical textbooks. Thus, his progressive views on education were still included in textbooks after World War II, while his laissez-faire liberalism and individualism remained marginalized in Scandinavia.

The variety of appropriations of Spencer in Scandinavia enlightens how philosophical and scientific ideas can be applied to serve multiple purposes even within relatively small and homogeneous regions such as Scandinavia. This study has shown that, besides the importance of national contexts, more local contexts such as university traditions and specific interests of publishing houses play a decisive role in the appropriation and circulation of ideas from abroad.

Table 11.1 *Translations of Herbert Spencer's works into the Scandinavian languages, 1870-1920*

Years	Language	Original	Title	Publisher	Translator	Comments
1876	Danish	*Education: Intellectual, Moral, and Physical* (1861)	*Om Opdragelse*	Høst & Søn, Copenhagen, 172 pages	Harald Høffding, philosopher	The first essay "What knowledge is of most worth?" is omitted. 2nd edition 1895
1878	Danish	"Progress: Its Law and Cause" (1857) "The Development Hypothesis" (1852) "The Physiology of Laughter" (1860) "The Use of Anthropomorphism" (1853) "Mill *versus* Hamilton – The Test of Truth" (1865)	*Mindre Afhandlinger*	Høst & Søn, Copenhagen, 160 pages	Harald Høffding, philosopher	Includes a biographical sketch originally published in *Popular Science Monthly*, March 1876. All essays are translated from *Essays: Scientific, Political and Speculative*, vol. I-III, 1857-1874.
1880-81	Swedish	*The Study of Sociology* (1873)	*Inledning till Samhällsläran*, vols. I-II	Björck, Stockholm, 177 and 162 pages	Victor Pfeiff	9th edition republished in instalments by Beiters in the book series "Kronbiblioteket" in 1909.
1882	Danish	"The Origin and Function of Music" (1857)	*Musikkens Vorden og Virken*	F.H. Eibes Forlag, Copenhagen, 56 pages	Anon.	
1883	Swedish	*Education: Intellectual, Moral, and Physical* (1861)	*Uppfostran i Intellektuellt, Moraliskt och Fysiskt Afseende*	Skoglund, Stockholm, 230 pages	Anon.	2nd edition 1890
1883	Swedish	*First Principles* (1862)	*Utvecklingsläran*	Almqvist & Wiksell, Uppsala, 503 pages	Victor Pfeiff	Foreword by Carl Yngve Sahlin.

1884	Swedish	*The Data of Ethics* (1879)	*Grunderna för Etiken*	Almqvist & Wiksell, Uppsala, 308 pages	Isidor Flodström, philosopher	
1887	Danish	*First Principles* (1862), Chapter I-V (§§ 1-34)	*Religion og Filosofi: En Bestemmelse af Erkendelsens Grænser og Forhold til Troen*	Wroblewsky, Copenhagen, 149 pages	Søren Hansen, philosopher	Foreword by Harald Høffding
1887	Norwegian	*Education: Intellectual, Moral, and Physical* (1861)	*Opdragelse*	Bibliothek for de tusin hjem, Fagerstrand, 240 pages	Hans Brekke	Published in a popular book series
1888	Swedish	"Gracefulness" (1852) "The Physiology of Laughter" (1860) "The Sources of Architectural Types" (1852) Extract of "The Origin and Function of Music" (1857) Extract of *The Classification of the Sciences* (1864) Extract of "Reasons for Dissenting from the Philosophy of M. Comte" (1864) "The Philosophy of Style" (1852)	*Behaget och andra Uppsatser*	Geber, Stockholm, 146 pages	Robinson	All essays are translated from *Essays: Scientific, Political and Speculative*, vol. I-III, 1857-1874.
1903	Danish	Extract of "Re-Barbarization" in *Facts and Comments* (1902)	"Tilbagefald til Barbari"	*Tilskueren*, pp. 667-671	A. Goldschmidt	Published in a leading highbrow monthly
1912	Swedish	*The Man 'versus' the State* (2nd ed. 1892 [1884])	*Individen gentemot Staten*	Beiters, Stockholm, 144 pages	Alban Keyser	Includes an introductory essay on political economy by Klas Teodor Gustaf Keyser

CHAPTER 12

A 'Spencerian Moment' in French Cultural History? Spencer in France (1870–1890)

Nathalie Richard

On July 25, 1878, a banquet in honor of Herbert Spencer was held in a famous Parisian restaurant. During the dinner, Spencer himself proposed a toast expressing his gratitude to the "French scientific audience" who had so favorably received his works. He commented that the natural eagerness of the French for general and methodically combined ideas had probably predisposed them to endorse his synthetic philosophy.[1] At that time, indeed, Spencer was a rising star in France, and contemporaries, such as Gaston Rageot in 1904 for example, could assert that "Spencer's fame, ultimately, was mainly due to the French."[2]

Interpreted with all necessary caution, statistics on the mention of Spencer's name in digitalized corpora show different trends in France and in Britain (Graphs 12.1 and 12.2). On the French side of the Channel, Spencer's success grew more rapidly, and decreased more quickly. It was also more exclusive: no other British author gained an equal reputation at that time. Not Charles Darwin, John Stuart Mill, or Thomas Huxley were nearly as famous, or at least as frequently quoted. In French sources, Spencer's name stood almost alone for all British liberal and evolutionary thought.

The years 1880–1885 marked the peak of Spencer's reputation in France. The first comments on his philosophy appeared shortly after 1860. Among the French positivists, Emile Littré was the first to review his ideas in *Auguste Comte et la philosophie positive* (1863). He focused on Spencer's dissent from Comte on the matter of the classification of sciences. The following year, a broader audience learned about Spencer's philosophy through an article in

1 "Peut-être le goût naturel des Français pour la généralisation des idées et leur enchaînement méthodique les disposait-il d'avantage à la bienveillance pour mon système de philosophie synthétique qui forme une sorte d'ensemble organique," Emile Alglave, "Herbert Spencer à Paris," *Revue des cours scientifiques*, June 1st, 1878: 1141. See also, Herbert Spencer, *An Autobiography* (London: William and Norgate, 1904), 311–312.

2 "Le renom de Spencer, en définitive, est surtout français," Gaston Rageot, "Herbert Spencer et la philosophie de la vie," *Revue des deux mondes*, August 1904/2: 808–809.

 | DOI 10.1163/9789004264007_014

the *Revue des deux mondes*, an influential review among educated elites. Auguste Laugel, a well known journalist who had a scientific column in the daily *Le Temps*, and who had reviewed Darwin's *Origin of Species* in the *Revue* in 1860, wrote a paper focused on *First Principles* and hailed the return of metaphysics to Britain, after a long (and in Laugel's mind unproductive) period of empiricism and utilitarianism, still illustrated by John Stuart Mill and Henry Thomas Buckle.[3] Spencer's major books were translated into French by the end of the 1870s, beginning with *First Principles* in 1871.[4] As this chapter will demonstrate, the political context of the time fostered and directed the appropriation of Spencer's philosophy. This context explains why Spencer's work was received in broad circles, and discussed in the general press as well as in political assemblies and in philosophical journals. The specific agenda of the founding of a new republican regime also explains why this reception is better understood in terms of 'appropriation,' and why the French invented their own 'Spencer,' different from the British one. The three sections of this paper are dedicated to the political background, to the academic responses to Spencer, and to the popular interpretations of Spencer in France, different aspects of a cultural importation which are distinguished for the sake of clarity, but which were in fact closely interlinked.

Spencer in a French Context: The Establishment of the Third Republic

Spencer's first translations coincided with the founding of the Third Republic. This new regime was proclaimed in September 1870 under difficult conditions, with the occupation of part of the country by Prussian troops, and the Paris Commune insurrection. The first National Assembly was elected in this troubled climate in February 1871. Voters chose the peace candidates who were for the most part Monarchists. A Monarchist parliament thus took charge of drafting a constitution for the new Republic. The process took a long time, until constitutional laws were finally passed in 1875. Only in 1879 were both

3 Auguste Laugel, "Les études philosophiques en Angleterre. M. Herbert Spencer," *Revue des deux mondes*, February 1864/2, 930–957.

4 Herbert Spencer, *Les Premiers principes*, traduit de l'anglais par M.E. Cazelles (Paris: Baillière, 1871); *Introduction à la science sociale* (Paris: Baillière, 1874); *Principes de psychologie*, trad. sur la nouvelle éd. anglaise par Th. Ribot et A. Espinas (Paris: Baillière, 1874–1875); *Essais de morale, de science et d'esthétique*, traduits de l'anglais par M.A. Burdeau (Paris: Baillière, 1877–1879); *Principes de biologie*, traduit de l'anglais par M.E. Cazelles (Paris: Baillière, 1877–1878); *De l'Éducation intellectuelle, morale et physique* (Paris: Baillière, 1878).

assemblies, the Chamber of Deputies and the Senate, dominated by a Republican majority. During these first years, continuing uncertainty surrounded the fate of the young regime. Its supporters sought to stabilize it by grounding it on a truly moral and political Republican philosophy. Spencer was often enlisted in these philosophical endeavors.

After 1879, the new Republican majority, dominated by the moderate 'Opportunist' party, engaged in broad reforms in civil rights, education, and the separation of church and state. The debates over some of these reforms directly impacted Spencer's interpretations. Such was the case regarding educational reforms in particular. Under the leadership of Jules Ferry, the Opportunists undertook to make primary schools free and mandatory. They believed this reform crucial to form more politically conscious citizen-voters. Debates took place mostly between 1879 and 1884, and soon focused on the sensitive issue of secularization. It was decided that the State (and not municipalities) would organize and control the new system. The State would see to the competence of teachers, and therefore also control the training of teachers in religious schools. The State would also ensure the respect of freedom of conscience for all pupils.

The moral content of primary education in the new secularized State system became a crucial problem. Before the reform, morality was equated to the religious (that is, mostly Catholic) education delivered in every school. After the reform, some argued that moral education had to be left out of the classroom and dealt with privately. But most participants in the debates believed that morality had to be dealt with in schools. Some argued that religious education should be maintained in primary schools. But all French were not Catholics, and this meant that non-Catholic children in the State system would attend these classes. Many wished therefore to promote a new, 'non-denominational' morality, and fiercely argued over what the exact content of such a morality would be. The proponents of such a secularized morality often quoted Spencer's works on ethics in order to claim that universal moral values could be defined in a scientific manner, according to the new "positive" knowledge on human psyche and societies, and independently of political opinions or religious creeds. Another issue was to introduce new pedagogy allowing primary schools to fulfill their main purpose, which was the education (and not the mere skills-training) of citizens. Along with primary schools, republicans also created non-compulsory nursery schools (*écoles maternelles*), in which pedagogical practices adapted to young children were discussed. In these debates Spencer's ideas on education were often mentioned to highlight the necessity of adapting educational practices in line with scientific psychology, thereby providing new knowledge on how to develop the aptitude of young students.

In the same years, higher education in France also underwent reforms, triggered by the new social demands linked to economic changes and fostered by the inferiority complex vis-à-vis Germany after the 1870 defeat. New teaching positions were created, and the number of scholarships for lower middle-class students increased. The reform process resulted in the 1896 "Liard Act," which created French modern universities. More specialized curricula were created, and the fields now called "human and social sciences" were established. Spencer's philosophy was used both in promoting new disciplines (sociology, and psychology), and in redefining older ones (philosophy). This process gave rise to a wider academic community, increasing the readership for academic books.

In the 1880s the republican majority also undertook reforms in civil rights. Freedom to form trade-unions was granted in 1884, and new legislation organized a more democratic municipal life. The 1881 Freedom of the Press Act removed all remaining barriers to the development of printed matters in France, leading to the creation of many new political, satirical, and popular periodicals. These reforms provided a stimulating environment for publishers, editors, journalists, and writers. New book series and new academic reviews were launched. New publishing companies specialized in academic and schoolbooks.[5] Spencer benefited from this economic, social and intellectual environment.

Spencer's appropriation in such a context had many, partly overlapping, facets.

Spencer in French Philosophical Controversies

The circulation and appropriation of Spencer's ideas in French academic circles has been well studied.[6] It was due in the first place to an active group of publishers and translators, who believed that his philosophy could be profitable on the printed market and instrumental in controversies over topical

5 Valérie Tesnières, *Le Quadrige. Un siècle d'édition universitaire (1860–1968)* (Paris: PUF, 2001).

6 Naomi Beck, *La Gauche évolutionniste. Spencer et ses lecteurs en France et en Italie* (Besançon: Presses universitaires de Franche-Comté, 2014); Daniel Becquemont and Laurent Mucchielli, *Le Cas Spencer. Religion, science et politique* (Paris: PUF, 1998); Daniel Becquemont and Dominique Ottavi (eds), *Penser Spencer* (Saint-Denis: Presses universitaires de Vincennes, 2011).

philosophical and political matters. Twenty books and brochures, and more than forty articles by Spencer himself were translated and published in France before World War I, as well as a great number of reviews and commentaries. Twelve translators and six publishers were involved (see Table 12.1 at the end of this chapter).[7]

Among them Gustave Germer Baillière was particularly important. His company published prestigious book series on contemporary philosophy and science and played a crucial role in the academic book market during the first decades of the Third Republic. Baillière was an anticlerical Republican, and he saw evolutionary theories as promising for new scientific developments, but also as a tool against clericalism. By 1870, he wanted to benefit from the internationalization of science. With European and American publishers, he took part in the creation of the "International Scientific Series" (*Bibliothèque scientifique internationale*).[8] Founded by the American Edward L. Youmans, this new series facilitated the translation and circulation of ideas throughout a broad market, including Germany, Italy, France, England, Russia and North America. It circulated syntheses on the latest scientific research intended for a wide audience, but the price of each volume (7 to 10 francs, while popular series volumes were about 1 franc) restricted the readership to middle- and upper-class readers. It cannot be considered as a 'popular science' collection, at least not in France. Spencer met Baillière for the first time in 1871, during a trip to Paris organized by Youmans in order to find a French publisher for the "International Scientific Series." Baillière became Spencer's main intermediary in France, but he was not the only one. Other important companies, such as Reinwald and Hachette, also published Spencer's work.

Baillière also owned influential scientific reviews. At that time, the *Revue scientifique* (or *Revue des cours scientifiques*, created in 1863) was the main source of general scientific information. The review editor, Emile Alglave, was also in charge of the International Scientific Series. A former law professor at Lille and Douai, he was a free-thought activist. Between 1870 and 1880, he published over ten papers by Spencer in his journal, mostly abstracts from books under press at Baillière. The *Revue philosophique de la France et de l'étranger*

7 Becquemont and Mucchielli, *Le Cas Spencer*, 215.

8 Leslie Howsam, "An experiment with science for the nineteenth-century book trade: the International Scientific Series," *The Bristish Journal for the History of Science*, 33/2 (2000): 187–207; Jean-Yves Mollier, *L'argent et les lettres. Histoire du capitalisme d'édition (1880–1920)* (Paris: Fayard, 1988), 295–297 ; see also, Tesnière, *Le Quadrige*.

also played an important role in this media organization.[9] Its editor was Théodule Ribot, a young philosophy teacher who espoused evolutionary theory as a cause to fight for. His doctoral thesis on heredity, *L'Hérédité, étude psychologique sur ses phénomènes, ses lois, ses causes et ses conséquences* (1873), was a manifesto in favor of evolutionary psychology. Ribot discovered *Principles of Psychology* in 1866, and became so enthusiastic that he undertook a translation, published in 1874.[10] Meanwhile, he reviewed Spencer's theory in a book on modern British psychology. He depicted the British philosopher as a genius with an exceptionally profound and unifying thought equal to Newton and Descartes.[11] Supported by Baillière, Ribot launched the *Revue philosophique* in 1874. Spencer soon became one of its main contributors. Over forty papers and reviews were published between 1876 and 1881.

Translators were also part of the activity of this group of book and media representatives. Among them, one had a prominent position and became friends with Spencer. Emile Honoré Cazelles was a Republican and a moderate free-thinker, a friend of Gambetta and other political leaders of the new regime. He had trained as a medical doctor, and held positions in State administration (as *préfet* and as State Counsellor) during the Third Republic. He was a well-known translator of modern English philosophy in the 1860s, and became Spencer's main translator for Baillière. In 1871, his translation of *First Principles* was the first of Spencer's books available in French. It included a long foreword, transforming Cazelles into an interpreter, as well as a translator.[12] Cazelles summarized and interpreted Spencer in the French context of the fierce philosophical controversies over metaphysics and transcendence since the 1850s, which opposed materialists and positivists on the one side, and spiritualists and catholic theologians on the other. He especially highlighted

9 Jacqueline Carroy, "Philosophie et psychologie au XIXe siècle en France. Taine, Ribot et *La Revue philosophique*," in *Unité et globalité de l'homme. Des humanités aux sciences humaines*, Marie-Louise Pelus-Kaplan (ed) (Paris: Syllepse, 2006), 79–91; Mara Meletti Bertolini, *Il pensiero e la memoria. Filosofia et psicologia nella* Revue philosophique *di Théodule Ribot (1876–1916)* (Milano: Franco Angeli, 1991).

10 *Principes de psychologie*. trad. sur la nouvelle éd. anglaise par Th. Ribot et A. Espinas (Paris: Baillière, 1874–1875).

11 Théodule Ribot, *La Psychologie anglaise contemporaine (école expérimentale)* (Paris: Baillière, 1870), 145–146.

12 Emile Cazelles, "Introduction" to *Les Premiers principes*, by Herbert Spencer, i-cix. Spencer praised this introduction in his *Autobiography*. It was considered important enough to be published separately in English: *Outline of the Evolution-Philosophy* (New York: Appleton, 1875).

the synthetic quality of Spencer's philosophy, seen as a system capable of integrating facts according to positive science without denying metaphysics.

Between the late 1860s and the mid 1880s, Spencer's philosophy not only became available to French academic readers, but it was also subjected to multiple, at times competing, uses. In an early phase, a spiritualist interpretation of Spencer predominated. The first reviews were published at the same time as the first volume of Littré's *Dictionary of the French Language* (*Dictionnaire de la langue française*), Ernest Renan's *Life of Jesus* (1863) and Hippolyte Taine's *History of English Literature* (1864). All raised fierce debates: Renan's book created a scandal, and Taine's introduction to his literary history was read as a materialist and determinist manifesto. Reactions by spiritualist and Catholic authors quickly followed. Bishop Dupanloup issued a "warning" to "young readers" and "fathers" against dangerous modern writers.[13] Elme Caro, a disciple of Victor Cousin, condemned these new critics of the "idea of God."[14]

These opposed philosophical trends were not as unified as they seemed. Positivists, such as Littré, differed from more materialist thinkers, such as Taine. Lay spiritualists, such as Caro, were not dogmatic Catholics like Dupanloup. Each group had its own interpretation of Spencer. As Spencer had done himself, the positivists focused on the differences between Comte's and Spencer's systems, especially on the classification of sciences. Materialists admired the systematic ambition of Spencer and the importance he gave to the law of the persistence of force. But they were ill at ease with Spencer's 'Unknowable.'[15] Orthodox Catholics equated Spencer to atheism. Lay spiritualists, such as Caro, thought that Spencer's *First Principles* could help reconcile positive science and religion.[16]

This spiritualist interpretation predominated around 1870. In the *Revue des deux mondes*, Elme Caro depicted Spencer as an ally for the spiritualist cause. At the Academy for Moral and Political Sciences (*Académie des sciences morales et politiques*), where they were influential, spiritualists such as Charles Lévêque held the same views.[17] One of the main speakers for this interpretation was Emile Cazelles. His introduction to *First Principles* stressed that, though

13 Félix Dupanloup, *Avertissement à la jeunesse et aux pères de famille sur les attaques dirigées contre la religion par quelques écrivains de nos jours, par M. l'évêque d'Orléans* (Paris: Douniol, Blanchard, 1863).

14 Elme Caro, *L'Idée de Dieu et ses nouveaux critiques* (Paris: Hachette, 1864).

15 Hippolyte Taine, "Th. Ribot, Bain, Herbert Spencer" (1874), reprinted in *Derniers essais de critique et d'histoire* (Paris: Hachette, 1903), 198–201.

16 Elme Caro, *Le Matérialisme et la science* (Paris: Hachette, 1868).

17 Charles Lévêque, "Principes de psychologie, par M. Herbert Spencer," *Séances et travaux de l'Académie des sciences morales et politiques*, 2 (1874): 405–406.

Spencer was an adept of positivism, his philosophy was congenial to metaphysics and to non-dogmatic religious beliefs.[18] Leading spiritualist philosophers later developed this balanced interpretation, focusing on Spencer as the author of *First Principles* and the prophet of the Unknowable. Paul Janet, the most influential spiritualist at that time, used Spencer's evolutionism to sustain his argument in favor of teleology in *Les causes finales* (1876). In his view, evolution demonstrated order in nature and pointed to supernatural final causes.

After 1870 however, a more militant view linking evolutionism, positivism and Republicanism became available. Such a view was held by a new generation of academics, who began their careers under the Third Republic. They were interested in *Principles of Psychology* and *Principles of Sociology* more than in *First Principles*, and involved Spencer in their Republican, sometimes anti-spiritualist crusade.[19] This interpretation was dominant in the *Revue philosophique*, formulated by Théodule Ribot himself. He advocated a new scientific psychology, called "experimental" though it was not practiced in laboratories, contrary to what was going on in Germany for example.[20] According to Ribot, psychological processes were due to evolution and heredity, and the formation of these processes could be understood by looking at cases of pathological regression. Mental processes had purely physiological causes, and scientific psychology had to get rid of explanations in terms of spiritual entities such as the "self" and the "will." Ribot argued that Spencer had set the agenda for this new psychology. Alfred Espinas, a close friend of Ribot, quoted Spencer's sociology in his doctoral thesis on "Animal Societies" in 1878 (*Les Sociétés animales*). He argued that a truly positive science of society could only be grounded on biology. But, he was not as radically anti-spiritualist as Ribot, and he believed that evolution and order in nature hinted at some kind of intelligent design.

Many Republican activists interpreted Spencer in such a way. They believed that Spencer's system made a new political philosophy possible, based on positive social and moral sciences, without falling into unbridled materialism.

---

18 "M.H. Spencer donne la main à la religion [...] et en même temps adhère à la doctrine des penseurs positifs," Cazelles, "Introduction" to *Les Premiers principes*, lxxx.

19 On Ribot and Espinas: John I. Brooks, *The Eclectic Legacy: Academic Philosophy and the Human Sciences in Nineteenth-Century France* (Newark: University of Delaware Press, 1998).

20 Jacqueline Carroy and Régine Plas, "The origins of French experimental psychology: experiment and experimentalism," *History of the Human Sciences*, 9/1 (1996): 73–84; Jacqueline Carroy and Henning Schmidgen, *Psychologies expérimentales: Leipzig-Paris (1890–1910)* (Berlin: Max Planck Institut für Wissenschaftsgeschichte, Preprint 206, 2002).

Alfred Fouillée, principal architect of the ideology of the Third Republic, commented on Spencer's sociology in a book on contemporary social science (*La Science sociale contemporaine*, 1880). His philosophy was intended to subsume the traditional opposition between positivism and idealism through the theory of "idea-forces" (*idées-forces*). He interpreted evolution as a partly spiritual process, assessing the power of ideas and will as factors of evolution in human societies.[21] Others focused on pedagogy and morality. Gabriel Compayré reviewed Spencer's work on education for the *Revue philosophique* and later summarized his ideas in a brochure (*Les Grands éducateurs: Herbert Spencer et l'éducation scientifique*, 1901). Reference to Spencer helped build an argument in favor of pedagogy as a science based on positive knowledge about the development of the human mind. Jean-Marie Guyau adopted an evolutionary point of view in his synthesis on modern moralists in Britain (*La Morale anglaise contemporaine*, 1879). He believed that evolution helped in the effort to better understand the birth of social and moral phenomena, but he rejected determinism. Education and action on the milieu could compensate for hereditary tendencies, and free will played a role in human affairs.

In the same period, supporters of the Republic from another philosophical school also commented on Spencer's system. *La Critique philosophique*, a periodical launched in 1872, published more than forty articles about Spencer in two decades. But contrary to the reviews in *La Revue philosophique* and in *La Revue scientifique*, these comments were critical. *La Critique philosophique* was the forum for Charles Renouvier's "neo-criticism,"[22] which provided a synthesis between Comte's positivism and Kant's criticism. After 1870, Renouvier focused on ethics and grounded his theory on a radical critique of evolutionism: Spencer was one of his main targets. He argued that humans were not the mere result of biological evolution. Even the most primitive human being had logical and spiritual faculties distinguishing him from animals.[23] Logical faculties allowed the human mind to forge general ideas, according to categories defined in a Kantian manner. Spiritual faculties gave rise to moral sense and free will. According to Renouvier, this inherently human moral sense was shaped by the social (that is, non-biological) environment. It was developed in "moral milieus" created by social groups, and the aim of "sociology" was to

21 Alfred Fouillée, *L'Evolutionnisme des idées-forces* (Paris: Alcan, 1890).

22 Marie-Claude Blais, *Au principe de la République. Le cas Renouvier* (Paris: Gallimard, 2000).

23 Renouvier clarified this argument in a series of papers on the "psychology of primitive man" published in *La Critique philosophique*. He frequently quoted Spencer's ideas as counterarguments (Becquemont and Mucchielli, *Le Cas Spencer*, 285–287).

study social organizations as such milieus, differentiating this science from biology. State intervention could be safely grounded on such knowledge: the Republic would be able to shape a new "social milieu," and give birth to a more moral society. Even though *La Critique philosophique* had few supporters, Renouvier's philosophy was very influential in the 1880s, especially through his followers. Some of his disciples gained important positions in State administration. Louis Liard, who was in charge of higher education at the Ministry of Education from 1884 to 1902, and who drafted the 1896 University reform Bill, was one of them.

These philosophers shaped the intellectual atmosphere of the last decade of the nineteenth century, in which Henri Bergson built his own spiritualist evolutionary philosophy, Théodule Ribot and Pierre Janet forged a new scientific psychology, and Emile Durkheim invented his new sociology, as a science of specifically social facts which could be reduced neither to biological nor psychological data.[24] But these philosophers also brought Spencer's reputation beyond the narrow circles of academia.

The Popular French Spencer

In the 1880s Spencer's name was introduced in crucial controversies over secularization, civic morality and education. His philosophy was frequently quoted during parliamentary debates, sometimes in support of competing political views. The general as well as popular press commented upon his system for various readers. In contrast to the politico-philosophical reception, the dissemination of Spencer's ideas among the general public, its main topics and its main channels, has not yet been studied by historians.

In the French Parliament, Spencer became a reference in debates over institutional reform and political liberties, notably over two priorities in the political agenda of the Opportunist party: school reform and the secularization of the State. On June 22, 1882, for instance, the Deputies discussed the reform of oaths taken before courts of justice. The former religious oath had to be replaced by a new one, respectful of the freedom of conscience, but powerful enough to induce respect for judicial decisions. The opportunist Frédéric Thomas advocated complete freedom of choice of the oath formula, according to religious and philosophical creeds. The radical Joseph Fabre spoke against

24 François Azouvi, *La Gloire de Bergson, essai sur le magistère philosophique* (Paris: Gallimard, 2007); Brooks, *The Eclectic Legacy*; Laurent Mucchielli, *La Découverte du social: naissance de la sociologie en France, 1870–1914* (Paris: La Découverte, 1998).

this proposal, in favor of a single, completely secularized oath upon honor and conscience. He mocked Thomas's proposal by asking his fellow MPs to imagine a court of justice before which Ernest Renan's followers would swear in the name of the "elusive Lord" ("*l'éternel insaisissable*"), while Hartman's followers swore in the name of the "Unconscious," and Spencer's disciples in the name of the "Unknowable."[25] In turn, the Catholic party opposed Fabre's proposal. Emile Freppel, bishop of Angers, argued that only fear of God could prevent witnesses and accused from perjury. These debates ended only in 1972.[26] Spencer's name had been introduced in a key episode of the complex history of secularization in France.

Jules Ferry himself often quoted Spencer in parliamentary debates, especially over education. On July 11, 1882, for instance, the deputies discussed State control over private schools. New legislation had been drafted to require a certificate of aptitude in pedagogy from all teachers, including (mostly Catholic) private-school teachers. A board of State administrators, excluding Church representatives, would deliver this certificate. Catholic deputies strongly opposed this reform, appealing to the freedom of education. Jules Ferry, then Ministry of Education, answered that pedagogy was a science, and that, therefore, pedagogical qualification could be assessed according to objective criteria, without political or religious bias. To sustain his argument, he quoted famous educationalists, including Spencer.[27] But Spencer could also be used by his opponents, who chose to stress his liberalism against what they saw as State encroachment upon the freedom of education. Such was the case, for instance, of Emmanuel Daguilhon-Pujol, a former Bonapartist who had joined the Catholic and Monarchist cause, during the debate over the composition of the State High Council of Education, in 1879–1880.[28] Therefore, in the early 1880s, Spencer's philosophy had become a shared, though polemical, source for grounding differing ideological statements in parliamentary debates.

Such was also the case among a broader audience, as his name became commonplace in newspapers. Since the 1860s, some periodicals targeted a

---

25 *Journal officiel de la République française. Débats parlementaires. Chambre des députés*, June 23, 1882, 993.

26 Jacqueline Lalouette, *La Libre-Pensée en France, 1848–1940* (Paris: A. Michel, 1997) ; Jean-Yves Pidoubès, "La liberté de conscience à l'épreuve du serment: Individu, religion et politique au XIXe siècle," in Anne-Emmanuelle Demartini and Dominique Kalifa (eds), *Imaginaire et sensibilités au XIXe siècle* (Paris: Créaphis, 2005), 157–167.

27 *Journal officiel de la République française. Débats parlementaires. Chambre des députés*, July 12, 1882, 1227.

28 For instance, "Chambre des députés. Séance du 17 juillet," *Journal des débats politiques et littéraires*, July 18, 1879.

wider, 'popular' audience of workers, young readers and women. Following *Le Petit Journal* (1863), new dailies focused on criminal affairs ("*faits divers*"), and news reports. They were sold by the copy (and not only by subscription), and were almost two-thirds cheaper than the higher national press. This economic and editorial model turned out to be very profitable, and *Le Petit Journal* was soon imitated. After 1870, the control over political content was lessened, and more politically oriented columns were introduced in the popular press. After the 1881 Freedom of the Press Act, political periodicals flourished.[29] Their numerous readers learned about Spencer in reports on parliamentary debates and in articles dedicated mostly to education, morality and patriotism.

Since 1870, Spencer was well known to the educated, middle-class, male readers of the high national press of all political views. The *Journal des débats*, a well-established, moderate periodical, which could be compared to the *Times*, gave book reviews and reported on references to Spencer in parliamentary debates. *Le Rappel*, founded in 1869 by Republican radicals, including Henri Rochefort (a famous journalist) and Victor Hugo, was among the most successful newspapers around 1880, selling up to 50,000 copies. It informed its readers that Jules Ferry frequently called upon Spencer in order to ground his arguments.[30] On January 25, 1880 *La Justice*, directed by Georges Clemenceau, quoted Spencer in an article on "schools and municipalities." This Republican newspaper opposed centralization and turned Spencer into a supporter of decentralization.[31] It had a relatively small audience (around 10,000 copies) but was influential in political spheres. On the anti-Republican side, *La Croix*, one of the main Catholic periodicals, created in 1880 by the religious order of the Assumptionists, commented on the "new laws of the month" in its August 1881 issue. It had been a particularly "sad month" said *La Croix*, marked by a "renewed anti-Christian fever" among deputies. Spencer was mentioned in a paragraph entitled "God in school" ("*Dieu à l'école*") on moral education.[32]

Spencer was also known to intellectual and social elites, male and female, enjoying lighter reading. *Gil Blas*, launched in 1879, was famous for its talented literary staff, including Théodore de Banville and Guy de Maupassant, and for

29 Christophe Charle, *Le Siècle de la presse (1830–1939)* (Paris: Seuil, 2004); Dominique Kalifa et al (eds), *La Civilisation du journal. Une histoire de la presse française au XIXe siècle* (Paris: Nouveaux Mondes, 2011).

30 For instance, "Au sénat," *Le Rappel*, January 29, 1880; "Chambre des députés, séance du 23 décembre," *Le Rappel*, December 25, 1880.

31 Jules Hoche, "L'école et la commune," *La Justice*, January 25, 1880.

32 "En France, le mois qui vient de s'écouler a été fécond en tristesses. La Chambre des députés s'est trouvée prise […] d'un redoublement de fièvre anti-chrétienne," "Histoire du mois. Les nouvelles lois," *La Croix*, August 1881, 315.

its ribald humor. It often quoted Spencer. His name appeared in a paper on "Suicide" on December 16, 1879,[33] in George Eliot's obituary on December 26, 1880, etc.[34] Guy de Maupassant (at times under the pen name of Maufrigneuse) regularly quoted Spencer in his columns. In an article on morality, dealing with selfishness and patriotism, he defined *The Study of Sociology* as "a kind of breviary of societies" ("*une sorte de bréviaire des peuples*").[35] The conservative *Le Gaulois* was widely read in the salons and among French nobility, selling between 20–30,000 copies around 1880. It helped to spread Spencer's reputation in the world Proust would describe a few decades later. On March 22, 1880, for instance, a paper about a conference given by a famous Catholic preacher, Father Henri Didon, associated Spencer with free-thought and atheism.[36]

In the same period, Spencer's name also became familiar to more numerous, less educated lower-class readers, sometimes hitting the headlines of popular dailies, such as *Le Petit parisien*. Created in 1876, this politically radical newspaper was renowned for its serial stories by famous novelists such as Emile Zola, Guy de Maupassant and Alphonse Daudet. By 1880 it was selling around 40,000 copies at 5 cents. On June 14, 1884 Jean Frollo's headline summarized a conference on "evolutionary moral" by the materialist Charles Letourneau who quoted Spencer in his conclusion.[37] On November 22, a piece on education by the same journalist entitled "Father and child" began by mentioning Spencer.[38] *La Lanterne*, founded in 1877, was then the most successful inexpensive political daily. It was characterized by its fierce anticlerical vein and sold around 100,000 copies per day. Spencer was mentioned in numerous reports on parliamentary debates.[39]

On the same themes (education, ethics and patriotism), women's newspapers and family-oriented periodicals also quoted Spencer, either positively or negatively, depending on which aspect of his philosophy they considered. *La Femme*, a bimonthly by the conservative National Friendly Union for Young Ladies (*Union nationale des amies de la jeune fille*), dealt with "women's rights" in its headline for January 15, 1882. Women had only one true right, which was to be responsible housewives and work for the greatness of their country. Spencer was quoted on children's education, as a useful reading for improving

33 Edouard Monin, "Propos du docteur. Le suicide," *Gil Blas*, December 12, 1879.

34 G. Dézieux, "George Eliot," *Gil Blas*, December 26, 1880.

35 Maufrigneuse, "Le fond du cœur," *Gil Blas*, October 14, 1884.

36 "Sixième et dernière conférence du Père Didon à la Trinité," *Le Gaulois*, March 22, 1880.

37 Jean Frollo, "La morale évolutionniste," *Le Petit Parisien*, June 14, 1884.

38 Jean Frollo, "Le père et l'enfant," *Le Petit Parisien*, November 22, 1884.

39 For example "Les Chambres. La journée parlementaire," *La Lanterne*, January 28, 1880.

domestic educational practices, by adapting them to each stage of a child's development.[40] Spencer's pedagogical work could therefore be considered suitable reading for women, even in conservative circles. But the expert on educational matters could also be condemned, in very similar circles, as a promoter of secularization. *La Semaine des familles*, a Catholic family-oriented weekly, gave a long critique of his views on education in two issues in March 1888. In a "Talk on Education," the editor Gaston Feugère criticized "utilitarian education" and "schools without God." He deplored that *On Education* had become the "new gospel" in pedagogical matters.[41] By the mid-1880s, Spencer was famous enough to become the butt of jokes. *L'Univers illustré*, an illustrated weekly magazine for middle-class families, spoke about him in its famous column "Courrier de Paris" signed by "Gérôme."[42] In March 1884, the columnist made fun of Spencer's sociology (and of Tylor's survivals theory) by giving a satirical explanation for the origins of business cards, referring to the alleged primitive custom of cutting one's arm to donate it to the tribe leader.[43]

These media reached numerous and diverse readers. They made Spencer's reputation mainly as a moralist and an educator, whose ideas had direct implications, mainly on pedagogical reform and on the question of secularization, and whose philosophy could be appropriated for different, even conflicting ideological aims. This partial image was reinforced by available books and by the school system. Among Spencer's work, *On Education* was the only one to be published in a "popular edition" (1880). In comparison with the original edition, the content was shortened, the format smaller (*in-16* instead of *in-8*), and it was sold for 60 cents instead of 5 francs. Sponsored by the French administration, the book was intended for educators and for libraries organized in primary schools.[44] But, as we have seen, it was also circulated among readers who did not agree with the increased state control on education and with the secularization of the school system, but who wished to promote the virtues of domestic, private education. By the end of the 1880s the program for the State examination of school teachers, directors and inspectors included *On Education* in the required reading list.[45] Spencer's books were also recom-

40 "Encore le droit des femmes," *La Femme*, January 15, 1882, 9–10.

41 Gaston Feugère, "Causerie sur l'éducation," *La Semaine des familles*, March 10,1888, 786–788; March 17, 1880, 806–808.

42 Gérôme was a shared penname. The column was sometimes written by Anatole France in the 1880s.

43 Gérôme, "Courrier de Paris," *L'Univers illustré*, March 29, 1884, 195.

44 Herbert Spencer, *De l'éducation, édition populaire* (Paris: Germer Baillière, 1880). See the foreword, iii. See also *Journal des instituteurs*, October 10, 1880, 640.

45 *Journal des instituteurs*, November 14, 1880, 705.

mended by the State Council of Education for high school students, and were given as prizes.[46] *Le Journal des instituteurs* ("Schoolteachers Journal") therefore quoted Spencer many times a year on various educational matters.[47] Ferdinand Buisson also mentioned Spencer often in the famous *Dictionary of Pedagogy*.[48] Among the growing professional milieu of primary school educators and administrators, Spencer became part of the canon, together with Jean-Jacques Rousseau, Heinrich Pestalozzi and others. Simplified versions of his ideas on evolutionary psychology, on pedagogy as a science adapted to this new developmental psychology, and on a scientifically based morality were adapted for the growing educational staff, in order to create a sense of professional identity and to turn educators into activists of a new Republican ideology.

Familiarized with Spencer's name and with some of his ideas through schoolbooks, the general press and popular editions, the general French reader could also come into contact with his philosophy through literature. The most prolific French novelist on Spencer was probably Guy de Maupassant. He discovered Spencer around 1880, at the same time as he found Schopenhauer, and interpreted the former in a pessimistic, determinist way. As a columnist for the general press, he often commented on Spencer's ideas in his papers, but he rarely mentioned his name in works of fiction.[49] Spencer also influenced Gustave Flaubert and Emile Zola, as the preparatory files for their works demonstrate, but they rarely mentioned him in their novels. Still, Spencer became a character in a best-seller. *The Disciple* (*Le Disciple*) by Paul Bourget was a sensation in 1889. It sold 22,000 copies in six weeks and triggered a public debate on the moral responsibility of scientists and writers. The novel centered around a middle-class student, Robert Greslou, who psychologically manipulated a naive young girl from an aristocratic family into falling in love with him, leading her to commit suicide. The student ended up in prison, where he wrote a confession sent to Adrien Sixte. In the novel, Sixte was depicted as "a modern

46 Spencer, *An Autobiography*, 326.

47 For instance, "Ecole maternelle. Enseignement simultané du dessin, de l'écriture et de la lecture par Melle Matrat," *Journal des instituteurs*, May 14, 1882, 187; "De l'éducation des femmes," *Journal des instituteurs*, March 25, 1883, 92.

48 Ferdinand Buisson (ed), *Dictionnaire de pédagogie*, Paris, Hachette, 1882–1893, 7 vol.

49 Antonia Fonyi argues that Maupassant made a poetic use of Spencer's Unknowable in his fantastic short stories, notably *Le Horla* (1887): "Présences de Spencer dans l'œuvre de Maupassant," *Arts et Savoirs*, 4 (2014), special issue Niklas Bender (ed), *Herbert Spencer en France: mise au jour d'une influence* <http://www.fabula.org/actualites/arts-et-savoirs-n-4herbert-spencer-en-france-mise-au-jour-d-une-influence_63308.php>.

philosopher" whose determinist philosophy had influenced Greslou, and was responsible for his lack of moral sense.

The success of the novel was partly due to two court cases in which students were indeed accused of having committed murders under the influence of dangerous, modern scientific ideas. The first case had taken place in Paris in 1878. One of the accused, Paul Lebiez, had given a public lecture on Darwinism shortly before the murder. The trial stressed that his moral sense was bent by his Darwinian creed, based on the survival of the fittest. The trial was widely broadcast in the press. Lebiez was compared to Raskolnikov from Dostoyevsky's *Crime and Punishment* (1866), and the case inspired writers such as Alphonse Daudet.[50] In 1888, another court case raised similar issues. Henri Chambige had intended to commit suicide together with his mistress, a married woman. She died and he survived, and the young man was accused of psychological abuse. Chambige's lack of moral sense was put forward by the accusation and related to the bad influence of his philosophical readings. This case, like the Lebiez trial, fascinated the general public,[51] and Bourget's novel was interpreted as a transposition. Adrien Sixte's real model was not revealed, but names of contemporary philosophers were dropped in the first chapter. Sixte was compared to Hippolyte Taine, and Théodule Ribot. He was also described as "he the British like to call the French Spencer" ("*celui que les Anglais appellent volontiers le Spencer français*").[52] However, Bourget distinguished Sixte from Spencer. Both philosophers were evolutionists, but the former, contrary to the latter, completely denied metaphysics. Bourget stressed that Spencer's philosophy, as presented in *First Principles*, allowed for the reconciliation of science and religious beliefs.[53] Through a short paragraph, readers were therefore introduced to the French spiritualist interpretation of Spencer, and warned against the danger of materialist, determinist systems such as Ribot's and Taine's.

---

50 Alphonse Daudet, *La Lutte pour la vie* (1889) in *Théâtre* (Paris: Charpentier, II, 1896).

51 Jacqueline Carroy and Marc Renneville, "Une cause passionnelle passionnante : Tarde et l'affaire Chambige," *Champ pénal*, 2005, 1 <http://champpenal.revues.org/document260.html>; "Le crime de Chambige (1888) : entre psychologie et littérature," in Jean-Louis Cabanès, Jacqueline Carroy and Nicole Edelman (eds), *Psychologies fin de siècle*, RITM, 2007, 38, 193–208.

52 Paul Bourget, *Le Disciple*, édition définitive (Paris: Plon, 1901), 18. Niklas Bender, "La théorie et ses abîmes: Herbert Spencer dans *Le Disciple* de Paul Bourget," in *Herbert Spencer en France* <http://www.fabula.org/actualites/arts-et-savoirs-n-4herbert-spencer-en-france-mise-au-jour-d-une-influence_63308.php>.

53 Bourget, *Le Disciple*, 35–36.

Conclusion

In the late 1880s, Spencer's name was well known among widely different circles. It is therefore legitimate to speak of a "Spencerian moment" in French cultural history. The main characteristics of this "moment" are related to the specific political context of the founding years of the Third Republic. In such a context, Spencer's philosophy helped to sustain hopes for a more efficient, scientifically based brand of politics, and helped to shape arguments on modernization,[54] on educational reform, on secularization and on public liberties. Taking place in a country which was considered by many liberals as the homeland of human rights, French debates influenced discussions in other nations facing similar issues in Europe and further, in Italy, Spain, Latin America and Russia for example. In some of these countries (in Italy for example), Spencer's philosophy was first introduced through French translations.[55] In this respect, the "French Spencer" directly triggered the globalization of Spencerism.

But the reception of Bourget's novel in France leads to a more complex appreciation. The readers of *The Disciple* had no difficulty in identifying Adrien Sixte. Even if Bourget denied it, they saw Hippolyte Taine as well as Spencer under the guise of his "modern philosopher." For Bourget's readers, "Spencer" was probably a representative of the synthesis of various materialist, determinist and evolutionary philosophical systems, including the influential philosophy of Taine. In France, Ernest Renan's, Littré's and Taine's reputations preceded and exceeded Spencer's. During the Second Empire, these philosophers developed their own branches of positivism, and campaigned for the creation of a positive science of man on which political reform could be grounded. In the 1860s Renan's *Life of Jesus* (1863) sold much more copies than any of Spencer's books and triggered a much wider scandal than Bourget's *Disciple*. It was interpreted as a manifesto for a positive science of religion, but also as an attempt to reconcile spirituality and reason in a modern religious synthesis.[56] In the same years, the first edition of Littré's famous *Dictionnary* (*Dictionnaire de la langue française*, 1863–1872) was seen as an attempt to introduce positive methods into the study of French language, depriving speech of its divine origins. In the 1870s and 1880s, Taine's *Origins of Modern France*

---

54 On the rising motto of "modernity" in late Nineteenth Century France, see Christophe Charle, *Discordance des temps. Une brève histoire de la modernité* (Paris: A. Colin, 2011).

55 Beck, *La Gauche évolutionniste.*

56 Ernest Renan, *Vie de Jésus* (Paris: Lévy frères, 1863). See also Nathalie Richard, *La* Vie de Jésus *de Renan. La Fabrique d'un best-seller* (Rennes: PUR, 2015).

attempted to interpret French political history since 1789 in the light of the scientific psychology developed in *On Intelligence*.[57] Spencer's system was understood as following and complementing these philosophies, exactly as Darwin's theory was read in France as a more modern, but quite similar, version of Lamarck's ideas. Even if Renan's and Taine's reputation spread beyond national borders, the overwhelming influence of their philosophy shaped the "French Spencer" in a manner specific to France. In other countries, the English philosopher was probably interpreted within other pre-existing intellectual frameworks, such as Krausism (*krausismo*), a spiritualist philosophy derived from the work of the German philosopher Friedrich Krause, in Spain and in Spanish speaking America.[58]

Spencer's ambiguous success soon faded in France. By the end of the 1880s, the Republic had undergone a series of crises, showing that civil rights, secularization and educational reforms were not sufficient to ensure its future, and raising the issue of welfare legislation. At the same time, scientific and philosophical agendas evolved with the renewal of spiritualism and the birth of more specialized social sciences. When Spencer's death became known in France, obituaries therefore portrayed an old-fashioned, almost forgotten glory. In the *Revue des deux mondes*, Gaston Rageot depicted the sad scene of Spencer's funeral, with its conventional, unemotional eulogies. He added that most obituaries had been critical, that Spencer's much expected *Autobiography* had mainly disappointed its readers, and that his philosophy was now relegated to history: *"Et enfin l'on a cessé de parler de l'illustre mort qui n'appartient plus aujourd'hui qu'à la morne actualité de l'histoire."*[59]

---

57 Hippolyte Taine, *Origines de la France contemporaine* (Paris: Hachette, 1875–1893); *De l'intelligence* (Paris: Hachette, 1870). Nathalie Richard, *Hippolyte Taine. Histoire, philosophie, littérature* (Paris: Garnier, 2013).

58 Pierre Bidart, "L'influence du philosophe allemand F. Krause dans la formation des sciences sociales en Allemagne," *Revue germanique internationale*, 21 (2004): 133–147. O. Carlos Stoetzer, *Karl Christian Friedrich Krause and His Influence in the Hispanic World* (Köln: Böhlau, 1998).

59 "And we stopped speaking of the famous deceased, who now only belongs to the boring past," Gaston Rageot, "Herbert Spencer et la philosophie de la vie," 808.

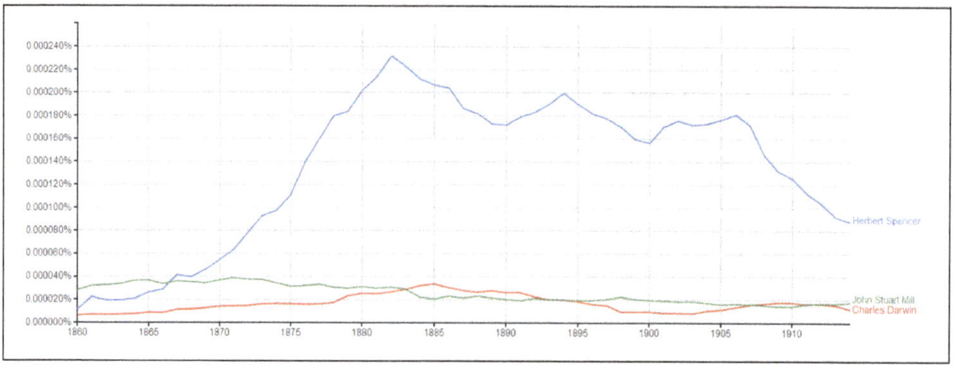

GRAPH 12.1 *Compared occurrence of "Herbert Spencer", "Charles Darwin", and "John Stuart Mill" in Google French Digitalized Corpus, 1860-1914* (*Ngram Viewer*).

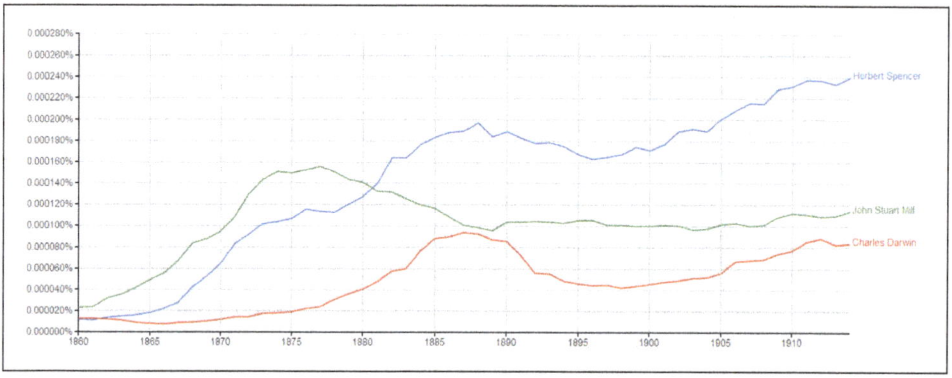

GRAPH 12.2 *Compared occurrence of "Herbert Spencer", "Charles Darwin", and "John Stuart Mill" in Google British English Digitalized Corpus, 1860-1914* (*Ngram Viewer*).

TABLE 12.1 *French translations of Spencer's works (1871-1914)*

Years	Title	Translator	Publisher	Comments
1871	*Les Premiers principes*	Emile Cazelles	Germer Baillières	
1874	*Introduction à la science sociale*		G. Baillières	Volume of the "Bibliothèque scientifique internationale"
1874-1875	*Principes de psychologie*	Théodule Ribot and Alfred Espinas	G. Baillières	
1877-1879	*Essais de morale, de science et d'esthétique*	Auguste Burdeau	G. Baillières	Collection of 30 papers by Spencer, abstracted from *Essays scientific, political and speculative*. 3 vol.
1877-1878	*Principes de biologie*	E. Cazelles	G. Baillières	
1878	*De l'Éducation intellectuelle, morale et physique*		G. Baillières	

Years	Title	Translator	Publisher	Comments
1878-1887	*Principes de sociologie*	E. Cazelles et J. Gerschel	G. Baillières	6 vol.
1880	*Les bases de la morale évolutionniste*		G. Baillières	*The Data of Ethics*. Volume of the "Bibliothèque scientifique internationale"
1880	*De l'éducation*		G. Baillières	Popular edition
1881	*Classification des sciences*	F. Réthoré	G. Baillières	
1885	*L'individu contre l'État*	J. Gerschel	Félix Alcan	Articles from the *Contemporary review*, 1884, collected under the title: *Man versus the State*.
1887	*L'éducation intellectuelle, morale et physique*	Alexis Bertrand	Vve E. Belin et fils	New translation, with notes and comments from the translator
1891	*Résumé de la philosophie de Herbert Spencer,* par F. Howard Collins, précédé d'une préface de M. Herbert Spencer	Henry de Varigny	F. Alcan	Foreword by Spencer
1893	*Justice*	Eloi Castelot	Guillaumin	Translation of Part IV of *Principles of Ethics*
1894	*Problèmes de morale et de sociologie*	Henry de Varigny	Guillaumin	Collected essays from various periodicals (1853-1893)
1895	*Le rôle moral de la bienfaisance*	Eloi Castelot	Guillaumin	Translation of Parts V and VI of *Principles of Ethics*
1895	*Le rôle moral de la bienfaisance*	Eloi Castelot and Etienne Martin Saint-Léon	Guillaumin	Revised translation
1895	*Le Principe de l'évolution, réponse à lord Salisbury*		Guillaumin	A paper published in *Le Journal des économistes*, 1895.
1896	*La morale des différents peuples et la morale personnelle*	Eloi Castelot and Etienne Martin Saint-Léon	Guillaumin	Translation of Parts II and III of *Principles of Ethics*
1898	*Les institutions professionnelles et industrielles, fin des Principes de sociologie*	Henry de Varigny	Guillaumin	Translation of Parts VII and VIII of *Principles of Sociology*
1903	*Faits et commentaires*	Auguste Dietrich	Hachette	
1907	*Une autobiographie*	Julie de Mestral-Combremont and Henry de Varigny	Guillaumin	Translated and adapted (abstracts from *An Autobiography*)
1909	*Qu'est-ce que la morale ?*	A. Desclos-Auricoste	Schleicher frères	New translation of *The Data of Ethics*

Afterword

Gowan Dawson and Gregory Radick

A long and still-living tradition depicts Herbert Spencer and Gilded Age America as made for each other, with Spencer cast as court philosopher to the robber barons. An early, classic expression dates from 1905, only two years after Spencer's death. Dissenting from the US Supreme Court's decision to overturn a New York law that had sought to limit the working hours of bakers, Oliver Wendell Holmes, Jr., waspishly complained that, contrary to the views of his conservative colleagues, "The Fourteenth Amendment does not enact Mr. Herbert Spencer's Social Statics." Four decades later, the progressive American historian Richard Hofstadter's *Social Darwinism in American Thought* (1944) likewise, and with enviable style and impressive scholarship, stressed what he saw as the singularly tight fit between Spencer's system and post-Civil War American culture – and the congruence of both with reform-resistant political conservatism. Nearer to our own day, we have Barry Werth's *Banquet at Delmonico's: Great Minds, the Gilded Age, and the Triumph of Evolution in America* (2008), the title referring to a famous, baron-studded dinner held in Spencer's honor in 1882, on his sole visit to the USA. As a work of popularized history, *Banquet at Delmonico's* is rightly absent from our editor Bernie Lightman's list, in his introduction, of recent Spenceriana. But Werth's book testifies to the continuing hold on the public imagination of the Holmes-Hofstadter version.[1]

At the same time, and with less fanfare, historians have been steadily undermining this version, ever more comprehensively. Thanks to their efforts, it is now plain that, geographically as well as politically, Spencerians could be found *everywhere* in the later nineteenth and early twentieth centuries. A representative example is Naomi Beck's contribution to a volume that does make Lightman's list, the 2004 collection *Herbert Spencer: The Intellectual Legacy*. There Beck analysed the progressive politics of Spencerian Italy, whose figurehead – as Paola Govoni shows in her chapter in this volume – was the socialist politician Enrico Ferri, author of an 1894 book setting out the case for seeing socialism as the logical consequence of Spencer's system. (Spencer interpreted his logic differently, and sent an angry letter of protest to an Italian

1 Richard Hofstadter, *Social Darwinism in American Thought* (Boston: Beacon Press, 1992; first published in 1944), ch. 2, esp. 31–2; Barry Werth, *Banquet at Delmonico's: Great Minds, the Gilded Age, and the Triumph of Evolution in America* (New York: Random House, 2008).

 | DOI 10.1163/9789004264007_015

newspaper.) A chapter in the same collection by John Laurent, mainly on Australia and New Zealand, noted how the heavy taxing of large properties in that part of the world in the early twentieth century could be traced back to the early Spencer's advocacy of "The Right to the Use of the Earth."[2]

Over and over again, *Global Spencerism* affirms this pluralizing re-direction in Spencer studies. The title of Mark Francis' chapter here on the US, "The Reforming Spencerians," could just as easily have been the title for any of the other chapters. So this new volume takes its place comfortably within the post-Hofstadter consensus on Spencer and politics. But it also throws new light on the question of what in Spencer's immense and varied oeuvre appealed to the reforming Spencerians, and how the local contexts of those reformers conditioned their interpretation and appropriation of his globally circulating work.

From New York to Damascus, and from Tokyo to Paris, people read Spencer. This simple fact is now something of a puzzle, as his writings have acquired a reputation for being dense and dull. Among other virtues, *Global Spencerism* helps us to recover – and understand – the enthusiasm with which Spencer's contemporaries read him. A few instances from the Anglophone world can stand for the wider, global class. After seeing the original prospectus for Spencer's system in a Boston bookstore, John Fiske wrote to a friend on June 24th, 1860, "my soul is on fire!" Later, when he read the "Laws of Organic Form," Fiske felt a "sense of awe" in the face of Spencer's omniscience.[3] Reading Spencer was almost a religious experience for Fiske. In 1895 the *Review of Reviews*, a popular British periodical, advised its readers that while they could use a recently published *Epitome* of Spencer's philosophy as "a refresher or index," they "must read [his work] through, not once or twice, but 'tearfully and prayerfully' many times over" in order to "assimilate its inner meaning."[4] Later in the same year, the American magazine *Arena* reported that a mother who "'became engrossed in Herbert Spencer's writings'" during pregnancy had

2 Naomi Beck, "The Diffusion of Spencerism and its Political Interpretations in France and Italy" and John Laurent, "Herbert Spencer's Influence on Economics," in Greta Jones and Robert A. Peel, eds., *Herbert Spencer: The Intellectual Legacy* (London: Galton Institute, 2004), chs. 3 & 4. For an overview of the volume, see the review by Radick in *The Galton Institute Newsletter* no. 52 (Sept. 2004): 5–6; available online at <http://www.galtoninstitute.org.uk/Newsletters/GINL0409/Herbert_ Spencer.htm>. Earlier critiques of Hofstadter's book include Robert C. Bannister, *Social Darwinism: Science and Myth in Anglo-American Social Thought* (Philadelphia: Temple University Press, 1979, reprinted with a new preface 1989) and Donald C. Bellomy, "'Social Darwinism' Revisited," *Perspectives in American History* 1 (1984): 1–129.

3 John Spencer Clark, *The Life and Letters of John Fiske*, 2 vols (Boston and New York: Houghton Mifflin Company, 1917), I, 230.

4 "Character sketch. Mr. Herbert Spencer," *Review of Reviews* 12 (1895), 399–407 (407).

given birth to a '"child [who] reflects the mother's mental condition in a most striking manner"' and is '"one of the finest reasoners...among children."'[5] If the past is a foreign country, a passport for entry into the era of Spencer is a willingness to take seriously the excitement that so many experienced in encountering his work.

Of course different audiences were interested in different aspects of that work. This leads us to a second striking insight from this volume. If we ask what, exactly, inspired the political reformers who embraced Spencer, the answer is in many cases a surprising one: his little book *Education: Intellectual, Moral, and Physical*, first published in 1860 in New York. For all that Spencer on evolution was and remains the best-known Spencer in the Anglophone world, Spencer on education was what attracted the most attention in such places as Russia, the Middle East, Japan, and China. This observation is so novel – and potentially so transformative of our picture of the global Spencer – that it is worth dwelling on at some length.

For anyone unfamiliar with it, *Education* consists of four chapters, each an essay published in journals in the late 1850s. The second, third and fourth chapters address the topics of Spencer's subtitle: intellectual aspects of education; moral aspects; physical aspects. "What knowledge is of most worth?" is the title of the first chapter, to which Spencer's gives a one-word and, in hindsight, predictable response: science. Still fascinating and gripping is how he builds to that answer, via a scathing critique of the British (but especially Anglican-Oxbridgian) tradition of classical education – in Spencer's portrait, a soul-and-body-destroying business of rote learning, enforced passivity, and unremitting focus on texts. Out of such a mis-education come students who know and, even worse, esteem pointlessly ornamental trivia, while remaining ignorant of the most basic facts about themselves and their world. Align teaching methods with science's discoveries about the mind and the laws of its development, Spencer advised, and teaching would look very different. Instead of stifling the spontaneous play of boys and girls, for example, teachers would encourage it, understanding that it was through such play that children learn at first hand some of the most important and enduring lessons the world has to teach them. For Spencer, experimental, experiential learning is the best kind; and when children are ready mentally for more formal instruction, it should be instruction in science.

Spencer made no claim to being original in what he recommended. He openly proclaimed the book as indebted to the work of the Swiss pedagogue

5 B.O. Flower, "The Right of the Child Considered in the Light of Heredity and Prenatal Influence," *Arena* 13 (1895), 243–62 (254).

and educational reformer J.H. Pestalozzi (1746–1827).[6] Yet derivative though it was – and in contrast to the difficult, abstract writing in much of the volumes that, from the 1860s onward, would accumulate into the synthetic philosophy – *Education* is wonderfully well written: vivid, iconoclastic, sympathetic, moving, and righteous. Spencer in these pages shows himself to be every bit as effective a writer of popular science as Robert Chambers in the wildly successful *Vestiges of the Natural History of Creation* (1844). To that extent, we need not wonder at *Education*'s status as a bestseller. *Education*, however, is not an evolutionary work, in anything like the metaphysically rich protozoa-to-politics, homogeneity-to-heterogeneity cosmic sense we now associate with "Spencerism." Yes, all of that comes up, and not unimportantly, as an explanation for why Pestalozzi's methods are sound.[7] But the surprising thing is how little the evolutionist element matters, which may be why the book has largely gone missing from history-of-science scholarship on Spencer.

We shall come back, under a different heading, to the global reading of Spencer on education. For now we wish to concentrate on the all-too-easily overlooked prerequisite for that global reading: the passage of the book, and Spencer's other books, into languages beyond English.[8] Needless to say, many of Spencer's readers were unable to read in English, so the existence of translations was crucial. In some cases, usually when the organs of publishing were well developed in a particular region, translations were produced and published in non-English speaking nations, with translators, sometimes famous in their own right and sometimes otherwise little known, operating as significant but generally overlooked "go-betweens" for the different national audiences for Spencer's writing.[9] For example, thirty of Spencer's works were translated into Japanese, while everything that Spencer published was translated into Russian. In the late seventies, the philosopher Harald Høffding translated a number of Spencer's works for Danish reading audiences. But even when translators indigenous to a region were not translating Spencer for the benefit

6 Spencer, *Education*, esp. 65–73. Spencer's own early education took place in the Pestalozzian school in Derby run by his father and, before him, his grandfather; see Piers J. Hale, *Political Descent: Malthus, Mutualism, and the Politics of Evolution in Victorian England* (Chicago: University of Chicago Press, 2014), 72.

7 Spencer, *Education*, e.g. 73–9.

8 On *Education*'s immense, international and, into the 1930s, enduring popularity, see F.A. Cavenagh's introduction to *Herbert Spencer on Education* (Cambridge: Cambridge University Press, 1932), xx.

9 On the role of so-called "go-betweens" in the global circulation of knowledge, see the essays in Simon Schaffer, Lissa Roberts, Kapil Raj, and James Delbourgo, eds., *The Brokered World: Go-Betweens and Global Intelligence, 1770–1820* (Sagamore Beach, MA: Watson, 2009).

of their fellow countrymen, Spencer's ideas could reach readers through translations prepared outside the country. Readers in French-speaking countries, for example, could obtain translations of Spencer's work produced by French publishers, although this could entail that the often complex power relations between centre and periphery operated even at the level of translations.[10] In fact, in some countries where French was neither the indigenous nor main alternative language – Italy for example – Spencer's philosophy was first introduced through French translations. Similarly, since few Spencer texts were available to Mexican readers in English or Spanish, they had no choice but to read Spencer in French.

It is well known that the process of translation can change meaning; so it was with the translation of Spencer's work. Not only was it sometimes difficult to find an exact parallel for the English words that Spencer used to discuss subjects such as evolution, but translators, as unheralded yet nonetheless very powerful mediators, could purposely modify his meaning to suit their aims or make his ideas more acceptable to their audience.[11] Missionaries, such as Yen Yung Kiung, could insert elements of natural theology into their translations of Spencer's work, while in Japan key terms used by Spencer could be translated to resonate powerfully with Confucian philosophy. Translators could also provide a substantial foreword or introduction, suggesting their particular interpretive approach to the philosopher of evolution. A case in point was Émile Cazelles, who included a long foreword to his French translation of Spencer's *First Principles* in 1871. The appearance of a translation of one of Spencer's works often determined when intellectuals began to engage with his ideas. Spencer did not become important in Italy until after the appearance of the first translation of *Education* in 1876. Tracking how long translations continued to be published gives us some indication of when it was seen to be advantageous to appropriate Spencer in a particular region. The publication of new translations of his work into the early twentieth century in Italy provides compelling evidence that his reputation remained strong there long after it had declined in other parts of the world.

---

10 On the complex relations between the centre and periphery in nineteenth-century science, see, for instance, Ruth Barton, "Haast and the Moa: Reversing the Tyranny of Distance," *Pacific Science* 54 (2000), 251–63; and Jim Endersby, *Imperial Nature: Joseph Hooker and the Practices of Victorian Science* (Chicago: University of Chicago Press, 2008), esp. 110.

11 On the new meanings given to scientific texts in the process of translation, see Nicolaas Rupke, "Translation Studies in the History of Science: The Example of *Vestiges*," *British Journal for the History of Science* 33 (2000), 209–22. See also Bernard Lightman, "Lost in Translation: Evolutionary Naturalists and Their Language Games," *History of Science* 53 (2015), forthcoming.

Close study of the translation of Spencer's writings can illuminate their many contexts. So can better understanding of the global dissemination and appropriation of Spencer's works. In Britain, many of Spencer's most avid supporters were professional men of science, such as his allies T.H. Huxley and John Tyndall: men who spoke with tremendous cultural authority, and whose backing helped to give Spencer scientific legitimacy (notwithstanding Huxley's cruel jibe that for Spencer a "tragedy was a deduction killed by a fact").[12] But the British situation now appears quite anomalous. In some national contexts, this volume teaches us, there was no well-defined scientific community to support Spencer. Instead, political and social thinkers, psychologists, journalists, and/or novelists became his champions, drawing on local cultural traditions to shape interpretation in desired directions.

Those directions were, of course, myriad. In fairness to the Holmes-Hofstadter version, some conservatives did find uses for Spencer, by selecting facets of his laissez-faire evolutionism to defend the notion of very slow change unimpeded by state interference. But to liberals and socialists who welcomed a break with the past, political or social upheaval provided the occasion to recruit Spencer's writings as legitimating the needed reform. And upheaval was ubiquitous. In Russia it was the era of Great Reforms in the 1860s; in Italy, the Italian Unification of 1861; in the United States, the Civil War (1861–1865); and in Japan the overthrow of the Tokugawa Shogunate in 1868. Political events later in the century could also show Spencer in a newly relevant light. The Sino-Japanese War in 1894, for example, led some Chinese intellectuals to look to Spencer for guidance in planning the future of their country.

By way of more concrete illustration of how Spencer's work could energize reform, we return to his *Education* and consider how it figures in a sample – small but representative – of the chapters in *Global Spencerism*. To start with Michael Gordin's chapter on imperial Russia, we learn there that *Education* is the only book of Spencer about which any of Gordin's characters had anything nice to say. And these were some characters! Dostoyevsky owned and read *Education*, maybe the 1866 translation later published in the Russian collected works. He made a non-satiric appropriation in an 1875 novel, *A Raw Youth*. He generally seems to have been such an admirer that he was much more tolerant, in both his correspondence and in his work as a "thick-journal" editor, of the rest of Spencer's oeuvre, given Dostoyevsky's generally conservative politics as well as the attractions of Spencer's evolutionary writings in Russia for radicals such as Mikhailovskii. Chekhov's engagement with *Education* was less visible artistically, but was nevertheless, Gordin reveals, prolonged and, again, very

12 Herbert Spencer, *An Autobiography*, 2 vols (London: Williams and Norgate, 1904), 1:109.

positive, leading him to advocate rejection of rote learning and equality of education between girls and boys.

Turning, with Marwa Elshakry's chapter, southward and westward to Egypt and the larger Arabic world of letters, she introduces us to the Grand Mufti Muhammed 'Abduh, a founder of Islamic modernism, but also a huge admirer of Spencer's, and above all of his *Education*, of which 'Abduh is thought to have done his own translation. Elshakry tells us that Spencer featured in 'Abduh's lectures on the Koran at Egypt's oldest mosque, and that 'Abduh even visited Spencer at his home on a visit to Britain (One wonders how it went…).[13] Her description of the 'Abduh-Spencer relationship helps us see why Spencer on education resonated so strongly with the intellectual elite as the British empire extended into the Ottoman sphere, and why that elite regarded *Education* as Spencer's "most important book," as 'Abduh wrote in 1903. The book spoke to new aspirations to break away from what men such as 'Abduh perceived as Ottoman backwardness and despotism, the better to put their people on the path to a gradually attained modernity. Yet it spoke equally to old ideals of education as character building, of moral upstandingness as the product of a good education, and of the family as the foundation of a good society. (Again, Spencer wrote much in *Education* about the need for parents to learn from science how to be good parents, to ensure that they brought up happy, healthy, good children.)

We encounter a Japanese counterpart to 'Abduh in G. Clinton Godart's chapter on Meiji-period Japan. Mori Arinori enjoyed a high political position from which he was able to enact his enthusiasm for Spencer more consequentially than 'Abduh. An early Meiji leader, Arinori was Japan's ambassador to Britain between 1879 and 1884, during which time he met Spencer, consulted with him, even socialized with him at the Athenaeum club (where Spencer attended his farewell party). When, later in his career, Arinori became Japan's Minister of Education, he began a programme of putting Spencer's *Education* into practice, notably by instilling an ideal of balance between three types of education, intellectual, moral and physical, with the aim of creating self-directed, self-fulfilled individuals whose goodness – that is, in Spencerian terms, their altruism, tempered with egoism – would hold society together progressively without the old, despotic military control. From Godart we learn that the form of military-style physical education Arinori sponsored was abandoned after 1945 (and it was arguably never exactly what Spencer himself would have

13 As 'Abduh does not come up either in Spencer's *Life and Letters* or in his *Autobiography*, the encounter does not appear to have been especially memorable on Spencer's side. We are grateful to Bernie Lightman for this point.

liked). Nevertheless, the "three educations" ideal lives on, as does memory of Spencer's greatest moment of potential political impact, when Mori sought Spencer's advice on Japan's constitution. In Spencer's view, as Godart relates the tale, the Japanese reformers were seeking too wrenchingly total and rapid a transition from despotism to constitutional government, and he advised them instead to go slower and increase freedoms only gradually, so as to guard against destabilizing the nation. Anyone doubting the extraordinary authority of Spencer for certain segments of his readership would do well to consider this episode closely.

The title of Ke Zunke and Li Bin's chapter, "Spencer and Science Education in China," accurately conveys its content. Like Mori Arinori, Yen Yung Kiung, the translator into Chinese of the first essay in *Education* – a translation which, from its first publication in 1881, was pretty much the only work of Spencer's available in Chinese until the very end of the nineteenth century – spent time as a young man in the USA among Asian-welcoming Christians. (It may be that, after all, Hofstadter was onto something in emphasizing the Americanness of Spencer.) As Ke Zunke and Li Bin show, Yen Yung Kiung taught natural sciences at a Christian college in Shanghai, where he came to despise Chinese traditional education, seeing in it a Confucian version of the English classical education Spencer railed against, and blaming it for what he saw as China's stagnant, backward condition. For the likes of Kiung, making Spencer on the value of science education available in Chinese was a pointed way of making the case that science education was what China needed if it was ever to join the modern world.

Francis' chapter on the USA, by contrast, presents us with what appears to be an exception to the emerging rule. On his telling, Spencer's *Education* is nowhere to be seen in the engagements of the American pragmatist philosophers James, Royce and Dewey with Spencer. Given what many of us think we know about Dewey and educational reform in the USA, this absence comes as a bit of a surprise at least in his case. But Francis is quite clear: "there is no support for the commonly held belief that Dewey's ideas were *deeply* rooted in his early reading of Spencer or of other scientifically-minded writers who had explored the habitual and instinctual origins of the human mind" (p. 116 of this volume). Other scholars have assessed the evidence for such rootedness differently. A notable recent example is Kieran Egan, in a 2002 book on progressivist education from Spencer to Dewey and beyond. Egan's *Getting It Wrong from the Beginning* signals in its title his own polemical intent, and clearly it is part of his discrediting strategy to father progressive education onto that biologizing amateur Herbert Spencer. One wonders how the situation would look if one could only strip away some of the distorting current-day politics.[14]

In sum, we find, in tracking the global spoor of Spencer's work, patterns of appropriation quite different from the repetitive processes that James Secord pointed to in his study of globalized Darwin (noted in this volume's introduction). According to Secord, in studies of Darwin's reception one usually finds stories that begin with opposing parties for and against, but that end in some form of accommodation.[15] Although for Spencer the stories all eventually end more or less the same, in some form or other of extinction and invisibility, they do not follow the same path to that conclusion. In some cases, such as Scandinavia, France, and Japan, we see a rise and fall pattern; in Russia, there is no rise and fall as Spencer is never influential in a straightforward way; and in the instance of Italy, the rise, fall, rise, and fall pattern resembles a roller-coaster ride. Again, the same printed objects bearing Spencer's authorial imprimatur could be interpreted in ways that suited the religious requirements of American audiences, as Lightman shows in his chapter, the socialist politics of readers in Italy, as Govoni makes clear, or the statist predilections of Scandinavians, as Hans Henrik Hjermitslev's chapter reveals. Spencer's authority, even within a single national context, could be (and was) used to support entirely contradictory positions: both optimism and pessimism in Argentina, as Adriana Novoa shows; different interpretations of the size of the original population of Brazil, as Heloisa Maria Bertol Domingues discusses; and both spiritualism and materialism in France, as Nathalie Richard demonstrates. Underpinning these divergences often are new texts: new translations, new editions, and/or new journalistic summaries. The process is similar to the one Secord has elsewhere termed "literary replication," in which "textual stability … [is] difficult to achieve," and reprinted texts such as translations or abstracts in newspapers become radically unstable and are denuded of the same range of meanings they previously held.[16]

In this light, Richard's observation that there was a distinctive French Spencer who largely differed from his English counterpart is especially suggestive. There is something rather uncanny and ghostly about the thought that the

---

14 Kieran Egan, *Getting it Wrong from the Beginning: Our Progressivist Inheritance, from Herbert Spencer, John Dewey, and Jean Piaget* (New Haven: Yale University Press, 2002), esp. 141–8.

15 James A. Secord, "Global Darwin," in *Darwin*, ed. by William Brown and Andrew C. Fabian (Cambridge: Cambridge University Press, 2010), 31–57.

16 See James A. Secord, *Victorian Sensation: The Extraordinary Publication, Reception, and Secret Authorship of "Vestiges of the Natural History of Creation"* (Chicago: University of Chicago Press, 2000), 126.

Gallic Spencer might ever have met the other Herbert across the Channel, but it nicely captures the way that authors exist as the creations, or projections, of their readers, who construct them in diverse ways, attuned to local concerns that are largely beyond the authors' own control (and many of these chapters reveal Spencer's frequent frustration with how his work was interpreted across the world). Rather than simply considering these varying local or national contexts in isolation, we should examine their interaction, as particular international crosscurrents evidently had an important impact on how Spencer's work was received in different locations and by particular groups. For instance, Richard's chapter reveals that the principal French translator of Spencer's work, Cazelles, interpreted it through the prism of his lay spiritualism. This becomes doubly important when we realize, as Novoa's and Domingues' chapters show, that in both Argentina and Brazil Spencer was read almost exclusively in French translations, with Spanish and Portuguese ones becoming available only much later.

Different national Spencers, then, have had occasion to meet and interact with one another. And to continue in this slightly spectral vein, we note too how the different timescales of the circulation of Spencer's ideas in different national contexts – with his influence in both France and Argentina developing quickly but then declining equally rapidly by the 1890s, while in Italy and Brazil he was a latecomer to the intellectual scene and, consequently, his influence lasted much later, and even well into the twentieth century – suggest the concept of an afterlife, or what has recently been called "undead science": a term used in social studies of contemporary science to signify the persistence of theories (such as cold fusion) that are supposedly rejected at the supposed closure of controversies.[17] Thanks to Global Spencerism, there may well have been – and perhaps may even now be – a spectral afterlife for Spencer in certain parts of the world long after his work was discarded in other places.[18] Thanks to *Global Spencerism*, historians can begin to investigate those global legacies and their interactions afresh.

17 Bart Simon, *Undead Science: Science Studies and the Afterlife of Cold Fusion* (New Brunswick: Rutgers University Press, 2002).

18 During the final editing of this Afterword, a new volume on Spencer's legacies appeared: Mark Francis and Michael W. Taylor, eds., *Herbert Spencer: Legacies* (London: Taylor and Francis, 2014).

Selected Scholarly Works on Global Spencerism

Alonso, Angela. *Ideias em Movimento – a Geração 70 e a Crise do Brasil Império*. São Paulo: Paz e Terra, 2002.

Andermann, Jens. *La Ciencia en la Argentina entre Siglos: Textos, Contextos e Instituciones*. Buenos Aires: Ediciones Manantial, 2000.

Andersen, Heine. "Modtagelsen af Spencer i Danmark – i tiden før 1920." *Dansk Sociologi* 19, no. 1 (2008): 103–131.

Azouvi François. *La Gloire de Bergson, Essai sur le Magistère Philosophique*. Paris: Gallimard, 2007.

Bannister, R.C. "'The Survival of the Fittest is our Doctrine': History or histrionics?" *Journal of the History of Ideas* 31 (1970): 370–98.

———. *Social Darwinism: Science and Myth in Anglo-American Social Thought*. Philadelphia, PA: Temple University Press, 1979.

———. "William Graham Sumner's Social Darwinism: A Reconsideration." *History of Political Economy* 5, no. 1 (1973): 89–109.

Bazhanov, V.A. *Vospriiatie Britanskoi sotsial'no-filosofskoi mysli v Rossii* (*XIX-nachalo XX vv*). Ul'ianovsk: ULGU, 2005.

Beck, Naomi. "Enrico Ferri's Socialism: A Marxist Interpretation of Herbert Spencer's Organic Analogy." *Journal of the History of Biology* 38 (2005): 301–25.

———. *La Gauche Évolutionniste. Spencer et Ses Lecteurs en France et en Italie*. Besançon: Presses Universitaires de Franche-Comté, 2014.

Becquemont, Daniel, and Mucchielli Laurent. *Le Cas Spencer. Religion, Science et Politique*. Paris: PUF, 1998.

Becquemont, Daniel, and Ottavi Dominique, eds. *Penser Spencer*. Saint-Denis: Presses Universitaires de Vincennes, 2011.

Beller, Walter, Bernardo Méndez, and Santiago Ramírez, eds. *El Positivismo Mexicano*. Mexico City: Universidad Autónoma Metropolitana Xochimilco, 1985.

Bender, Niklas, ed. *Herbert Spencer en France: Mise au Jour d'une Influence*, special issue *Arts et Savoirs*, 2014, 4, online periodical <http://www.fabula.org/actualites/arts-et-savoirs-n-4herbert-spencer-en-france-mise-au-jour-d-une-influence_63308.php>.

Berman, Milton. *John Fiske: The Evolution of a Popularizer*. Cambridge: Harvard University Press, 1961.

Biagini, Hugo E., ed. *El Movimiento Positivista Argentino*. Buenos Aires: Editorial de Belgrano, 1985.

Billington, James H. *Mikhailovsky and Russian Populism*. New York: Oxford University Press, 1958.

Blais, Marie-Claude. *Au Principe de la République. Le Cas Renouvier*. Paris: Gallimard, 2000.

Bliksrud, Liv, Geir Hestmark, and Tarald Rasmussen. *Norsk idéhistorie, bind IV: Vitenskapens Utfordringer*. Oslo: Aschehoug, 2002.

Breymann, Walter N. "The Científicos: Critics of the Díaz Regime, 1892–1903." *Proceedings of the Arkansas Academy of Science* 7 (1954): 91–97.

Brooks, John I., *The Eclectic Legacy. Academic Philosophy and the Human Sciences in Nineteenth-Century France*. Newark: University of Delaware Press, 1998.

Burns, E. Bradford. *The Poverty of Progress: Latin America in the Nineteenth Century*. Oakland, California: University of California Press, 1980.

Carneiro, Robert L. "Herbert Spencer as an Anthropologist." *Journal of Libertarian Studies* 5, no. 2 (1981): 153–210.

Castro, Nayelli, and Clara Foz. "La Circulación de las Ideas Positivistas en Argentina y en México: Editores y Traductores (1850–1950)." *MonTI* 5 (2013): 365–388.

Charle, Christophe. *Discordance des Temps. Une Brève Histoire de la Modernité*. Paris: A. Colin, 2011.

Clasen, Mathias, Stine S. Grumsen, Hans H. Hjermitslev, and Peter C. Kjærgaard. "Translation and Transition: The Danish Literary Response to Darwin." in *The Literary and Cultural Reception of Charles Darwin in Europe*, vol. III, edited by Thomas F. Glick and Elinor Shaffer, 128–59. London: Bloomsbury Publishing, 2014.

Collichio, Terezinha, *Miranda de Azevedo e o Darwinismo no Brasil*. São Paulo: EDUSP, 1988.

Conry, Yvette, *L'Introduction du Darwinisme en France au XIXe* Siècle. Paris: Librairie Philosophique J. Vrin, 1974.

Crook, D.P. *Darwin's Coat-Tails: Essays on Social Darwinism*. New York: Peter Lang Publishing Inc., 2007.

Cruz Costa, João. *Contribuição à História das Ideias no Brasil*. Rio de Janeiro: Livraria José Olympio Editora, 1956.

Cunha, Dulce F. Fernandes da. *A Biblioteca do Museu Nacional do Rio de Janeiro*. Rio de Janeiro: Museu Nacional, 1966.

Danielsson, Ulf. "Darwinismens inträngande i Sverige." *Lychnos* (1963–64): 157–210; (1965–66): 262–334.

Dantes, Maria Amélia M. "Institutos de Pesquisa Científica no Brasil." in *História das Ciências no Brasil*, edited by Mario G. Ferri and Motoyama, Shozo, 2º vol., 341–380. São Paulo: EPU, Editora da USP, 1979/81.

Davis, Winston. *The Moral and Political Naturalism of Baron Kato Hiroyuki*. Berkeley: University of California Press, 1995.

de Gortári, Eli. *La Ciencia en la Historia de México*. Mexico City: Editorial Grijalbo, 1980.

de María y Campos, Alfonso. "Porfirianos prominentes: orígenes y años de juventud de ocho integrantes del grupo de los Científicos, 1846–187." *Historia Mexicana* (1985): 610–661.

Domingues, Heloisa M. Bertol, Magali R. Sá, and Thomas S. Glick. *A Recepção do Darwinismo no Brasil*. Rio: Ed. Fiocruz, 2003.

Egan, K. *Getting it Wrong from the Beginning: Our Progressive Inheritance from Herbert Spencer, John Dewey, and Jean Piaget*. New Haven, CT: Yale University Press, 2002.

Eisen, Sydney. "Herbert Spencer and the Spectre of Comte." *Journal of British Studies* 7 (1967): 48–67.

Elman, Benjamin A. *On Their Own Terms: Science in China, 1550–1990*. Cambridge, Mass.: Harvard University Press, 2005.

Elshakry, Marwa. *Reading Darwin in Arabic, 1860–1950*. Chicago: University of Chicago Press, 2013.

Ferrarotti, Franco. "Introduzione," in Herbert Spencer, *Principi di sociologia*, ed. Franco Ferrarotti, 2 vols. Turin: UTET, [1967] 1988, 7–61.

Flower, Elizabeth. "The Mexican Revolt Against Positivism." *Journal of the History of Ideas* 10, no. 1 (1949): 115–129.

Francis, Mark. *Herbert Spencer and the Invention of Modern Life*. Durham: Acumen, 2007.

———."Herbert Spencer." in *The Wiley-Blackwell Companion to Major Social Theorists*, edited by G. Ritzer & J. Stepinsky, vol. 2, 165–84. Oxford: Wiley-Blackwell, 2011.

———. "The Evolutionary Turn in Positivism." in *The Oxford Handbook of British Philosophy in the Nineteenth Century*, edited by W.J. Mander, 285–305. Oxford: Oxford University Press, 2014.

Frängsmyr, Tore. *Svensk idéhistoria: Bildning och vetenskap under tusen år*, vol. II. Stockholm: Natur och Kultur, 2000.

Freeman, Derek, Carl Jay Bajema, John Blacking, Robert L. Carneiro, U.M. Cowgill, Santiago Genovés, Charles C. Gillispie *et al*. "The Evolutionary Theories of Charles Darwin and Herbert Spencer [and Comments and Replies]." *Current Anthropology* (1974): 211–237.

Garin, Eugenio. "Il Positivismo Italiano alla fine del secolo XIX." *Giornale Critico della Filosofia* 1 (1980): 1–27.

Glick, Thomas F., Miguel Angel Puig-Samper, and Rosaura Ruiz, eds. *The Reception of Darwinism in the Iberian World: Spain, Spanish America, and Brazil*. New York: Springer, 2001.

———. "The Comparative Reception of Darwinism: A Brief History." *Science & Education* 19, no. 6–8 (2010): 693–703.

Godfrey-Smith, Peter. "Spencer and Dewey on Life and Mind." in *Artificial life IV*, edited by R. Brooks and P. Maes, 80–89. Cambridge, MA: MIT Press, 1994.

González Ascencio, Gerardo. "Positivismo y organicismo en México a fines del siglo XIX." *Alegatos* (2010): 693–724.

Gray, Tim. *The Political Philosophy of Herbert Spencer: Individualism and Organicism*. Aldershot: Avebury, 1996.

Greene, John C. "Biology and Social Theory in the Nineteenth Century: Auguste Comte and Herbert Spencer." in *Critical Problems in the History of Science*, edited by M. Clagett, 419–46. Madison: University of Wisconsin Press, 1959.

Gusdorf, George. *Les Príncipes de la Pensée au Siècle des Lumières*. Paris: Payot, 1971.

Haines, Valerie A. "Is Spencer's Theory an Evolutionary Theory?" *American Journal of Sociology* 93, no. 5 (1988): 1200–1223.

———. "Spencer, Darwin, and the Question of Reciprocal Influence." *Journal of the History of Biology* 24, no. 3 (Fall 1991): 409–431.

Hale, Charles Adams. *Justo Sierra: un Liberal del Porfiriato*. Mexico City: Fondo de Cultura Económica, 1997.

———. *La Transformación del Liberalismo en México a fines del siglo XIX*. Mexico City: Fondo de Cultura Económica, 2002.

Hallewell, Laurence. *O livro no Brasil*. São Paulo: EDUSP, 1985.

Hawkins, Mike. *Social Darwinism in European and American Thought, 1860–1945: Nature as Model and Nature as Threat*. Cambridge: Cambridge University Press, 1997.

Hjermitslev, Hans H. "Danes Commemorating Darwin: Apes and Evolution at the 1909 Anniversary." *Annals of Science* 67, no. 4 (2010): 485–525.

———. "Protestant Responses to Darwinism in Denmark, 1859–1914." *Journal of the History of Ideas* 72, no. 2 (2011): 279–303.

Hofstadter, Richard. *Social Darwinism in American Thought*. Boston: Beacon Press, 1955.

Howell, David. "Society Reified: Herbert Spencer and Political Theory in Early Meiji Japan." *Comparative Studies in Society and History* 42, no. 1 (2000): 67–86.

Howland, Douglas. *Translating the West: Language and Political Reason in Nineteenth Century Japan*. Honolulu: University of Hawai'i Press, 2002.

Howsam, Leslie. "An Experiment with Science for the Nineteenth-century Book Trade: the International Scientific Series." *The British Journal for the History of Science* 33 (2000): 187–207.

Hung-chun, Zen, trans. *Jiaoyu lun*. Shanghai: Shangwu yinshuguan, 1923.

Jones, Greta. *Social Darwinism and English Thought: The Interaction Between Biological and Social Theory*. New York: Humanities Press, 1980.

Jones, Greta, and R.A. Peel, eds. *Herbert Spencer: The Intellectual Legacy*. London: Galton Institute, 2004.

Kazamias, A. *Herbert Spencer on Education*. New York: Teachers College Press, 1966.

Knapp, Shoshana. "Herbert Spencer in Čexov's 'Skučnaja Istorija' and 'Duél': The Love of Science and the Science of Love." *Slavic and East European Journal* 29, no. 3 (Autumn 1985): 279–296.

Koch, Carl Henrik. *Dansk filosofi i positivismens tidsalder 1880–1950*. Copenhagen: Gyldendal, 2004.

Lalouette, Jacqueline. *La Libre-Pensée en France, 1848–1940*. Paris: A. Michel, 1997.

Lanaro, Giorgio. *L'evoluzione, il progresso e la società industriale. Un profilo di Herbert Spencer*. Florence: Nuova Italia, 1997.

Lang, Xiao. "Kejiao xingguo de qianyin: Sibinsai jiaoyu sixiang zai jindai Zhongguo." *Huadong shifan daxue xuebao(jiaoyu kexue ban)* no.2 (2008): 88–96.

Lauwerys, J.A. "Herbert Spencer and the Scientific Movement." in *Pioneers of English Education*, edited by A.V. Judges, 160–93. London: Faber and Faber, 1952.

La Vergata, Antonello. "Darwinismo, Evolução e Guerra." in *Darwinismo, Meio Ambiente, Sociedade*, edited by Heloisa M. Bertol Domingues, Magali R. Sá, Miguel Angel Puig-Samper, and Rosaura Gutiérrez, 237–260. São Paulo: Via Lettera, Rio, MAST, 2009.

———. "Herbert Spencer: Biology, Sociology, and Cosmic Evolution." in *Biology as Society, Society as Biology: Metaphors*, edited by Sabin Maasen, Everett Mendelsohn, and Peter Weingart, 193–229. Boston: Kluwer Academic Publishers, 1995.

Levine, Alex, and Adriana Novoa. *¡Darwinistas! The Construction of Evolutionary Thought in Nineteenth Century Argentina*. Leiden: Brill, 2012.

Lie, Thore. "The Introduction, Interpretation and Dissemination of Darwinism in Norway during the period 1860–1890." in *The Reception of Charles Darwin in Europe*, edited by Eve-Marie Engels and Thomas F. Glick, 156–174. London: Continuum, 2008.

Liedman, Sven-Eric. "Sweden and Victorian Britain." *The European Legacy: Toward New Paradigms* 4, no. 6 (1999): 72–83.

Lightman, Bernard. "Ideology, Evolution and Late-Victorian Agnostic Popularizers," in *History, Humanity and Evolution: Essays for John C. Greene*, edited by James R. Moore, 285–309. Cambridge: Cambridge University Press, 1989.

———. "The International Scientific Series and the Communication of Darwinism," *Journal of Cambridge Studies* 5 (December 2010): 27–38.

Liru, Zhao. "Youguan *Xinlingxue* yishu de yanjiu." *Xinli xuebao* no. 4 (1983): 380–388.

Matute, Álvaro. "Notas sobre la historiografía positivista mexicana." *Secuencia. Revista de Historia y Ciencias Sociales* 21 (1992): 49–64.

Mead, Karen. "Gendering the Obstacles to Progress in Positivist Argentina, 1880–1920." *Hispanic American Historical Review* (1997): 645–675.

Micelli, Sergio, Org. *História das Ciências Sociais no Brasil*. São Paulo: Vértice, Editora dos Tribunais, IDESP, 1989.

Michio, Nagai. "Herbert Spencer in Early Meiji Japan." *The Far Eastern Quarterly* 14, no. 1 (1954): 55–64.

Miller, Nicola. *In the Shadow of the State: Intellectuals and the Quest for National Identity in Twentieth-Century Spanish America*. London: Verso, 1999.

Montserrat, Marcelo. "La presencia evolucionista en el positivismo argentino." *Quipu* 3, no. 1 (1986): 91–101.

Moore, James R. "Herbert Spencer's Henchman." in *Darwinism and Divinity: Essays on Evolution and Religious Belief*, edited by John R. Durant, 76–100. Oxford: Basil Blackwell, 1985.

Novoa, Adriana, and Alex Levine. *From Man to Ape: Darwinism in Argentina, 1870–1920.* Chicago: University of Chicago Press, 2010.

Offer, J. *Herbert Spencer and Social Theory.* Basingstoke: Palgrave Macmillan, 2010.

———, ed. *Herbert Spencer: Critical Assessments of Leading Sociologists.* 4 vols. London: Routledge, 2000.

Papa, Emilio R., ed. *Il positivismo e la cultura italiana.* Milan: Angeli, 1985.

Peel, J.D.Y. *Herbert Spencer: the Evolution of a Sociologist.* London: Heinemann, 1971.

———. "Spencer and the Neo-Evolutionists." *Sociology,* 3, no. 2 (1969): 173–191.

Perrin, Robert G. *Herbert Spencer: A Primary and Secondary Bibliography.* New York and London: Garland Publishing Inc., 1993.

———. "Herbert Spencer's Four Theories of Social Evolution." *American Journal of Sociology* 81 (1976): 1347–1359.

Plumyene, Jean. *Les Nations Romantiques.* Paris: Fayard, 1979.

Podgorny, Irina, and María Margaret Lopes. *El desierto en una vitrina: museos e historia natural en la Argentina, 1870–1890.* Rosario: Editorial LIMUSA, 2008.

Raat, William Dirk. "Los intelectuales, el positivismo y la cuestión indígena." *Historia Mexicana* 20 (1971): 412–427.

Ramos-Escandón, Carmen. "Genaro García, historiador feminista de fin de siglo." *Signos Históricos* 5 (2001): 87–107.

Rem, Tore. "Darwin and Norwegian Literature." in *The Literary and Cultural Reception of Charles Darwin in Europe,* vol. III, edited by Thomas F. Glick and Elinor Shaffer, 160–180. London: Bloomsbury Publishing, 2014.

Richard, Nathalie. *Hippolyte Taine. Histoire, philosophie, littérature.* Paris: Garnier, 2013.

Richards, R.J. "The Moral Foundations of the Idea of Evolutionary Progress: Darwin, Spencer and the Neo-Darwinians," in *Evolutionary Progress,* edited by M. Nitecki, 129–48. Chicago: Chicago University Press, 1988.

Rodríguez, Julia. *Civilizing Argentina: Science, Medicine, and the Modern State.* Chapel Hill: University of North Carolina Press, 2006.

Rossi, Paolo, ed. *L'età del positivismo.* Bologna: il Mulino, 1986.

Ruiz Gutiérrez, Rosaura. *Positivismo y evolución: introducción del darwinismo en México.* Mexico City: Ed. Limusa: 1991.

Saez, Carmen. "'La Libertad,' periódico de la dictadura porfirista." *Revista Mexicana de Sociología* (1986): 217–236.

Santucci, Antonio, ed. *Scienza e filosofia nella cultura positivistica.* Milan: Feltrinelli, 1982.

Shkurinov, P.S. *Pozitivizm v Rossii XIX veka.* Moscow: Izd. Moskovskogo universiteta, 1980.

Shuchang, Yan. "Yan Yongjing dui xifang xinlixue yinru ji qi hanyu xinlixue shuyu chuangzhi." *Nanjing shida xuebao(shehui kexue ban),* no.4 (2012): 116–120.

Shwarcz, Lilia K.M. "Usos e abusos da mestiçagem e da raça no Brasil: uma história das teorias raciais em finais do século XIX." *Afro-Ásia* 18 (1996): 77–101.

Silberman, Richard. "Herbert Spencer on Education – Prophet or False Prophet." *Journal of Education* 184, no. 2 (2003): 85–122.

Skovgaard-Petersen, Vagn. "De blev lærere – læreruddannelse mellem 1860 og 1945." in *For at blive en god lærer: Seminarier i to århundreder*, edited by Karen B. Braad, Christian Larsen, Ingrid Markussen, Erik Nørr, and Vagn Skovgaard-Petersen, 179–312. Odense: Syddansk Universitetsforlag, 2005.

Smith, C.U.M. "Evolution and the Problem of Mind: Part I – Herbert Spencer." *Journal of the History of Biology* 15, no. 1 (Spring 1982): 55–88.

Sodré, Nelson W. *História da Imprensa no Brasil*. Rio de Janeiro: Ed. Civlização Brasileira, 1966.

Soler, Ricaurte. *El positivismo argentine*. Buenos Aires: Paidós, 1968.

Sora, Gustavo. *Brasilianas: José Olympio e a gênese do mercado editorial brasileiro*. São Paulo: EDUSP, Com Arte, 2010.

Stoetzer, O. Carlos. *Karl Christian Friedrich Krause and His Influence in the Hispanic World*. Köln: Böhlau, 1998.

Swale, Alister. *The Political Thought of Mori Arinori: A Study of Meiji Conservatism*. London: Routledge, 2000.

Taylor, Michael W. *Herbert Spencer and the Limits of the State: The Late Nineteenth Century Debate between Individualism and Collectivism*. Bristol: Thoemmes Press, 1996.

———. *Men versus the State: Herbert Spencer and Late Victorian Individualism*. Oxford: Clarendon Press, 1992.

———. *The Philosophy of Herbert Spencer.* London: Continuum, 2007.

Terán, Oscar. *Vida intelectual en el Buenos Aires fin-de-siglo (1880–1910): Derivas de la cultura científica*. Mexico: Fondo De Cultura Economica, 2000.

Terán, Oscar, and José María Ramos Mejía. *Positivismo y nación en la Argentina*. Buenos Aires: Puntosur, 1987.

Tesnières Valérie. *Le Quadrige. Un siècle d'édition universitaire (1860–1968)*. Paris: PUF, 2001.

Thomas, Julia. *Reconfiguring Modernity: Concepts of Nature in Japanese Political Ideology*. Berkeley: University of California Press, 2002.

Tiangen, Wang. "Yan Fu yu Zhang Taiyan shehuixue sixiang de duizhi yu jiaoliu." *Guangxi daxue xuebao(zhexue shehui kexue ban)* no. 1 (2003): 49–54.

Tomlinson, Stephen. "From Rousseau to Evolutionism: Herbert Spencer on the Science of Education." *History of Education* 25, no. 3 (September 1996): 235–54.

Tort, Patrick. *Pour Darwin*. Paris: PUF, 2001.

Toscano, Mario A. *Malgrado la storia: Per una lettura critica di Herbert Spencer*. Milan: Feltrinelli, 1980.

Turner, Jonathan H. *Herbert Spencer: A Renewed Appreciation*. Beverly Hills: Sage, 1985.

Utkina, N.F. *Pozitivizm, antropologicheskii materializm i nauka v Rossii (vtoraia polovina XIX veka)*. Moscow: Nauka, 1975.

Vucinich, Alexander. *Social Thought in Tsarist Russia: The Quest for a General Science of Society, 1861–1917*. Chicago: University of Chicago Press, 1976.

Walicki, Andrzej. *A History of Russian Thought: From the Enlightenment to Marxism*, tr. Hilda Andres-Rusiecka. Stanford, CA: Stanford University Press, 1979 [1973].

Weinstein, D. *Equal Freedom and Utility: Herbert Spencer's Liberal Utilitarianism*. Cambridge: Cambridge University Press, 1998.

Werth, Barry. *Banquet at Delmonico's: Great Minds, the Gilded Age, and the Triumph of Evolution in America*. New York: Random House, 2009.

West, David A. *A Naturalist in Brazil*. Blacksburg, Virginia: Pocahontas Press, 2003.

Wiltshire, David. *The Social and Political Thought of Herbert Spencer*. Oxford: Oxford University Press, 1978.

Xianming, Wang. "Yanyi mingzhu yu Zhongguo wenhua de xiandaihua: yi Yan Fu yi *Qunxue yiyan* weili de kaocha." *Fuzhou daxue xuebao(shehui kexue ban)* no. 2 (2008): 24–30.

———. "*Gezhi Huibian* yu xifang jindai keji zhishi zai qingmo de chuanbo." *Zhongguo keji shiliao* no. 1 (1996): 36–47.

———, ed. *Jindai kexue zai Zhongguo de chuanbo*. Jinan: Shandong jiaoyu chubanshe, 2007.

Yuozhi, Xiong. *Xixue dongjian he wanqing shehui*. Beijing: Zhongguo renmin daxue chubanshe, 2011.

Zea, Leopoldo. *El positivismo en México: nacimiento, apogeo y decadencia*. Mexico: FCE, 1968.

———, ed. *Pensamiento positivista latinoamericano*. Caracas: Fundación Biblioteca Ayacucho, 1980.

Zimmermann, Eduardo A. "Racial Ideas and Social Reform: Argentina, 1890–1916." *Hispanic American Historical Review* (1992): 23–46.

Zebrowski, Robin L. "Mind is Primarily a Verb; An Examination of Mistaken Similarities Between John Dewey and Herbert Spencer." *Educational Theory* 58, no. 3 (2008): 305–320.

Zini, Zino. *Spencer*. Milan: Athena, 1926.

Zohrab, Irene. "Dostoevsky and Herbert Spencer." *Dostoevsky Studies* 7 (1986): 45–72.

Index

www.ingramcontent.com/pod-product-compliance
Lightning Source LLC
Chambersburg PA
CBHW071249180926
51613CB00002B/178
* 9 7 8 9 0 0 4 2 6 3 9 9 4 *